PROGRESS

IN

HETEROCYCLIC CHEMISTRY

Volume 2

Related Pergamon Titles of Interest

Books

Organic Chemistry Series

CARRUTHERS: Cycloaddition Reactions in Organic Synthesis
DAVIES: Organotransition Metal Chemistry: Applications to Organic Synthesis
DEROME: Modern NMR Techniques for Chemistry Research
DESLONGCHAMPS: Stereoelectronic Effects in Organic Chemistry
GIESE: Radicals in Organic Synthesis: Formation of Carbon–Carbon Bonds
HANESSIAN: Total Synthesis of Natural Products: The 'Chiron' Approach
PAULMIER: Selenium Reagents and Intermediates in Organic Synthesis
PERLMUTTER: Conjugate Addition Reactions in Organic Synthesis
WILLIAMS: Synthesis of Optically Active α-Amino Acids
WONG & WHITESIDES: Enzymes in Synthetic Organic Chemistry

Major Works of Reference

ALLEN & BEVINGTON: Comprehensive Polymer Science
BARTON & OLLIS: Comprehensive Organic Chemistry
HANSCH *et al*: Comprehensive Medicinal Chemistry
KATRITZKY & REES: Comprehensive Heterocyclic Chemistry
TROST & FLEMING: Comprehensive Organic Synthesis
WILKINSON *et al*: Comprehensive Organometallic Chemistry

Also of Interest

KATRITZKY: Handbook of Heterocyclic Chemistry
NICKON & SILVERSMITH: Organic Chemistry—The Name Game
PERRIN & ARMAREGO: Purification of Laboratory Chemicals, 3rd edition
RIGAUDY & KLESNEY: Nomenclature of Organic Chemistry ('The Blue Book')

Journals

TETRAHEDRON (rapid publication primary research journal for organic chemistry)
TETRAHEDRON: ASYMMETRY (new international journal for rapid publication on all aspects of asymmetry in organic, inorganic, organometallic, physical chemistry and bio-organic chemistry)
TETRAHEDRON LETTERS (rapid publication preliminary communications journal for organic chemistry)
TETRAHEDRON COMPUTER METHODOLOGY (new international electronic journal for rapid publication of original research in computer chemistry)

Full details of all Pergamon publications, and a free specimen copy of any Pergamon journal, are available on request from any Pergamon office

PROGRESS IN HETEROCYCLIC CHEMISTRY

Volume 2

A critical review of the 1989 literature preceded by one chapter on a current heterocyclic topic

Editors

H SUSCHITZKY

Department of Chemistry and Applied Chemistry, University of Salford, UK

and

E F V SCRIVEN

Reilly Industries Inc., Indianapolis, Indiana, USA

PERGAMON PRESS

Member of Maxwell Macmillan Pergamon Publishing Corporation

OXFORD · NEW YORK · BEIJING · FRANKFURT
SÃO PAULO · SYDNEY · TOKYO · TORONTO

U.K.	Pergamon Press plc, Headington Hill Hall, Oxford OX3 0BW, England
U.S.A.	Pergamon Press, Inc., Maxwell House, Fairview Park, Elmsford, New York 10523, U.S.A.
PEOPLE'S REPUBLIC OF CHINA	Pergamon Press, Room 4037, Qianmen Hotel, Beijing, People's Republic of China
FEDERAL REPUBLIC OF GERMANY	Pergamon Press GmbH, Hammerweg 6, D-6242 Kronberg, Federal Republic of Germany
BRAZIL	Pergamon Editora Ltda, Rua Eça de Queiros, 346, CEP 04011, Paraiso, São Paulo, Brazil
AUSTRALIA	Pergamon Press Australia Pty Ltd., P.O. Box 544, Potts Point, N.S.W. 2011, Australia
JAPAN	Pergamon Press, 5th Floor, Matsuoka Central Building, 1-7-1 Nishishinjuku, Shinjuku-ku, Tokyo 160, Japan
CANADA	Pergamon Press Canada Ltd., Suite No. 241, 253 College Street, Toronto, Ontario, Canada M5T 1R5

First edition 1990

Library of Congress Cataloging in Publication Data

Progress in heterocyclic chemistry: a critical review of the 1989 literature preceded by one chapter on a current heterocyclic topic/editors, H. Suschitzky and E.F.V. Scriven.—1st ed.
p. cm.
Includes index.
1. Heterocyclic compounds.
I. Suschitzky, H. II. Scriven, Eric F.V.
QD399.P76 1990 547'.59—dc20 89-8531

British Library Cataloguing in Publication Data

Progress in heterocyclic chemistry.
Vol. 2
1. Heterocyclic compounds.
547'.59

ISBN 0-08-037069-1 (Hardcover)
ISBN 0-08-037070-5 (Flexicover)

Printed in Great Britain by BPCC Wheatons Ltd, Exeter

Contents

Foreword

Progress in Heterocyclic Chemistry (PHC) Volume 2 reports on important original material abstracted from the 1989 literature of heterocyclic chemistry in a systematic way. It also contains an article on a heterocyclic topic hitherto unreviewed, written by Dr Aitken who is active in the relevant field. We much regret that the advertized review on 'Asymmetric Synthesis of Azetidinones' by Dr Bateson was not ready in time. However, it will appear fully updated in PHC Volume 3.

The arrangement of the chapters follows the scheme adopted for Volume 1 and this, aided by a subject index, should facilitate the search for new developments in a particular area. The reference system is that used in 'Comprehensive Heterocyclic Chemistry' (Pergamon, 1984) where a full explanation of the system can be found.

We thank all authors for delivering most readable and camera-ready reports with well drawn diagrams. We are also grateful to Dr Colin Drayton, editorial director of Pergamon Press, for his expert advice and especially for editorial improvements of the chapter presentation. Like its predecessor, PHC Volume 2 is designed to enable academic and industrial chemists and advanced students of chemistry to keep abreast of heterocyclic developments in an effortless way.

H. SUSCHITZKY
E. F. V. SCRIVEN

Advisory Editorial Board Members

CHAPTER 1

Oxidation of Five-Membered Ring Heterocycles Containing N and S

R ALAN AITKEN, DAVID P ARMSTRONG & SHAUN T E MESHER
University of St Andrews, UK

1.1 INTRODUCTION

Five-membered ring heterocycles containing both nitrogen and sulphur show an interesting pattern of reactivity towards various oxidising agents. The most important processes are formation of N- or S-oxides and, for partly saturated systems, dehydrogenation, but various ring-opening and oxidative coupling reactions have also been observed. In this article the most important reactions of five-membered ring heterocycles containing (only) N and S with oxidising agents are described.

1.2 THIAZOLE BASED SYSTEMS

1.2.1 Thiazoles

The first successful oxidation of this relatively resistant ring system was reported in 1947. Treatment of 2,4-dimethylthiazole (**1**) with hydrogen peroxide in acetic acid gave the N-oxide (**2**) in 60% yield [47JPJ34]. With thiazole itself the N-oxide was formed in low yield due to extensive ring cleavage. Monoperphthalic acid was also used to convert (**1**) to (**2**). Hydrogen peroxide in acetic acid readily converts sulphathiazole (**3**) to its N-oxide (**4**) [59AK225]. A number of 2,5-diarylthiazoles are readily oxidised to their N-oxides using monoperoxymaleic acid [77CPB3270] while, in studies on novel antibacterial agents, peracetic acid was used to obtain the N-oxides of thiazoles (**5**) [72JPP42662].

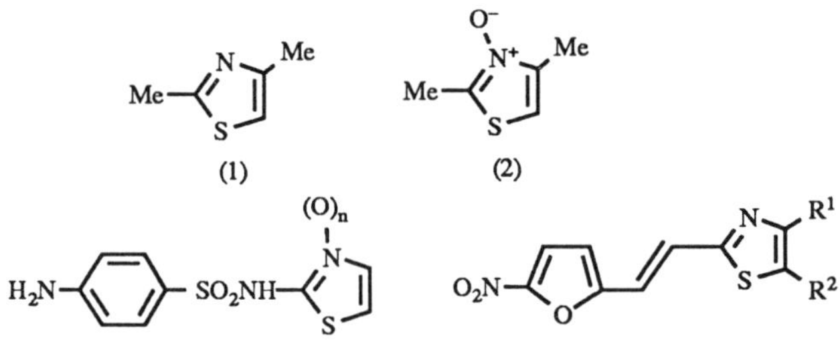

(1) (2) (3) n=0 (4) n=1 (5)

Photo sensitised oxidation of 2,4,5-triphenylthiazole (**6**) proceeds by initial cycloaddition of $^1O_2$ either in a 1,4-sense to give (**7**) or 1,2 to give (**11**). Irradiation with O_2 in MeOH in the presence of rose bengal gives benzamide and benzil which arise from extrusion of sulphur from (**7**) to give (**8**) which is then hydrolysed [69BCJ2973]. With methylene blue in $CHCl_3$ the main product is dibenzoyl(thiobenzoyl)amine (**10**). This may arise either from rearrangement of (**7**) to (**9**) followed by a 1,3-thiobenzoyl shift, or, alternatively, by cycloreversion of (**11**) to (**12**) followed by a 1,3-benzoyl shift. A later report confirmed the pathway in MeOH but reported the formation of 2,4,5-triphenyloxazole as a minor product and in $CHCl_3$ it was found to be the only

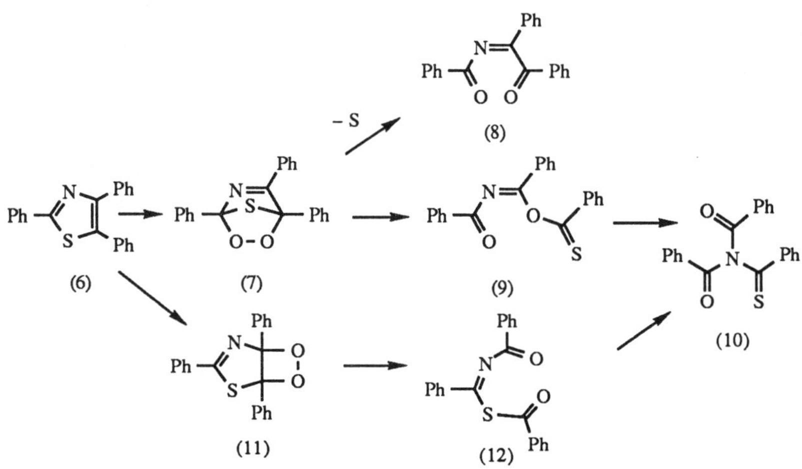

(6) (7) (8) (9) (10) (11) (12)

product [77HCA284]. Its formation is rationalised by loss of sulphur from either (**7**) or (**11**) followed by rearrangement to the oxazole O-oxide which is deoxygenated by $^1O_2$.
The photo-oxidation of 2-amino-4-phenylthiazole to thiobenzoylurea (**14**) with methylene blue in $CHCl_3$ involves loss of CO from in initial 1,4-adduct (**13**) [81IJC(B)870].

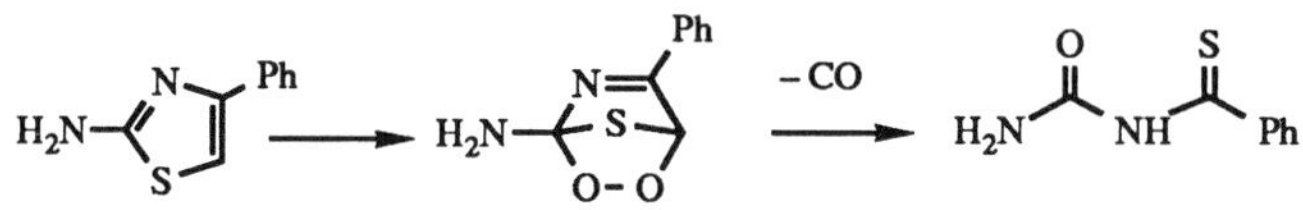

(13) (14)

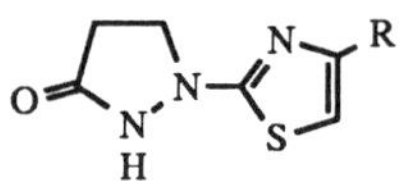

(15)

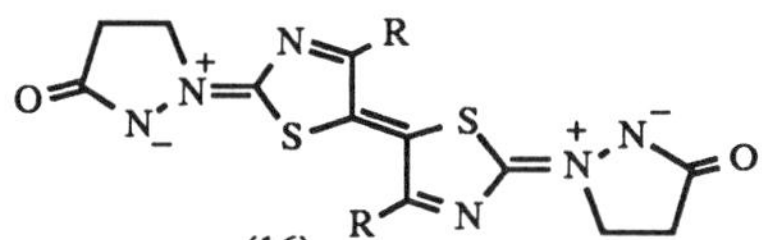

(16)

Oxidative coupling occurs on treatment of thiazoles (**15**) with $NaIO_4$ to give bis-azomethine imines (**16**) [82JPR873].

1.2.2 2-Thiazolines

MCPBA has been used to oxidise (**17**) to the corresponding S,S-dioxide [76TL1137] as well as (**18**) to a mixture of diastereomeric S-oxides [77G289]. The thiazoline ring of the N-terminal ThzIle residue of bacitracin A is oxidised to its S-oxide by $NaIO_4$ and to the S,S-dioxide by $KMnO_4$ [82JBI25].

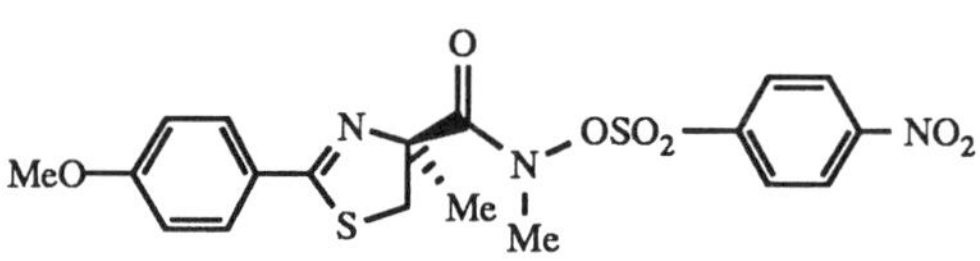

(17)

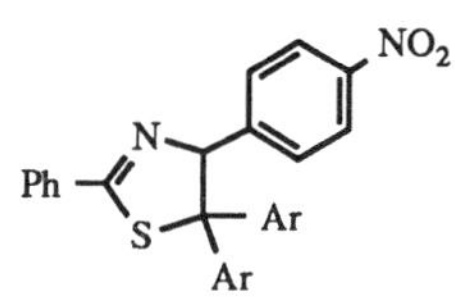

(18)

In our own laboratory we have found that chiral thiazolines (**19**) are generally oxidised to mixtures of S,S-dioxide (**20**), sulphonic acid (**21**) and disulphide (**22**) with such reagents as MCPBA, peracetic acid, performic acid, $NaIO_4$ and $KMnO_4$. However the use of N_2O_4 or peroxytrifluoroacetic acid gives exclusively the previously unknown N-oxides (**23**) [90UP1]. An early report describes oxidation of 2-phenylthiazoline (**19**, R^1=Ph, R^2=H) to N-benzoyltaurine (**21**, R^1=Ph, R^2=H) using bromine water [1890CB157].

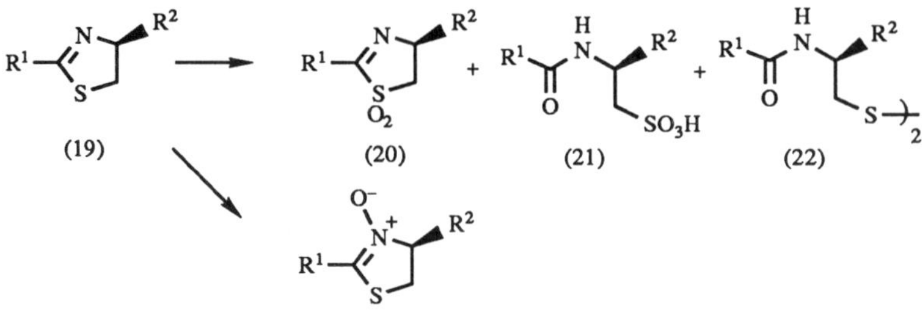

(19) (20) (21) (22) (23)

Various transition metal oxidants dehydrogenate 2-thiazolines to the corresponding thiazoles. Manganese dioxide is effective in oxidising thiazolines (**24**) in this way, a transformation also possible using phenanthraquinone [66JCS(C)1061]. Nickel peroxide is similarly effective in a wide range of examples [77JA8078] [78JOC1624] [79JOC497], including the thiazoylthiazoline moiety (**25**) in the anti tumour compound phleomycin A^2.

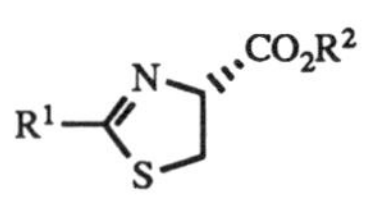

(24)

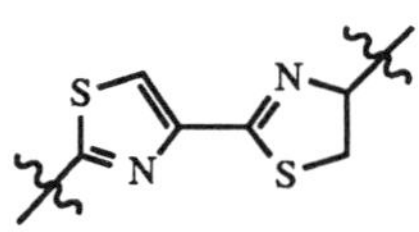

(25)

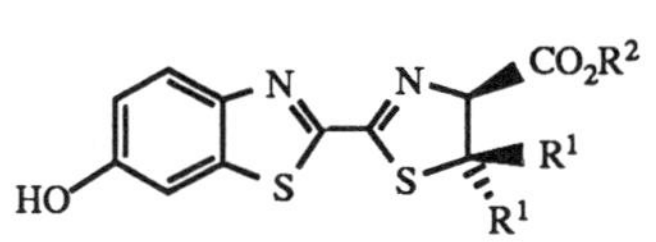

(26)

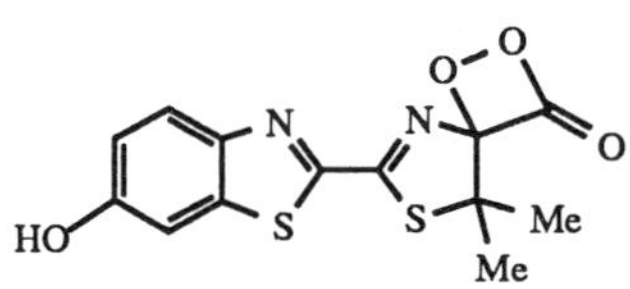

(27)

Potassium ferricyanide in basic solution oxidises compounds of type (**24**) to the thiazoles although there is competing hydrolysis to give disulphides [68JCS(C)1526]. Basic solutions of firefly luciferin (**26**, R^1=R^2=H) are dehydrogenated by $K_3Fe(CN)_6$ as well as by atmospheric oxygen [63JA337]. Mercuric acetate in acetic acid has also been used for (**24**) [68JCS(C)1526].

Products from the vapour phase oxidation of 2-methyl-2-thiazoline over a variety of mixed vanadium/molybdenum oxide catalysts include 2-methylthiazole, thiazole-2-carboxaldehyde and thiazole[82MI718].

Photo-oxidation of luciferin analogue (**26**, R^1=Me, R^2=Ph) in DMSO proceeds by addition of O_2 at C-4, displacement of PhO^- to give (**27**), which then loses CO_2 with luminescence to give the thiazolin-4-one [68CC22].

1.2.3 3-Thiazolines

Oxygen converts 3-thiazolines such as (**28**) and (**29**) to the corresponding S-oxides at room temperature [64LA(672)103].

A number of agents have been successfully used to bring about dehydrogenation of 3-thiazolines. Ferric chloride is effective for (**30**, R=H, Me, Et, Ph) [58LA(611)121]. For di- and trisubstituted examples such as (**31**)

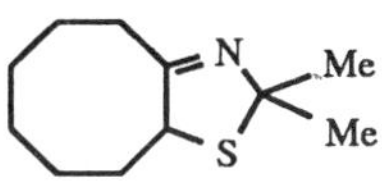

(28)

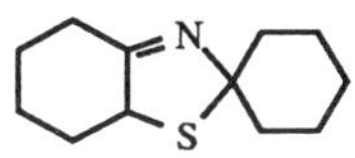

(29)

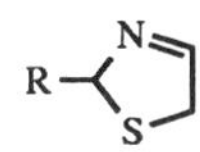

(30)

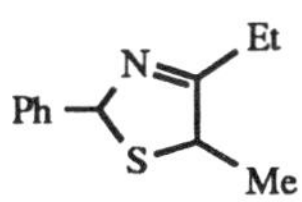

(31)

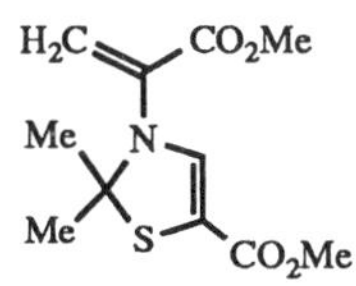

(32)

sulphur is most effective for conversion to the thiazoles while $FeCl_3$, hydrogen peroxide, potassium ferricyanide and potassium dichromate can all be used with somewhat lower yields [57LA(610)49].

1.2.4 4-Thiazolines

Treatment of (**32**) with one equivalent of MCPBA at –78°C affords the corresponding S-oxide [76JCS(P1)2540].

1.2.5 2-Thiazolin-5-ones

Oxidative coupling through the 4-position is the main mode of oxidation for these compounds. Thus treatment of (**33**) with iodine in the presence of triethylamine gives (**34**) [72MI29]. Aeration of (**35**) in aqueous dioxane gives the oxidised dimer in addition to α-ketoglutaric acid from hydrolysis

[78TL2063], and photo-oxidation of (**36**) in CH_2Cl_2 also leads to dimerisation [87H1313] [80CL717].

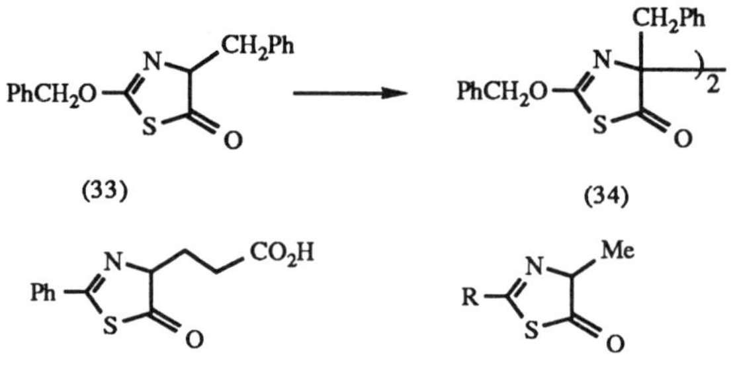

(33) (34) (35) (36) R=H or Ph

1.2.6 4-Thiazoline-2-thiones

Many of the oxidation reactions of these compounds can be rationalised in terms of the 2-mercaptothiazole tautomer. Thus treatment of (**37**) with iodine [64JOC2148] or ammonium persulphate [49JOC1111] leads to the disulphide (**38**). This can also be obtained with H_2O_2 in neutral solution [41JOC764] [48JA3419] but under acidic conditions, as well as with nitric acid, sulphur is

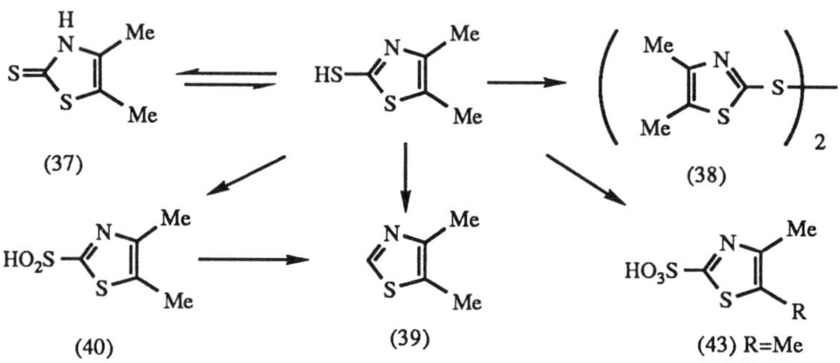

(37) (38) (40) (39) (43) R=Me (44) R=H

lost to give the thiazole (**39**) [49JOC1111] [41JOC764]. This reaction most likely occurs by oxidation to the sulphinic acid (**40**) which can be isolated as its sodium salt [50USP2509453] [50USP2509454] but decomposes with loss of SO_2 on acidification. With H_2O_2 in acetic acid, sulphur is also lost from (**41**) and the resulting thiazole can be isolated as its perchlorate (**42**) [74CJC3021]. Oxidation to thiazole-2-sulphonic acids is also possible as illustrated by the conversion of (**37**) to (**43**) with N_2O_4 [49JOC1111] and the formation of (**44**) from 4-methyl-4-thiazoline-2-thione with H_2O_2 under basic conditions [39JPJ43].

A final mode of oxidation of these compounds is the replacement of the exocyclic sulphur by oxygen. Both mercuric acetate [73BSF270] and

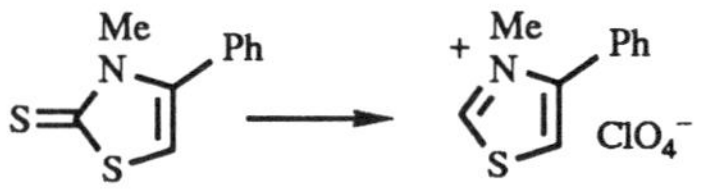

(41) (42)

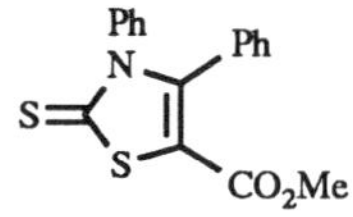

(45)

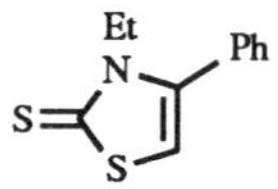

(46)

H_2O_2/KOH [56JPJ1153] convert thiazolinethiones such as (**45**) and (**46**) to the corresponding thiazolinones.

Further examples of the different oxidation processes in this section are to be found in a review [76PS185].

1.2.7 Thiazolidines

Thiazolidines, particularly those with a carboxylic acid function in the 4-position derived from cysteine, have been extensively studied with a variety of oxidising agents. As early as 1937 the oxidation of (**47**) to its S-oxide with H_2O_2 in acetone and to its S,S-dioxide with excess H_2O_2 in acetic acid was reported [37JA200]. With the non-acetylated derivative (**48**), on the other hand, H_2O_2 or iodine resulted in ring cleavage to cystine while bromine produced

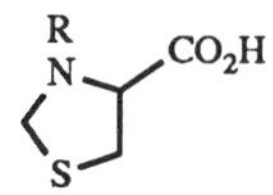

(47) R=Ac
(48) R=H

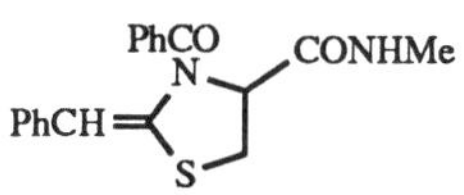

(49)

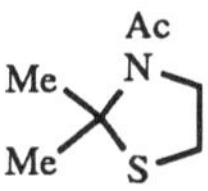

(50)

cysteic acid [37JA200]. Thiazolidine itself was similarly converted to a disulphide with iodine and a sulphonic acid with bromine. MCPBA has been used to obtain S-oxides from thiazolidines such as (**49**) [75JA5010] and (**50**) [78S744]. The penicillin derivative (**51**) is converted with peracetic acid to an S-oxide which can be further oxidised to the S,S-dioxide with $KMnO_4$

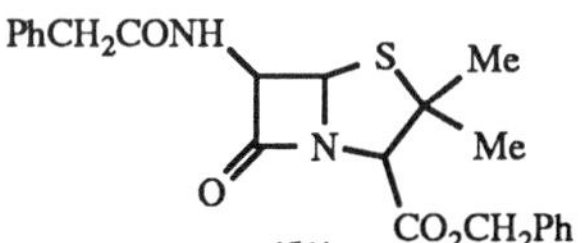

(51)

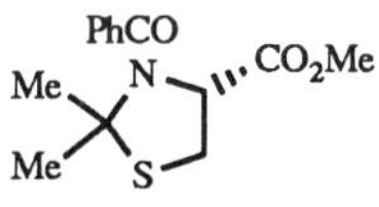

(52)

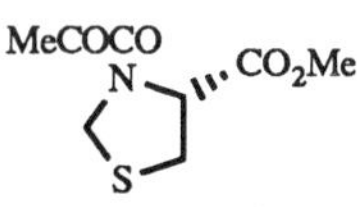

(53)

[50JOC815]. A detailed study of the stereoselectivity of S-oxide formation from substrates such as (**52**) using MCPBA, $NaIO_4$ and $Bu^tOOH/Ti(OPr^i)_4$ has appeared [86TL3391]. The methyl ester of (**47**) is converted to its S-oxide using $NaIO_4$ [80BSB749] and the S-oxide of (**53**), obtained similarly, is further oxidised to the S,S-dioxide using $KMnO_4$ [73TL1103].

A single example of dehydrogenation of a thiazolidine to the 3-thiazoline is provided by the enzymatic reaction of (**48**) with D-amino acid oxidase [79MI252].

The final important mode of oxidation of this ring system is oxygenation at the 5-position. Lead tetra-acetate oxidation of (**54**) produced (**55**) as part of the first synthesis of cephalosporin C [66JA852].

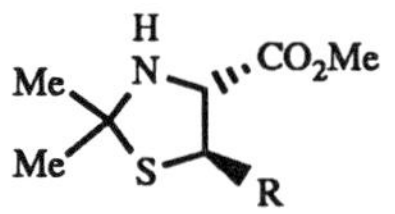

(54) R=N(CO_2Me)NHCO_2Me
(55) R=OH

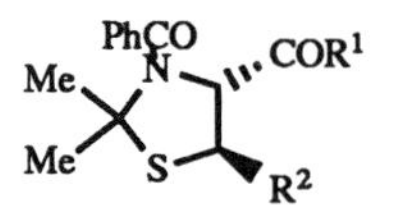

(56) R^1=OMe, R^2=H
(57) R^1=OMe, R^2=OH
(58) R^1=NHCH(Pr^i)CO_2Me, R^2=H
(59) R^1=NHCH(Pr^i)CO_2Me, R^2=OCOPh
(60) R^1=NHCH(Pr^i)CO_2Me, R^2=OH

Photo-oxidation in THF in the presence of tetraphenylporphyrin results in conversion of (**56**) to (**57**). With methylene blue in DMSO this is accompanied by formation of the thiazolidin-5-one as well as the S-oxide of (**56**) [84TL4767] while in methanol the S-oxide is the sole product [85T2133]. Oxidation of (**58**) with perbenzoic acid at 70°C gives (**59**) which can be hydrolysed to (**60**) [77GEP2615621].

1.2.8 Thiazolidin-2-ones

A few cases of S-oxidation have been reported for this ring system. Thus MCPBA can be used to convert thiazolidinones such as (**61**) and (**62**) to either the S-oxide or the S,S-dioxide [81JAP87574] while for (**63**) peracetic acid

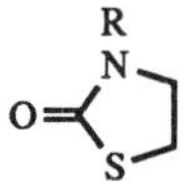

(61) R=n-C_6H_{13}
(62) R=Bu^n
(63) R=$CH_2CH_2CONH_2$

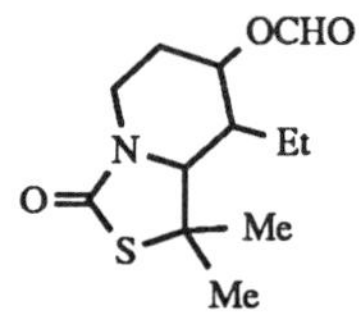

(64)

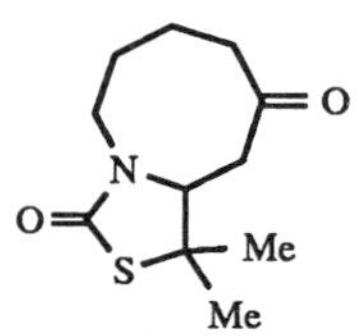

(65)

produces the S,S-dioxide [61JOC5103]. S,S-Dioxides are also formed on treatment of bicyclic thiazolidinones such as **(64)** with MCPBA and **(65)** with performic acid [85T2861].

1.2.9 Thiazolidin-4-ones

The oxidation of this ring system has been extensively studied and has been reviewed in detail elsewhere [61CRV463] [79AHC83]. Peracetic acid, sodium periodate and chloramine-T [81JHC633] have been used to form S-oxides, while $KMnO_4$, CrO_3 and excess peracetic acid give the S,S-dioxides.

1.2.10 Thiazolidine-2-thiones

The most facile oxidative process in this system is replacement of the exocyclic sulphur by oxygen. A variety of substrates such as **(66)** [49JCS2367] and **(67)** [83JAP5829775] are converted to the corresponding thiazolidinones

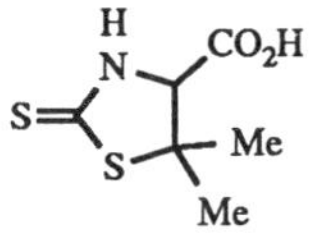

(66)

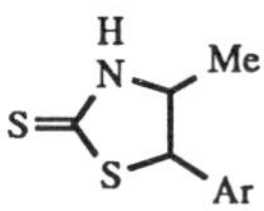

(67)

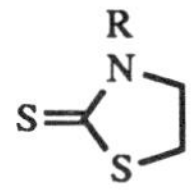

(68) R=p-MeO-C_6H_4-
(69) R=CH_2CH_2CN
(70) R=Me
(71) R=H

using H_2O_2 under basic conditions. The reactions involve initial S-oxidation to a sulphine, isolable as an iron complex [66LA(695)77], which cyclises to the oxathiirane and loses S. Mercuric acetate also brings about this oxidation of **(68)** [47HCA1336] and **(69)** [61JOC5106] as does the more unusual oxidant propylene oxide with **(70)** [70BCJ168]. In a process analogous to that for the 4-thiazoline-2-thiones (Section 1.2.6), iodine in the presence of triethylamine causes oxidative dimerisation of the parent thiazolidinethione **(71)** to **(72)** [80JCS(P1)665].

Further examples of these types of oxidation have been reviewed [76PS185].

A single paper describes the further oxidation of thiazolidine-2-thiones **(73)** to thiazolidinone S,S-dioxides using peracetic acid [61JOC5103].

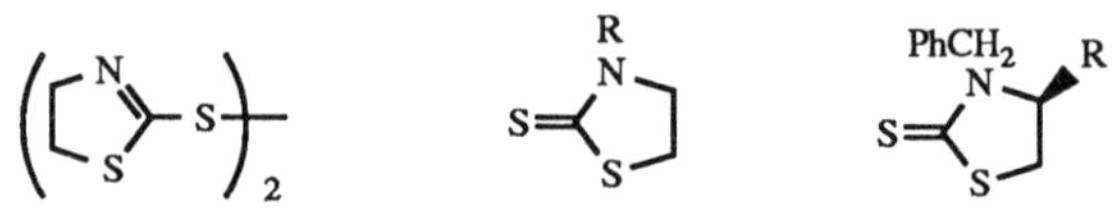

(72) (73) (74)

The rapid conversion to the thiazolidinone is apparently followed by slower oxidation of the ring S. In our own laboratory we have similarly obtained thiazolidinone S,S-dioxides from a variety of chiral 4-substituted thiones (**74**) [90UP2].

1.2.11 Thiazolidine-4-thiones

A single example of this type is provided by the conversion of (**75**) to the corresponding thiazolidinone using H_2O_2 [71KGS145].

1.2.12 Thiazolidine-2,4-diones and Thiazolidine-2-thione-4-ones

Thiazolidinedione (**76**) can be oxidised by MCPBA either to the S-oxide or the S,S-dioxide [86LA1787] and $KMnO_4$ converts (**77**) to its S,S-dioxide [79IJC(B)377]. Selenium dioxide introduces a third carbonyl group into (**78**) to produce the 2,4,5-trione [76S190].

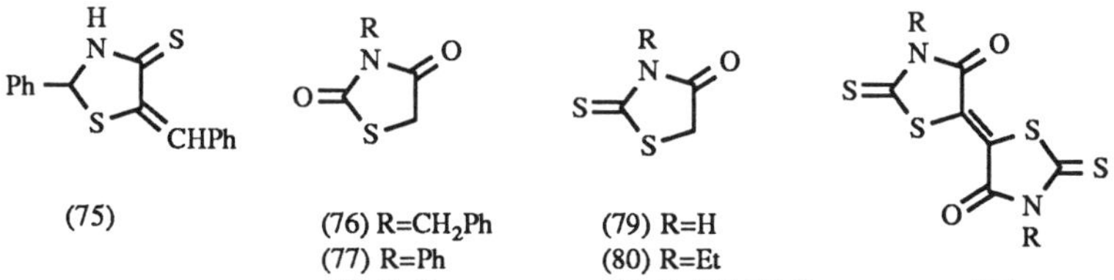

(75)

(76) R=CH_2Ph
(77) R=Ph
(78) R=H

(79) R=H
(80) R=Et
(81) R=CH_2CO_2Et

(82)

Rhodanine (**79**) is oxidised to the corresponding 2,4-dione by H_2O_2 in basic solution [38JPJ8] by way of an isolable sulphine [65LA(681)55] but with excess H_2O_2 ring opening gives sulphoacetamide.

Oxidative coupling to give (**82**) can be achieved for (**79**) and (**80**) using SeO_2 [57AG138] and for (**81**) using bromine in the presence of Et_3N [55JCS927].

1.3 ISOTHIAZOLE BASED SYSTEMS

1.3.1 Isothiazoles

Examples of isothiazole oxidation are rare. Nitric acid in either sulphuric or acetic acid converts compounds (**83**) to the S-oxides which are further oxidised

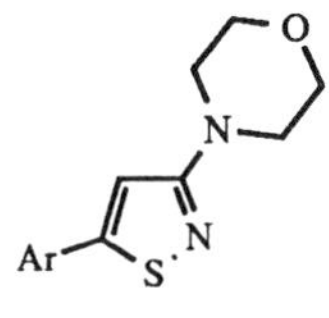

(83)

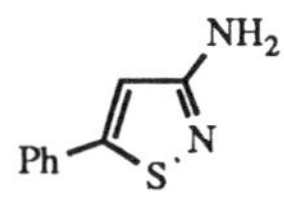

(84)

to S,S-dioxides by peracetic acid [66G1009]. Peroxysulphuric acid reacts with (**84**) to give the corresponding S-oxide [72JCS(P1)1247].

1.3.2 4-Isothiazolin-3-ones

The formation of S-oxides and S,S-dioxides has been reported. The parent ring system (**85**) as well as various substituted derivatives are converted to S-oxides with nitric acid, N_2O_4 or chromic acid, [71JHC591]. Peracetic acid

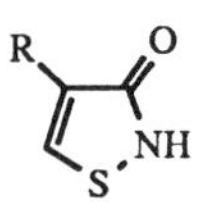

(85) R=H
(86) R=Ph

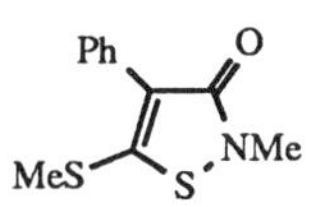

(87)

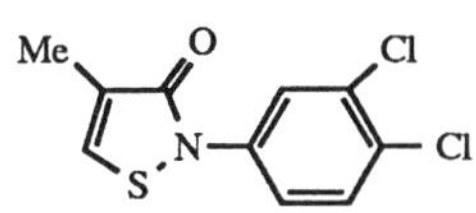

(88)

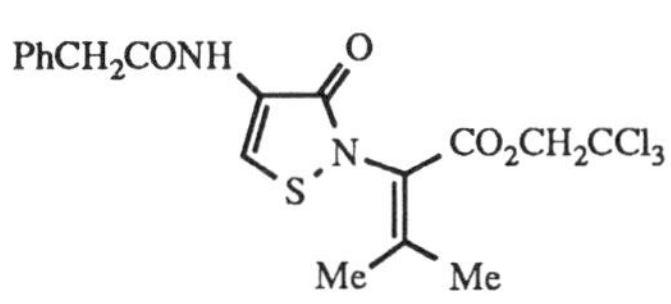

(89)

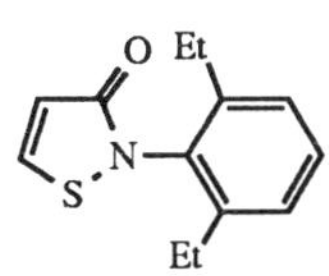

(90)

oxidises isothiazolinones such as (**86**) and (**87**) to the corresponding S,S-dioxides [77AJC1815]. MCPBA can be used to obtain either S-oxides or S,S-dioxides in a variety of cases illustrated by **(88)** [71JHC591], **(89)** and **(90)** [75JA5020] [81JAP81573].

1.3.3 3-Isothiazoline-5-thiones and 4-Isothiazoline-3-thiones

In a process analogous to that observed for the 4-thiazoline-2-thiones (Section 1.2.6), treatment of compounds such as **(91)** and **(92)** with perchloric

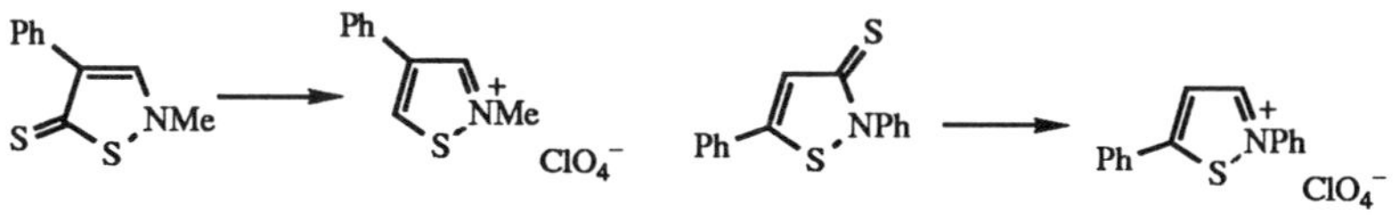

(91) (93) (92) (94)

acid results in loss of sulphur to give isothiazolium perchlorates **(93)** and **(94)** respectively [76PS185].

1.3.4 Isothiazolidin-3-ones

Although the S-oxides and S,S-dioxides of this ring system are well known they have generally been prepared by direct ring closure routes in the oxidised form. MCPBA has however been used to convert compounds such as **(95)** to their S-oxides [74JOC1210] and **(96)** to its S,S-dioxide [73TL5213].

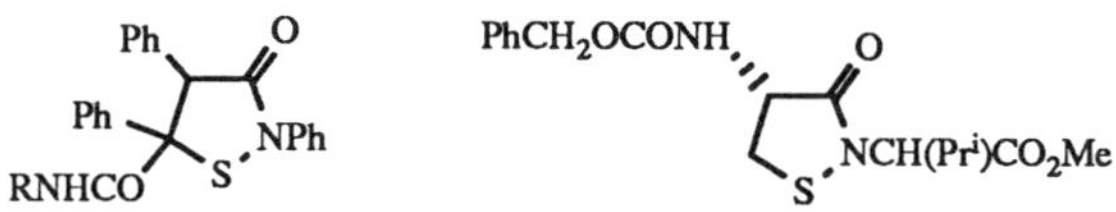

(95) (96)

1.4 THIADIAZOLE BASED SYSTEMS

1.4.1 1,2,3-Thiadiazoles

These compounds undergo both N and S-oxidation. Treatment of thiadiazoles such as **(97)** [72T5655] and **(98)** [75T637] with either MCPBA or peracetic acid first gives the 2-oxides. With excess peracetic acid these are further

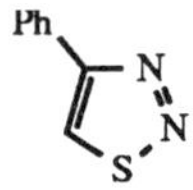

(97)

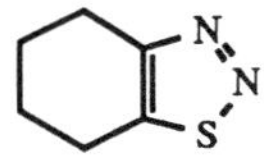

(98)

oxidised in low yield to the 1,1,2-trioxides. Upon photolysis the 2-oxide derived from (**97**) rearranged to the 3-oxide [75LA1257] whose X-ray structure unambiguously established the identity of the two isomers [78ZN(B)316].

1.4.2 1,2,5-Thiadiazoles

The 3,4-diethoxythiadiazole (**99**) readily forms an S-oxide on treatment with MCPBA [81H1561]. For the dihydroxy compound (**100**) reaction with DDQ produces the S-oxide which is better regarded as the thiadiazolidinedione structure [690PP255].

A fascinating recent example of this ring system is (**101**) which also contains a 1,3,2-dithiazole ring, thus allowing an internal comparison between

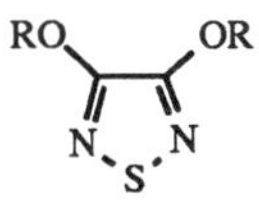

(99) R=Et
(100) R=H

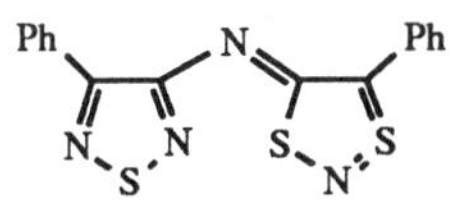

(101)

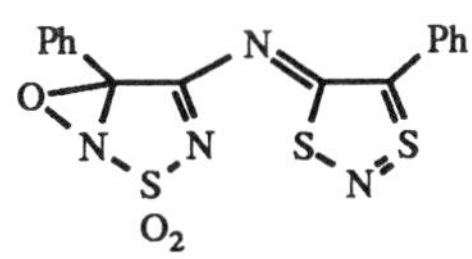

(102)

the oxidation behaviour of the two heterocycles [89CC1134]. All the oxidation occurs on the thiadiazole ring. Treatment with N_2O_4 at –20°C gives the S-oxide which is further oxidised to the S,S-dioxide with excess N_2O_4 at 20°C. Treatment of this with MCPBA then gives the trioxide (**102**) which can also be obtained directly from (**101**) with excess MCPBA.

1.4.3 Δ^2-1,3,4-Thiadiazolines

The highly substituted thiadiazolines (**103**) can be converted to S-oxides and S,S-dioxides with monoperphthalic acid [73T3569] or MCPBA [78JCS(P1)1218]. Where there is a hydrogen at the 5-position, as in (**104**) or (**105**), oxidation may be accompanied by N-deacylation to give the thiadiazoles. Thus while (**104**) gives an S-oxide (with concomitant oxidation of MeS to $MeSO_2$) on treatment with MCPBA [86JCS(P1)1357], the use of $KMnO_4$

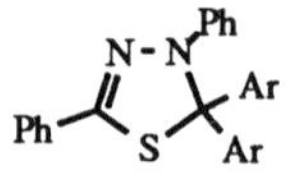

(103)

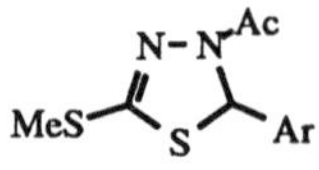

(104)

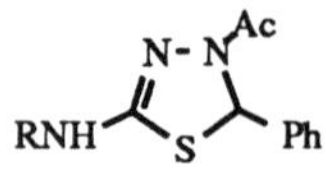

(105) R=H
(106) R=Ac

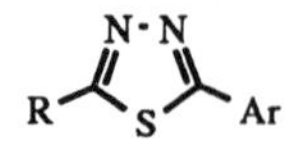

(107) R=$MeSO_2$
(108) R=NH_2, NHAc

[83JCS(P1)967] or peracetic acid [83JCS(P1)2957] gives mainly (**107**). Similarly reaction of (**106**) with MCPBA gives the S-oxide [82CC901], but with $KMnO_4$ the main product is (**108**) [83JCS(P1)967]. Conversion of (**105**) to (**108**) is possible using ferric chloride [80JOC1473].

1.4.4 Δ^3-1,3,4-Thiadiazolines

Peracetic acid is effective in converting thiadiazoline (**109**) to both the S-oxide and the S,S-dioxide [72JCS(P1)305].

1.4.5 1,3,4-Thiadiazolidines

The main oxidative process in this system is dehydrogenation to Δ^3-1,3,4-thiadiazolines. Compound (**109**) and similar examples have been obtained

(109)

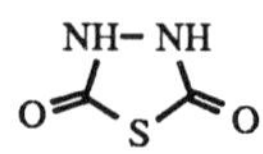

(110)

in this way from the corresponding thiadiazolidines using lead tetra-acetate, DDQ and MnO_2 [72JCS(P1)305] [78JCS(P1)45]. A further interesting oxidant used to prepare (**109**) and the parent thiadiazoline is diethyl azodicarboxylate [72JOC4045]. The thiadiazolidinedione (**110**) has been oxidised to the corresponding thiadiazolinedione, a highly reactive dienophile, using lead tetra-acetate, t-butyl hypochlorite, and cupric chloride [73JOC3622] [74JOC2951].

1.5 OTHER RING SYSTEMS

1.5.1 Δ^3-1,2,3-Dithiazolines

The unusually strained dithiazoline (**111**) is oxidised to a mixture of the 1-oxide, and the 2-oxide (**112**) which has lost one molecule of isobutene, by treatment with MCPBA [79BCJ3615].

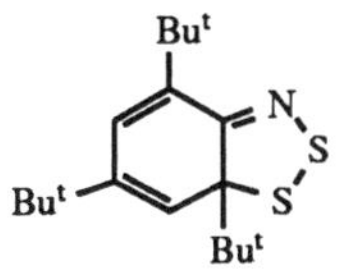
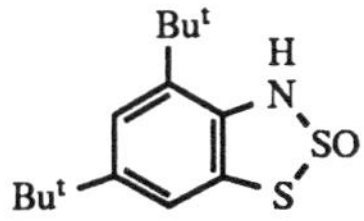
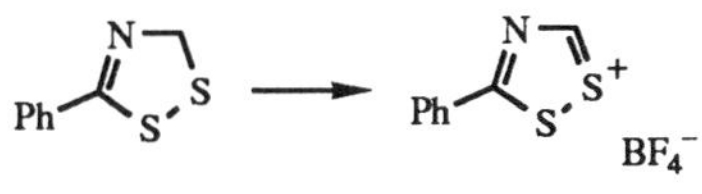

(111) (112) (113) (114)

1.5.2 Δ⁴-1,2,4-Dithiazolines

The hydride abstraction from (**113**) to give dithiazolium salt (**114**) using trityl fluoroborate is formally an oxidation [74AP828].

1.5.3 Δ⁴-1,2,4-Dithiazoline-3-thiones and Δ⁴-1,3,4-Dithiazoline-5-thiones

Replacement of the exocyclic sulphur atom by oxygen occurs on treatment with $KMnO_4$ [67TL1013], mercuric acetate [66BSF1183], peracetic acid and chlorine [70CJC2142] as exemplified by conversion of (**115**) to (**116**) and (**117**) to (**118**).

(115) (116) (117) (118)

1.5.4 1,2,3,4-Thiatriazoles

Although the thiatriazole (**119**) is very resistant to oxidation, being unaffected by chlorine, nitric acid, hydrogen peroxide or performic acid, it is

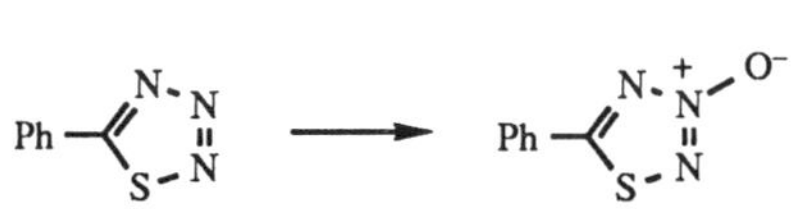

(119) (120)

converted to the 3-oxide (**120**) with peroxytrifluoroacetic acid [75T1783]. Although the yield is low due to competing fragmentation of (**119**), the product is stable [78JOC4816].

1.6 CONCLUSION

Five membered ring heterocycles containing N and S show a highly varied and interesting pattern of behaviour towards oxidation. While some systems have been investigated in detail and are well understood, many others have scarcely been examined and provide the opportunity for the discovery of many new useful reactions in the future.

1.7 REFERENCES

1890CB157 S. Gabriel and P. Heymann, *Ber. Dtsch. Chem. Ges.*, 1890, **23**, 157.

37JA200 S. Ratner and H.T. Clarke, *J. Am. Chem. Soc.*, 1937, **59**, 200.

38JPJ8 R. Kitamura, *J. Pharm. Soc. Jpn.*, 1938, **58**, 8. [*Chem. Abstr.* 1938, **32**, 3759].

39JPJ43 E. Ochiai, *J. Pharm. Soc. Jpn.*, 1939, **59**, 43. [*Chem. Abstr.*, 1939, **33**, 3791].

41JOC764 E.R. Buchman, A.O. Reims, and H. Sargent, *J. Org. Chem.*, 1941, **6**, 764.

47HCA1336 H. Erlenmeyer, H. Schulthess, and H. Bloch, *Helv. Chim. Acta*, 1947, **30**, 1336.

47JPJ34 E. Ochiai and E. Hayashi, *J. Pharm. Soc. Jpn.*, 1947, **67**, 34. [*Chem. Abstr.*, 1951, **45**, 9533].

48JA3419 J.J. Ritter and H. Sokol, *J. Am. Chem. Soc.*,1948, **70**, 3419.

49JCS2367 A.H. Cook, J.A. Elvidge, and G. Shaw, *J. Chem. Soc.*, 1949, 2367.

49JOC1111 F.D. Stewart and R.A. Mathes, *J. Org. Chem.*, 1949, **14**, 1111.

50JOC815 C.J. Cavallito and J.H. Harley, *J. Org. Chem.*, 1950, **15**, 815.

50USP2509453 *U.S. Pat.* 2,509,453 (1950). [*Chem. Abstr.*, 1950, **44**, 7885].

50USP2509454	*U.S. Pat.* 2,509,454 (1950). [*Chem. Abstr.*, 1950, **44**, 7885].
55JCS927	E.B. Knott and R.A. Jeffreys, *J. Chem. Soc.*, 1955, 927.
56JPJ1153	H. Hirano, *J. Pharm. Soc. Jpn.*, 1956, **76**, 1153 [*Chem. Abstr.*, 1957, **51**, 3606].
57AG138	U. Schmidt and R. Schuering, *Angew. Chem.*, 1957, **69**, 138.
57LA(610)49	F. Asinger, M. Thiel, and L. Schröder, *Liebigs Ann. Chem.*, 1957, **610**, 49.
58LA(611)121	M. Thiel, F. Asinger, and K. Schmiedel, *Liebigs Ann. Chem.*, 1958, **611**, 121.
59AK225	H. von Euler and H. Hasselquist, *Arkiv för Kemi*, 1959, **17**, 225.
61CRV463	F.C. Brown, *Chem. Rev.*, 1961, **61**, 463.
61JOC5103	R.J. Gaul and W.J. Fremuth, *J. Org. Chem.*, 1961, **26**, 5103.
61JOC5106	R.J. Gaul, W.J. Fremuth, and M.N. O'Connor, *J. Org. Chem.*, 1961, **26**, 5106.
63JA337	E.H. White, F. McCapra, and G.F. Field, *J. Am. Chem. Soc.*, 1963, **85**, 337.
64JOC2148	W.J. Humphlett and R.W. Lamon, *J. Org. Chem.*, 1964, **29**, 2148.
64LA(672)103	F. Asinger, W. Schäfer, M. Baumann, and H. Römgens, *Liebigs Ann. Chem.*, 1964, **672**, 103.
65LA(681)55	W. Walter and G. Randau, *Liebigs Ann. Chem.*, 1965, **681**, 55.
66BSF1183	J.-L. Derocque and J. Vialle, *Bull. Soc. Chim. Fr.*, 1966, 1183.
66G1009	A. Bruno and G. Purello, *Gazz. Chim. Ital.*, 1966, **96**, 1009.
66JA852	R.B. Woodward, K. Heusler, J. Gosteli, P. Naegeli, W. Oppolzer, R. Ramage, S. Ranganathan, and H. Vorbrüggen, *J. Am. Chem. Soc.*, 1966, **88**, 852.
66JCS(C)1061	M.A. Barton, G.W. Kenner, and R.C. Sheppard, *J. Chem. Soc. (C)*, 1966, 1061.
66LA(695)77	W. Walter, J. Voss, and J. Curts, *Liebigs Ann. Chem.*, 1966, **695**, 77.
67TL1013	H. Behringer and D. Deichmann, *Tetrahedron Lett.*, 1967, 1013.
68CC22	F. McCapra, Y.C. Chang, and V.P. Francois, *J. Chem. Soc., Chem. Commun.*, 1968, 22.

68JCS(C)1526	N.A. Fuller and J. Walker, *J. Chem. Soc. (C)*, 1968, 1526.
69BCJ2973	T. Matsuura and I. Saito, *Bull. Chem. Soc. Jpn.*, 1969, **42**, 2973.
69OPP255	M. Carmack, I.W. Stapleton, and R.Y. Wen, *Org. Prep. Proced. Int.*, 1969, **1**, 255.
70BCJ168	Y. Ueno, T. Nakai, and M. Okawara, *Bull. Chem. Soc. Jpn.*, 1970, **43**, 168.
70CJC2142	M. Ahmed and D.M. McKinnon, *Can. J. Chem.*, 1970, **48**, 2142.
71JHC591	S.N. Lewis, G.A. Miller, M. Hausman, and E.C. Szamborski, *J. Heterocycl. Chem.*, 1971, **8**, 591.
71KGS145	S.N. Baranov and B.E. Zhitar, *Khim. Geterosikl. Soedin.*, 1971, **3**, 145. [*Chem. Abstr.* 1973, **78**, 71985].
72JCS(P1)305	D.H.R. Barton and B.J. Willis, *J. Chem. Soc., Perkin Trans. 1*, 1972, 305.
72JCS(P1)1247	R.J.A. Walsh and K.R.H. Wooldridge, *J. Chem. Soc., Perkin Trans. 1*, 1972, 1247.
72JOC4045	J. Buter, S. Wassenaar, and R.M. Kellogg, *J. Org. Chem.* 1972, **37**, 4045.
72JPP42662	S. Ueno, T. Kawasaki, D. Inmaru, and Y. Oosaka, *Jap. Pat.* 42,662 (1972) [*Chem..Abstr.*,1973,**78**, 72123].
72MI29	I.Z. Siemion, W. Steglich, and L. Wilschowitz, *Rocz. Chem.*, 1972, **46**, 29.
72T5655	P. Braun, K.-P. Zeller, H. Meier, and E. Muller, *Tetrahedron*, 1972, **28**, 5655.
73BSF270	C. Gueden and J. Vialle, *Bull. Soc. Chim. Fr.*, 1973, 270.
73JOC3632	E.J. Corey and B.B. Snider, *J. Org. Chem.*, 1973, **38**, 3632.
73T3569	B.F. Bonini, G. Maccagnani, L. Thijs, and B. Zwanenburg, *Tetrahedron Lett.*, 1973, 3569.
73TL1103	K.R. Henery-Logan and C.G. Chen, *Tetrahedron Lett.*, 1973, 1103.
73TL5213	R.B. Morin, E.M. Gordon, and J.R. Lake, *Tetrahedron Lett.*, 1973, 5213.
74AP828	H. Böhme and K.H Ahrens, *Arch. Pharm.*, 1974, **307**, 828.
74CJC3021	J.L. Charlton, S.M. Loosmore, and D.M. McKinnon, *Can. J. Chem.*, 1974, **52**, 3021.
74JOC1210	T. Minami and T. Agawa, *J. Org. Chem.*, 1974, **39**, 1210.
74JOC2951	S.W. Mojé and P. Beak, *J. Org. Chem.*, 1974, **39**, 2951.

75JA5010	S. Nakatsuka, H. Tanino, and Y. Kishi, *J. Am. Chem. Soc.*, 1975, **97**, 5010.
75JA5020	T. Kamiya, T. Teraji, M. Hashimoto, O. Nakaguchi, and T. Oku, *J. Am. Chem. Soc.*, 1975, **97**, 5020.
75LA1257	H.P. Braun, K -P. Zeller, and H. Meier, *Liebigs Ann. Chem.*, 1975, 1257.
75T637	H.P. Braun and H. Meier, *Tetrahedron*, 1975, **31**, 637.
75T1783	A. Holm, L. Carlsen, S.-O. Lawesson, and H. Kolind-Andersen, *Tetrahedron*, 1975, **31**, 1783.
76JCS(P1)2540	A.G.W. Baxter and R.J. Stoodley, *J. Chem. Soc., Perkin Trans. 1*, 1976, 2540.
76PS185	S.M. Loosmore and D.M. McKinnon, *Phosphorus Sulfur*, 1976, **1**, 185.
76S190	M. Pergal, K. Popov-Pergal, and N. Kobilarov, *Synthesis*, 1976, 190.
76TL1137	A.I. Scott, S.E. Yoo, S.-K. Chung, and J.A. Lacadie, *Tetrahedron Lett.*, 1976, 1137.
77AJC1815	M. Davis, M.C. Dereani, J.L. McVicars, and I.J. Morris, *Aust. J. Chem.*, 1977, **30**, 1815.
77CPB3270	T. Matusuda, N. Honjo, M. Yamazaki, and Y. Goto, *Chem. Pharm. Bull.*, 1977, **25**, 3270.
77G289	B.F. Bonini, G. Maccagnani, G. Mazzanti, and B. Zwanenburg, *Gazz. Chim. Ital.*, 1977, **107**, 289.
77GEP2615621	*Ger. Pat.* 2,615,621 (1976). [*Chem. Abstr.*, 1977, **86**, 72628].
77HCA284	G. Vernin, S. Treppendahl, and J. Metzger, *Helv. Chim. Acta.*, 1977, **60**, 284.
77JA8078	D.A. McGowan, U. Jordis, D.K. Minster, and S.M. Hecht, *J. Am. Chem. Soc*,, 1977, **99**, 8078.
78JCS(P1)45	D.J. Humphreys, C.E. Newall, G.H. Phillips, and G.A. Smith, *J. Chem. Soc., Perkin Trans. 1*, 1978, 45.
78JCS(P1)1218	B.F. Bonini, G. Maccagnani, G. Mazzanti, L. Thijs, G.E. Veenstra, and B. Zwanenburg, *J. Chem. Soc. Perkin Trans. 1*, 1978, 1218.
78JOC1624	D.K. Minster, U. Jordis, D.L. Evans, and S.M. Hecht, *J. Org. Chem.*, 1978, **43**, 1624.
78JOC4816	A. Holm, L Carlsen, and E. Carlsen, *J. Org. Chem.*, 1978, **43**, 4816.
78S744	F. Chioccara, L. Oliva, and G. Prota, *Synthesis*, 1978, 744.

78TL2063	G.C. Barrett, L.A. Chowdhury, and A.A. Usmani, *Tetrahedron Lett.*, 1978, 2063.
78ZN(B)316	W. Winter, U. Plucken, and H. Meier, *Z. Naturforsch., B*, 1978, **33**, 316.
79AHC83	G.R. Newkome and A. Nayak, in 'Advances in Heterocyclic Chemistry', ed. A.R. Katritzky and A.J. Boulton, Academic Press, 1979, vol.25, p.83.
79BCJ3615	Y. Inagaki, R. Okazaki, and N. Inamoto, *Bull. Chem. Soc. Jpn.* 1979, **52**, 3615.
79IJC(B)377	R.P. Rao, S.R. Singh, N.B. Zaidi, and B. Singh, *Indian J. Chem., Sect. B*, 1979, **18**, 377.
79JOC497	D.L. Evans, D.K. Minster, U. Jordis, S.M. Hecht, A.L. Mazzu, and A.I. Meyers, *J. Org. Chem.*, 1979, **44**, 497.
79MI252	R. Coccia, C. Blarzino, C. Foppoli, and C. Cini, *Ital. J. Biochem.*, 1979, **28**, 252.
80BSB749	W.A. Nachtergaele and M.O. Anteunis, *Bull. Soc. Chim. Belg.*, 1980, **89**, 749.
80CL717	K. Hiroshi, T. Kazue, H. Kurumisawa, and Y. Tamura, *Chem. Lett.*, 1980, **6**, 717.
80JCS(P1)665	A.G.M. Barrett, D.H.R. Barton, and R. Colle, *J. Chem. Soc., Perkin Trans. 1*, 1980, 665.
80JOC1473	S. Kubota, Y. Ueda, K. Fujikane, K. Toyooka, and M.Shibuya, *J. Org. Chem.*, 1980, **45**, 1473.
81H1561	S. Karady, J.S. Amato, D. Dortmund, and L.M. Weinstock, *Heterocycles*, 1981, **16**, 1561.
81IJC(B)870	M.H. Bahar and B.K. Sabata, *Indian J. Chem, Sect. B*, 1981, **20**, 870.
81JAP81573	*Jap. Pat.*, 81,573 (1981) [*Chem. Abstr.*, 1981, **95**, 187238].
81JAP87574	*Jap. Pat.*, 87,574 (1981). [*Chem. Abstr.*, 1981, **95**, 187236].
81JHC633	M.T. Omar, A.E. El-Khamry, and F.A. Sherif, *J. Heterocycl. Chem.*, 1981, **18**, 633.
82CC901	S. Kubota, K. Toyooka, N. Yamamoto, M. Shibuya, and M. Kido, *J. Chem. Soc., Chem Commun.*, 1982, 901.
82JBI25	S Rambhav, *J. Biosci.*, 1982, **4**, 25.
82JPR873	A.M. Richter and V. Mojjoukhine, *J. Prakt. Chem.*, 1982, **324**, 873.
82MI718	I. Iovels and M.V. Shimanskaya, *Latv. PSR Zinat. Akad. Vestis, Khim. Ser.*, 1982, 718. [*Chem. Abstr.*, 1983, **98**, 125951].

83JAP5829775	Jap. Pat. 5829,775 (1983). [*Chem. Abstr.*, 1983, **99**, 70718].
83JCS(P1)967	S. Kubota, K. Toyooka, J. Ikeda, N. Yamamoto, and M. Shibuya, *J. Chem. Soc., Perkin Trans. 1*, 1983, 967.
83JCS(P1)2957	S. Kubota, K. Toyooka, H.K. Misra, M. Kawano, and M. Shibuya, *J. Chem. Soc., Perkin Trans. 1*, 1983, 2957.
84TL4767	T. Takata, K. Hoshino, E. Takeuchi, Y. Tamura, and W. Ando, *Tetrahedron Lett.*, 1984, **25**, 4767.
85T2133	T. Takata, Y. Tamura, and W. Ando, *Tetrahedron*, 1985, **41**, 2133.
85T2861	J.A.M. Hamersma and W.N. Speckamp, *Tetrahedron*, 1985, **41**, 2861.
86JCS(P1)1357	S. Kubota, K. Toyooka, M. Shibuya, and Z. Taira, *J. Chem. Soc., Perkin Trans. 1*, 1986, 1357.
86LA1787	W. Hanefeld and M.A. Jalili, *Liebigs Ann. Chem.*, 1986, 1787.
86TL3391	W. Ando and L. Huang, *Tetrahedron Lett.*, 1986, **27**, 3391.
87H1313	H. Kato, K. Tani, H. Kurumisawa, and Y. Tamura, *Heterocycles*, 1987, **26**, 1313.
89CC1134	P.J. Dunn, C.W. Rees, A.M.Z. Slawin, and D.J. Williams, *J. Chem. Soc., Chem. Commun.*, 1989, 1134.
90UP1	R.A. Aitken and D.P. Armstrong, in preparation.
90UP2	R.A. Aitken, D.P. Armstrong, and S.T.E. Mesher, in preparation.

CHAPTER 2
Three-Membered Ring Systems

ALBERT PADWA AND FREDERICK R. KINDER
Emory University, Atlanta, GA, USA

Three-ring heterocycles play an important role in many aspects of organic chemistry. These small membered rings are found in a significant number of natural products as well as in synthetic compounds of importance in biological studies. Aziridines and epoxides, in particular, have found growing utility in organic synthesis as precursors to stereodefined cyclic and acyclic molecules. In this chapter, selective reports from 1989 have been chosen which illustrate some of the recent developments in this field.

2.2 EPOXIDES

2.2.1 Preparation

The growing importance of epoxides as versatile synthetic intermediates has stimulated the development of more efficient methods for asymmetric epoxidations. Recent developments in the titanium-tartrate mediated asymmetric epoxidation of prochiral alcohols [80JA5976] have been aimed at gaining insight into the nature and composition of the catalyst [89JOC4016]. The titanium complexes derived from the linked-ligand **3** induce high levels of asymmetric induction, but do not catalyze efficient kinetic resolution of secondary allylic alcohols [89JOC2826].

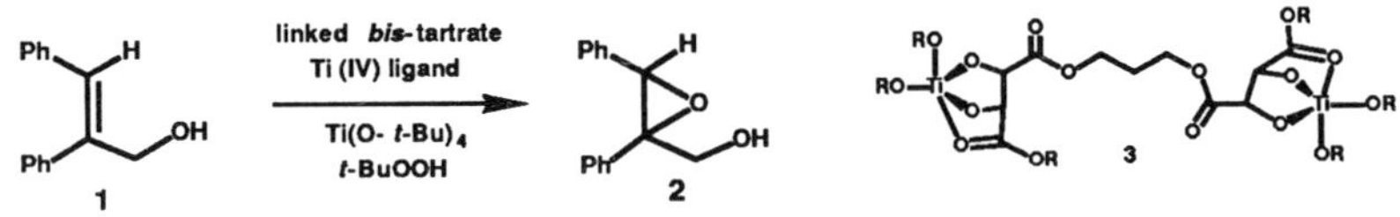

The reaction of arsonium ylides with carbonyl compounds has attracted a great deal of interest since the initial report by Henry and Wittig [60JA563]. Unstabilized arsonium ylides generally give rise to E-epoxides [81JA1283]. Several interesting reports on arsonium ylide epoxidations have appeared in 1989. Heteroatom-substituted ylides have been used for the homologation of aldehydes [89TL6023]. For example, phenylthiomethylene arsorane **5** gave epoxide **6** upon treatment with benzaldehyde. This material rearranged to **7** when subjected to silica gel.

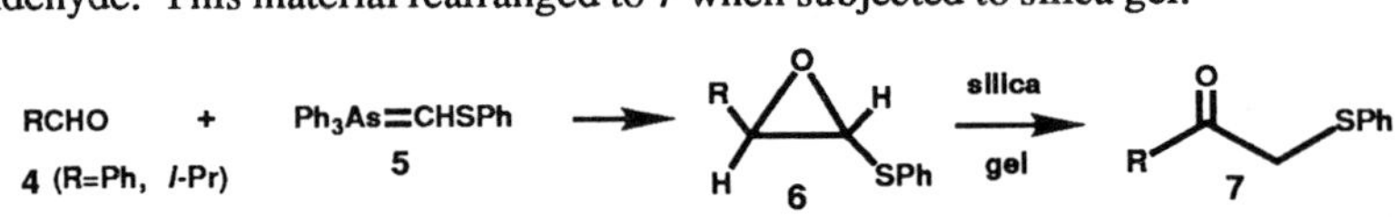

Amino acid-derived doubly protected α-amino aldehydes were treated with arsonium ylides to give the corresponding aminoalkyl substituted epoxides with high diastereofacial selectivity [89TL5425]. The epoxides were further utilized as chiral building blocks.

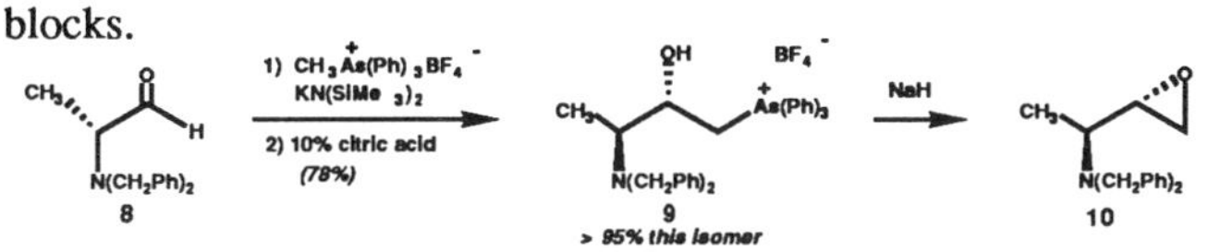

(+)-Hepoxilin A_3, a biologically active metabolite of arachidonic acid, was prepared by a convergent strategy that featured an arsonium ylide condensation to produce a *trans*-epoxide in the key synthetic step [89TL2545]. Semistabilized arsonium ylides are known to give mixtures of olefins and epoxides [65CI(L)1726]. In a recent study, allyltriphenylarsonium tetrafluoroborate was treated with aldehydes. In this case, the selectivity for the formation of either the epoxide or olefin was found to be dependent upon the choice of base used for the generation of the arsonium ylide [89JOC3229].

The synthetic utility of sulphur ylides to prepare epoxides is well established [75MI1]. During this past year, optically active epoxides were prepared utilizing chiral sulphides [89JOC4222]. Thus, the camphorsulphonic acid-derived sulphide **13** was treated with benzyl bromide in the presence of potassium hydroxide and benzaldehyde under phase transfer conditions. This resulted in a quantitative yield of **12** with a modest degree of enantioselectivity.

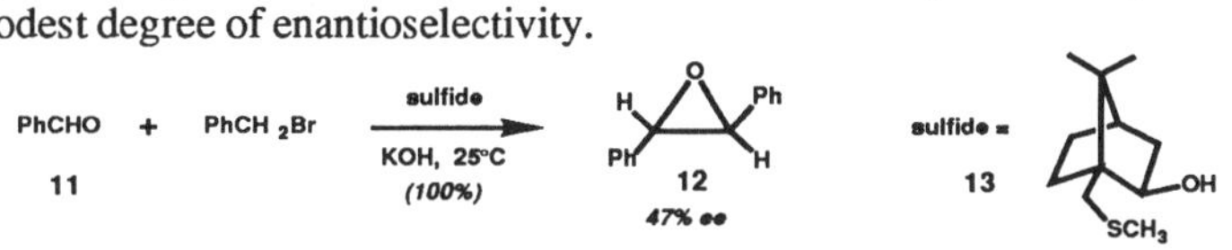

Optically active sulphinyl groups have been used to prepare chiral epoxides [89CPB184]. For example, the addition of the anion derived from optically pure chloromethyl sulphide **14** to aldehydes and ketones afforded chlorohydrins with 1,2-asymmetric induction in good yields [89JOC3130]. Treatment of these chlorohydrins with base gave optically active α,β-epoxy sulphoxides, which may be stereospecifically desulphinylated by n-butyllithium in THF.

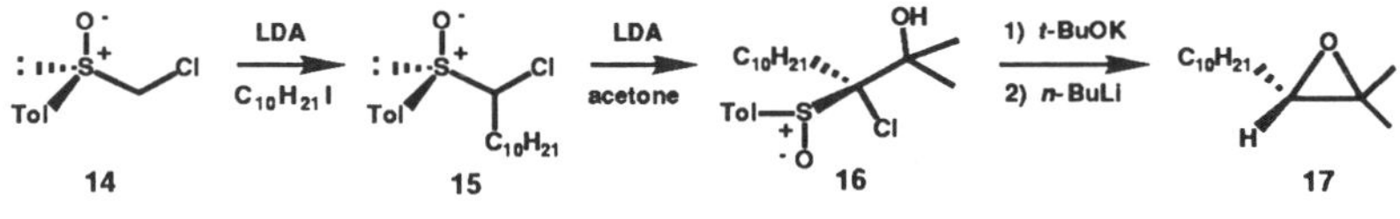

Dioxiranes have been used as mild epoxidizing agents which react readily at low temperatures [86JA2470]. For unstable epoxides, such as those derived from enol ethers, dioxirane epoxidation is the method of choice. Several reports during the year have exploited this reaction in order to obtain the previously unisolable acyloxo, alkoxo and silyloxooxiranes. An example of this approach involves the preparation of epoxide **19** from enol lactone **18** via the dimethyldioxirane oxidation.

Dioxirane epoxidation of silyl enol ethers afforded the corresponding epoxides in excellent yields. These substrates readily undergo an acid-catalyzed rearrange-

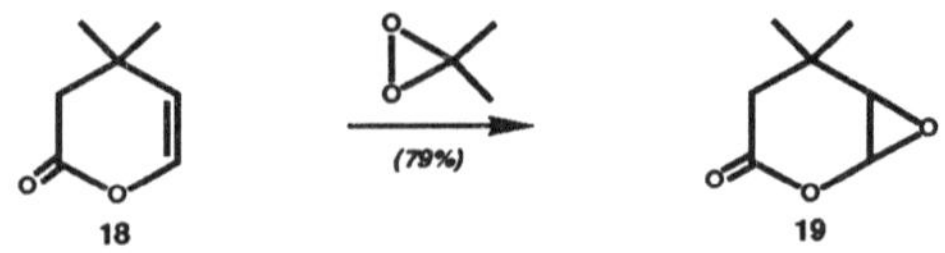

ment to α-trimethylsiloxy carbonyl derivatives [89TL6497] [89JOC4249]. Danishefsky described the use of dimethyldioxirane for the direct epoxidation of glycals. The 1,2-anhydrosugars produced were employed in the stereospecific construction of β-linked oligosaccharides [89JA6661].

The photooxidation of olefins in the presence of transition-metal complexes derived from titanium has been used for an efficient "one-pot" synthesis of epoxy alcohols. Use of singlet oxygen transformed olefin **20** into its allylic hydroperoxide. This material was subsequently converted via a transition-metal-catalyzed oxygen transfer reaction to epoxy alcohol **21** [89JA203].

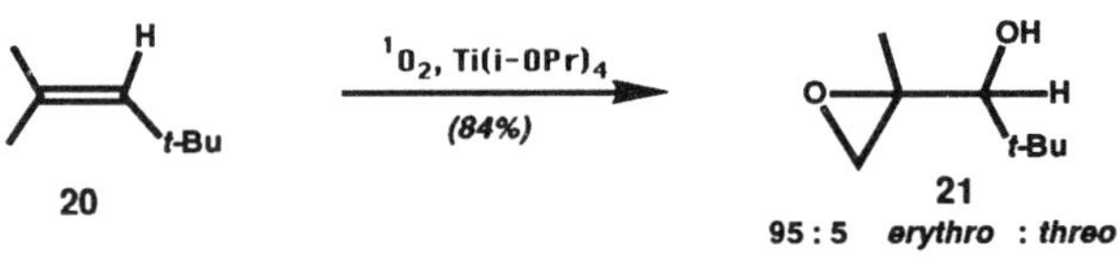

Metalloporphyrin-catalysed epoxidation of olefins has been used to mimic the activity of the cytochrome P-450 enzyme. Manganese porphyrins catalyzed the epoxidation of various alkenes by magnesium monoperoxyphthalate. Alkene conversion was complete within several minutes at low temperatures and high yields of epoxides (60-93%) were obtained [89CC889]. Oxygen transfer reactions from 2,6-disubstituted pyridine *N*-oxides to olefins were efficiently catalyzed by ruthenium porphyrins under mild conditions. A variety of olefins were converted into the corresponding epoxides in high yield. For example, styrene was converted to styrene oxide in 100% yield via oxidation with 2,6-dichloropyridine *N*-oxide and a catalytic amount of ruthenium porphyrin [89TL6545]. Chiral porphyrin-metal epoxidation catalysts have also been investigated this past year. When manganese (III) was incorporated into a chiral phorphyrin ligand, extremely high catalytic efficiency and moderate enantioselectivity was observed with a variety of aryl olefins [89JA9116].

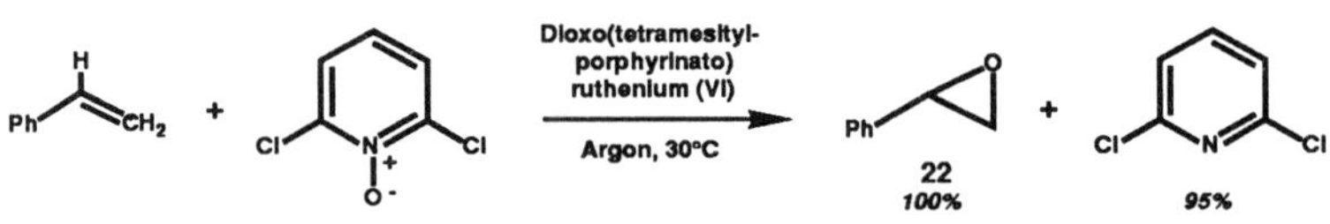

The epoxidation of α,β-unsaturated ketones using dialkylzinc-oxygen reagents was also described in 1989 [89CL1149]. Autooxidation of diethylzinc occurred prior to attack by the diethylzinc on the enone chromophore. The resulting alkylperoxozinc species acts as an efficient epoxidizing reagent.

2.2.2 Reactions of Epoxides

2.2.2.1 Ring Opening-Anion Induced

Epoxides are widely recognized as extremely versatile synthetic intermediates. The enhanced reactivity of these species, attributable to a high degree of ring strain, permits a range of nucleophilic ring openings, Lewis acid catalyzed rearrangements and isomerization reactions. The reaction of vinyl epoxides with isocyanates [88JA7933] provides a facile entry to oxazolidin-2-ones. Treatment of either vinyl epoxide **23** or **26** with aryl isocyanate **24** in the presence of Pd(0) afforded the thermodynamically less stable *cis*-4,5-disubstituted oxazolidin-2-one **25** [89TL3893]. It was suggested that the nitrogen atom is initially coordinated to the palladium in the intermediate π-allyl complex and that this species then collapses to the product.

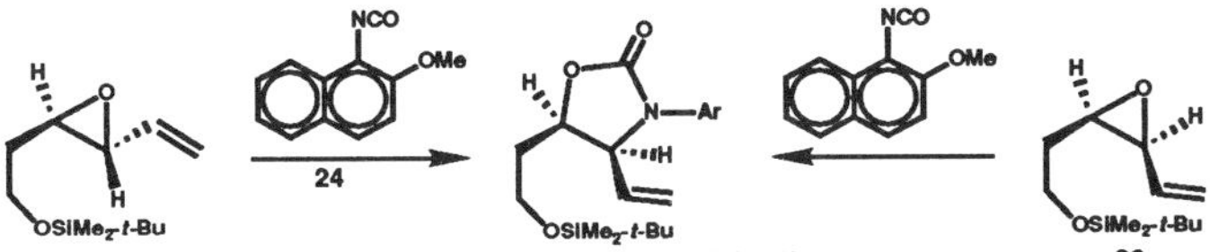

23 25 26

Epoxides undergo ring opening with organometallic nucleophiles to give larger oxygen heterocycles. For example, the addition of 2-(phenoxymethyl)-2-propenylmagnesium chloride (**27**) to epoxide **28** afforded 3-methylenetetrahydropyran **30** in high yield [89TL4863].

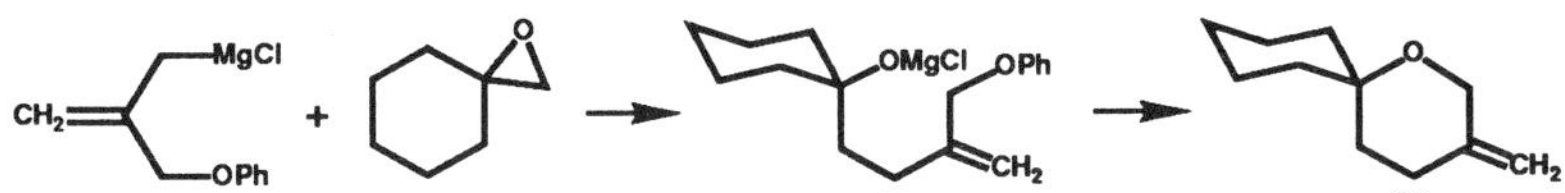

27 28 29 30

Regiospecific nucleophilic substitution reactions involving attack of the nitrogen atom of oximes on epoxides were used to generate nitrones which were then trapped in 1,3-dipolar cycloaddition reactions. Four variations of this process were investigated, whereby epoxide opening was either intermolecular or intramolecular. For the case where both processes were intramolecular, epoxide **31** produced tricyclic isoxazolidine **32** upon heating [89TL5489].

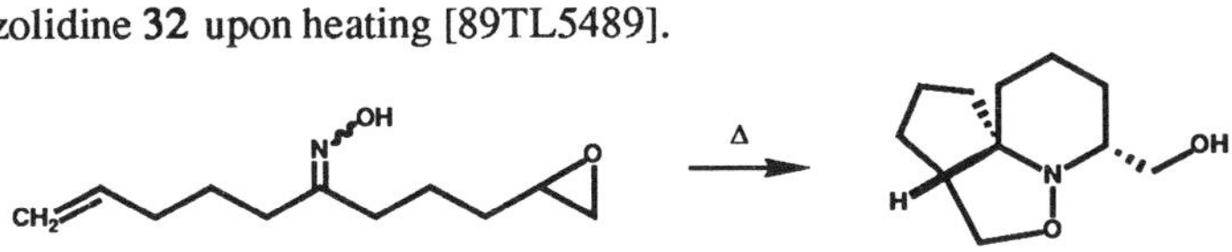

31 32

The lithiation of substituted tosylated epoxides (i.e. **33**) derived from allylic sulphones with methyllithium led to lithium (*E*)-3-lithio-3-tosyl-2-propenolates in a stereoselective manner. The further reaction of these intermediates with different electrophiles (such as methyl iodide) in the presence of a copper (I) salt afforded the functionalized tosylated allylic alcohol in a regio- and stereoselective manner [89JOC1491].

33 34 35

Organometallic addition reactions have long been exploited for the stereospecific transformations of epoxides to β-alkyl alcohols. Recent studies by Flippen and coworkers have dealt with substituent-directed alkyl addition reactions of epoxides. High regioselectivity was observed when a trialkyl aluminum-*n*-butyllithium mixture was reacted with epoxides. The major product corresponded to β-alkyl alcohols which contained remote oxygen functional groups [89JOC3588].

Epoxide ring opening by fluoride ion to give fluorohydrins has generated interest due to the importance of organic fluorides in biologically active compounds [87T3123]. The traditional anhydrous hydrogen fluoride method for the preparation of the fluorohydrins has been improved upon through the use of a hydrogen fluoride salt and porous aluminum fluoride as a solid reagent. These reagents, when sonicated in the presence of simple aliphatic epoxides, gave rise to the corresponding fluorohydrins in good yield [89CC1848].

2.2.2.1.2 Lewis Acid Promoted Ring Openings

Nicolaou and coworkers have described a well-defined and predictable route to tetrahydrofurans and tetrahydropyrans [89JA5330]. The method relies on the stereo- and regioselective opening of hydroxy epoxides by acid catalysis. The presence of a saturated chain at the remote secondary epoxide position led to tetrahydrofuran systems (5-*exo* epoxide opening). Placement of an electron-rich double bond at that position led to tetrahydropyran formation (6-*endo* epoxide opening).

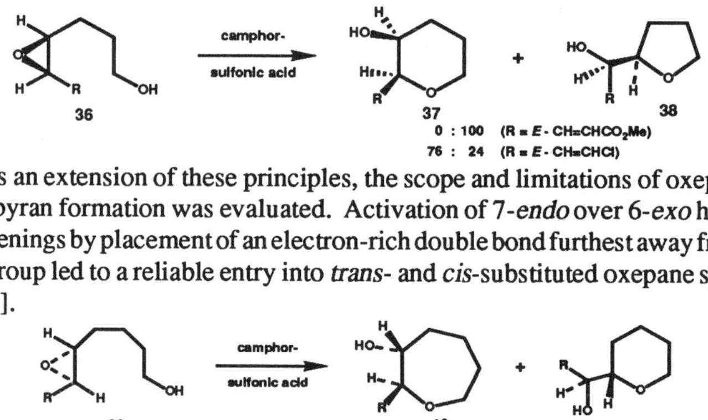

36 → 37 + 38

0 : 100 (R = *E*-CH=CHCO$_2$Me)
76 : 24 (R = *E*-CH=CHCl)

As an extension of these principles, the scope and limitations of oxepane *vs* tetrahydropyran formation was evaluated. Activation of 7-*endo* over 6-*exo* hydroxy epoxide openings by placement of an electron-rich double bond furthest away from the hydroxyl group led to a reliable entry into *trans*- and *cis*-substituted oxepane systems [89JA5335].

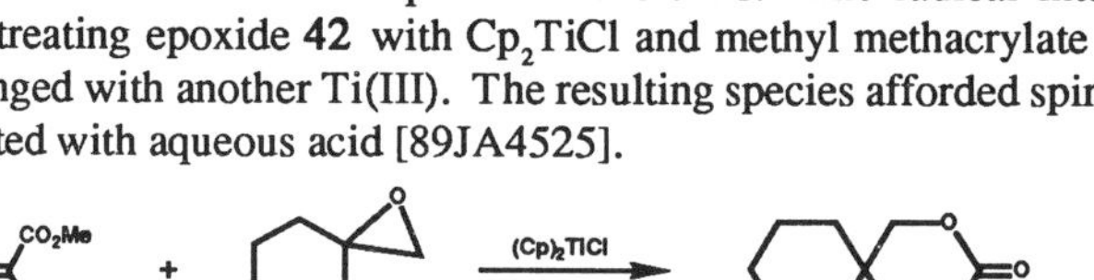

39 → 40 + 41

0 : 100 (R = CH$_2$CH$_2$CO$_2$Me)
92 : 8 (R = *E*-CH=CHCl)

The titanium (III) mediated epoxide ring opening has recently been established as an efficient means to trap activated olefins. The radical intermediate produced by treating epoxide **42** with Cp$_2$TiCl and methyl methacrylate was efficiently scavenged with another Ti(III). The resulting species afforded spiro-lactone **43** when treated with aqueous acid [89JA4525].

42 → 43

Molander and Andrews have described the preparation of functionalized oxepanes via the Lewis acid-directed regioselective intramolecular addition of allylstannanes (or allylsilanes) to 2,3-epoxy ethers. Excellent regiocontrol and diastereoselectivity was achieved by titanium (IV) chloride promoted cyclization of substituted epoxy ethers [89JOC3114]. Cyclizations occur in high yields, providing direct routes to functionalized oxepanes.

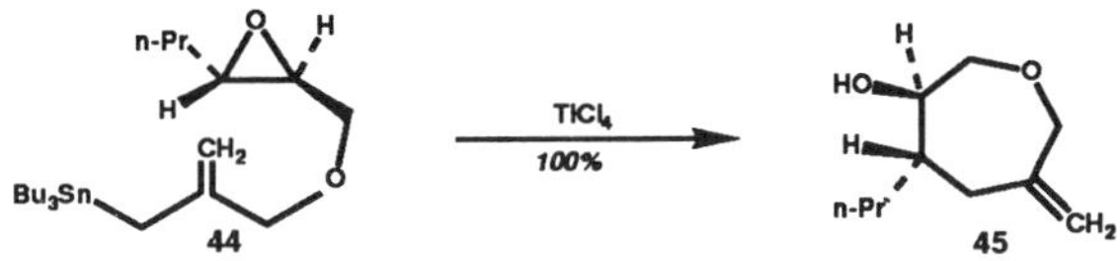

2.2.2.2 Rearrangements of Epoxides

A recent synthesis of an oxygenated derivative of the plant hormone gibberellin, utilizes an epoxide rearrangement as a key synthetic step [89TL6899]. Epoxide **46** was treated with three equivalents of a Ti(IV) amide which induced a Lewis acid-catalyzed rearrangement to the hydroxyolefin **47**. These titanium reagents may have considerable value as Lewis acid rearrangement promoters in gibberellin chemistry.

A new approach to the synthesis of optically active β-hydroxy aldehydes via the organoaluminium-promoted rearrangement of epoxy silyl ethers has been described. α-Silyloxy epoxides such as **48** and **51** were treated with 2 equivalents of methylaluminium *bis*-(4-bromo-2,6-di-*tert*-butylphenoxide) at -78°C producing the corresponding β-silyl aldehydes in good yield with high optical activity [89JA6431]. The stereochemistry at the migrating siloxy carbon is rigorously retained in the rearrangement.

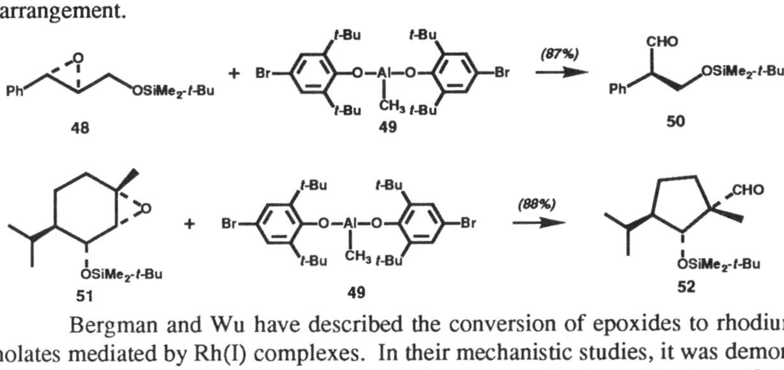

Bergman and Wu have described the conversion of epoxides to rhodium enolates mediated by Rh(I) complexes. In their mechanistic studies, it was demonstrated that intermediate **54** was generated photochemically from dihydride **53** by oxidative addition of rhodium into the three-membered ring C-H bond. The epoxyrhodium complex then rearranged to enolate **55** by a 1,2-hydrogen shift [89JA7628].

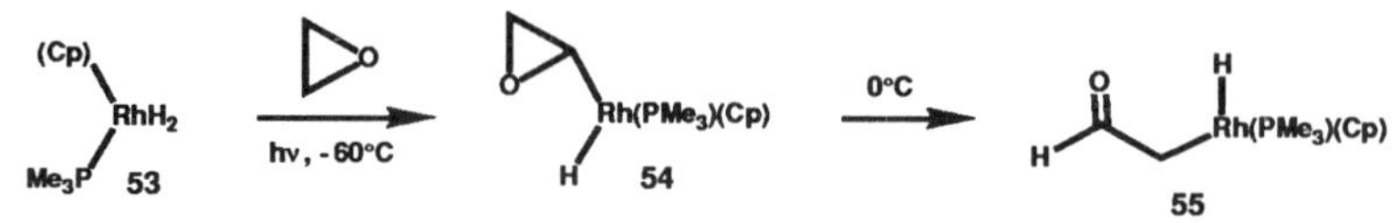

An efficient general method for the construction of optically active 3-hydroxyalkynes has been developed using chiral α,β-epoxy alcohols readily accessible from non-chiral allyl alcohol precursors. For example, epoxychloride **56**, prepared via Sharpless epoxidation of the allylic alcohol precursor, was treated with n-butyllithium. The initial reaction involves epoxide ring opening via a β-elimination to produce a vinyl chloride intermediate. A subsequent elimination gave the optically active alkyne **57** [89CC1344].

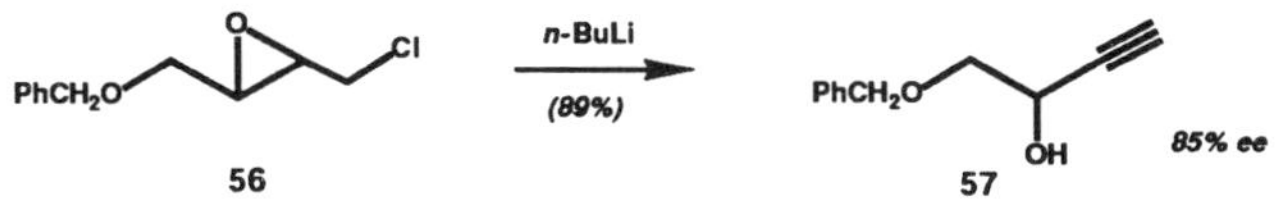

2.2.2.3 Oxiranyl Anions

Numerous reports implicating metalated oxiranes intermediates have been recorded [83OR345]. During the past year, Molander and Mautner have described an improved synthesis of trimethylsilyl-substituted oxiranyl anions and their reactions with aldehydes to give the corresponding epoxy alcohols. Although low diastereoselectivity with alkyl aldehydes was observed, good to excellent diastereofacial addition to ketones and conjugated aldehydes was achieved [89JOC4042]. Most of the anions exhibited strict retention of configuration at the metalated carbon stereocenter.

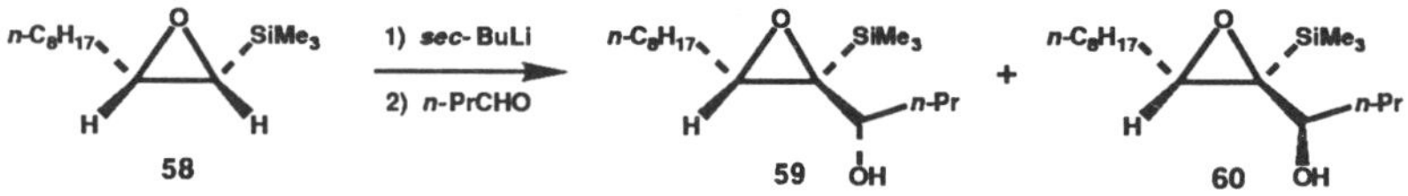

Oxiranyl phenyl sulphones were also reported to function as acyl anion equivalents. Epichlorohydrin was readily converted to phenylsulphonyl epoxide **61**. Deprotonation of the α-sulphonyl hydrogen followed by addition of an electrophile afforded good yields of the substituted oxiranes with retention of configuration. The oxiranyl anions derived from *cis*-epoxides were much more reactive than the anions generated from the corresponding *trans*-isomers. Treatment of the sulfonyl substituted epoxide **62** with magnesium bromide gave good yields of the corresponding bromoketones [89JCS(P1)835].

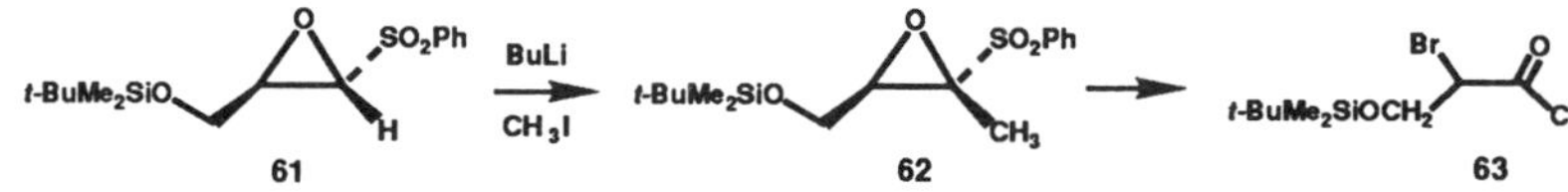

2.2.2.4 Reductions of Epoxides

Selective hydrogenolysis of alkenyloxiranes to give homoallylic alcohols was carried out using formic acid in the presence of a palladium-phosphine catalyst

[89JA6280]. The reaction proceeded stereoselectively, because the hydride derived from formic acid attacked the allyl groups intramolecularly from the palladium side of the π-allylpalladium hydride intermediate. The stereoselectivity of hydride attack, which induces the ring opening of alkenyloxiranes, can be controlled by the olefin geometry.

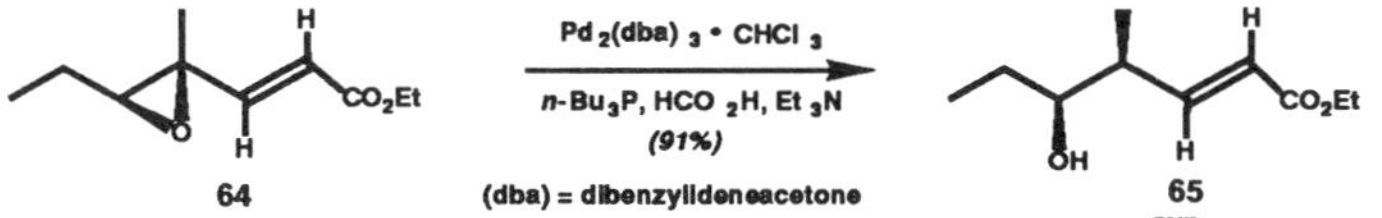

64 65 *syn*

Other investigations involving the reduction of epoxides during the past year include the deoxygenation reaction of epoxides to olefins using the dianion complex $[Fe_4S_4(SPh)_4]^{2-}$ in the presence of sodium borohydride. The reduced form $[FeS_4(SPh)_4]^{3-}$ was suggested as the active reductant in this reaction [89TL6387].

2.3 DIOXIRANES

Aside from the dioxirane epoxidation of olefins described in Section 3.2.1.4, several other dioxirane-mediated oxidative processes were also reported. Oxidation of prochiral sulphides by dioxiranes in buffered aqueous solutions, using bovine serum albumin as a chiral auxiliary, afforded the corresponding sulphoxides in up to 89% enantiomeric excess [89TL6233].

2.4 AZIRIDINES

2.4.1 Preparation of Aziridines

The growing importance of functionalized aziridines in organic synthesis and their presence in bioactive molecules continues to provide strong impetus for the development of new synthetic methods for the preparation of these heterocycles. Optically active aziridine-2-carboxylic acids have been synthesized from chiral glycidic esters [89TL4881]. Allylic alcohols were subjected to the Sharpless epoxidation and the resulting hydroxyperoxide was oxidized to the acid followed by esterification. These glycidic esters were treated with sodium azide to give a mixture of isomeric azido alcohols. In a subsequent reaction with triphenyl phosphine, the azido alcohols were converted into aziridine-2-carboxylic esters of high optical purity in good yields.

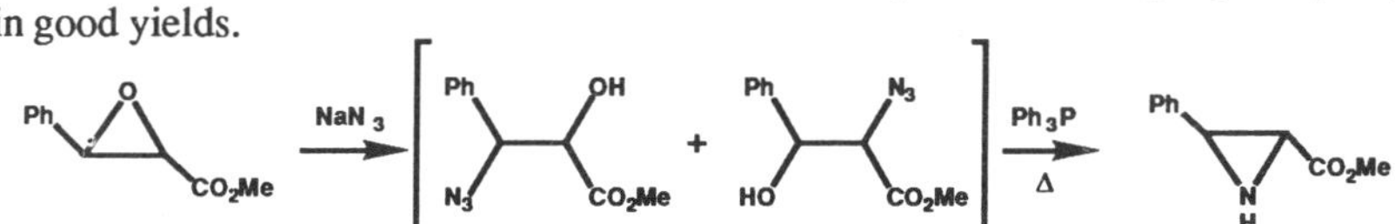

66 67 68 69

Acetoxyaminoquinazolone **71** has been reported to function as an effective aziridinating agent for alkenes [87CC1362]. Atkinson and coworkers had previously investigated the similarities of aziridination of allylic alcohols using **71** and the epoxidation of allylic alcohols using peracids [88CC624]. During the past year, this group demonstrated that cyclohex-3-en-1-ol (**70**) underwent stereospecific aziridination with **71** as compared to the corresponding peracid epoxidation which proceeded with poor stereoselectivity. Aziridination of the acetate of **70**, on the other hand, afforded a mixture of two isomeric acetates with little stereoselectivity and in low yield [89CC562].

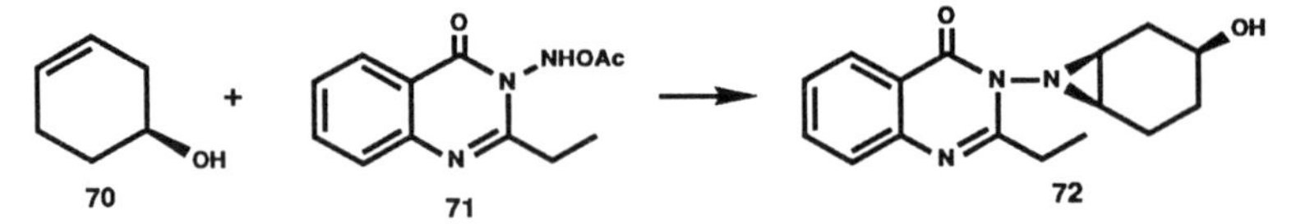

Sharpless and coworkers have recently reported the synthesis of homochiral N-substituted aziridines from homochiral 1,2-cyclic sulphates [89TL2623]. Diol **73**, prepared from the catalytic asymmetric dihydroxylation of the olefin was converted into cyclic sulphate **74** in two steps [89JA1123]. Treatment of **74** with an excess of primary amine afforded β-aminosulphate **75** which, following deprotonation by n-butyllithium cyclized to the aziridine **76** in excellent yield and in high enantiomeric excess.

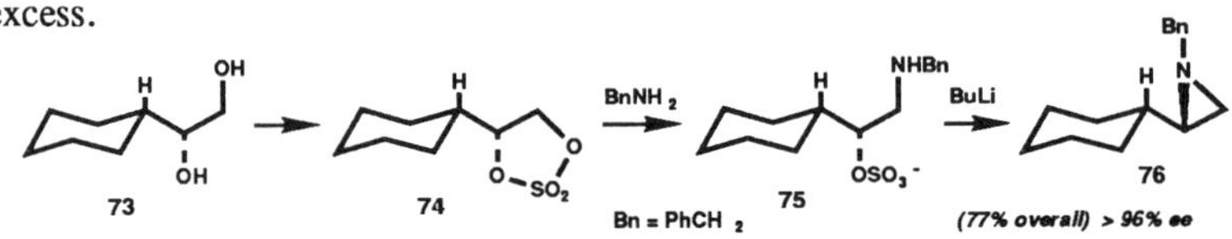

An unusual *N*-acylation of glycine Schiff bases has recently been described. Treatment of azaallyl anions, generated from glycine Schiff bases such as **77**, with acyl halides furnished *N*-acylaziridines via an intramolecular cyclization [89TL4717]. The results were rationalized in terms of a AR_1-5 transition state for the cyclization.

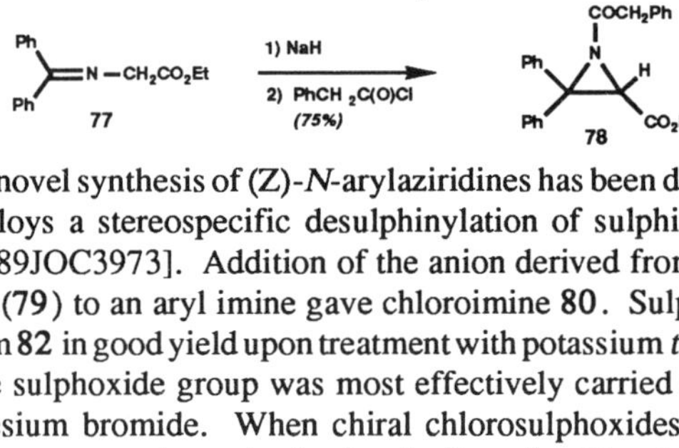

A novel synthesis of (Z)-*N*-arylaziridines has been described by Yamakawa which employs a stereospecific desulphinylation of sulphinylaziridines using alkylmetals [89JOC3973]. Addition of the anion derived from 1-chloromethyl tosyl sulphoxide (**79**) to an aryl imine gave chloroimine **80**. Sulphinylaziridine **81** was formed from **82** in good yield upon treatment with potassium *tert*-butoxide. Displacement of the sulphoxide group was most effectively carried out using an excess of ethylmagnesium bromide. When chiral chlorosulphoxides were used as starting materials, an asymmetric synthesis of optically active *N*-aryl-aziridines was realized.

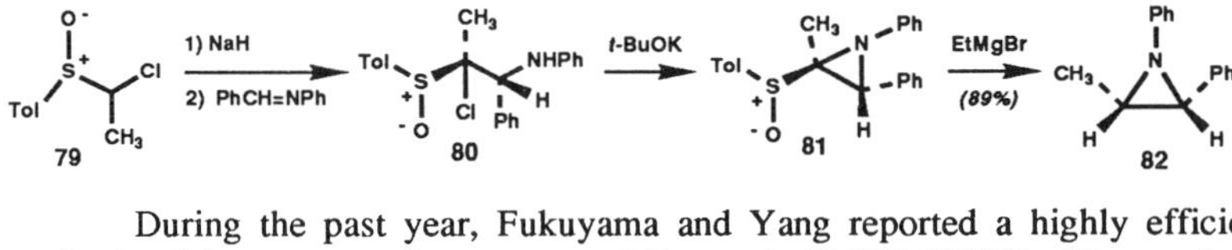

During the past year, Fukuyama and Yang reported a highly efficient synthesis of the potent antitumor agent, Mitomycin C [89JA8303]. The synthesis featured an intramolecular azide-mediated aziridination to give intermediate **84**. Following a series of synthetic reactions that ultimately afforded aziridine **85**, the key step involved a remarkable rearrangement which was induced by treatment of **85** with methanolic ammonia giving the natural product **86** in 10% overall yield from commercially available 2,6-dimethoxytoluene.

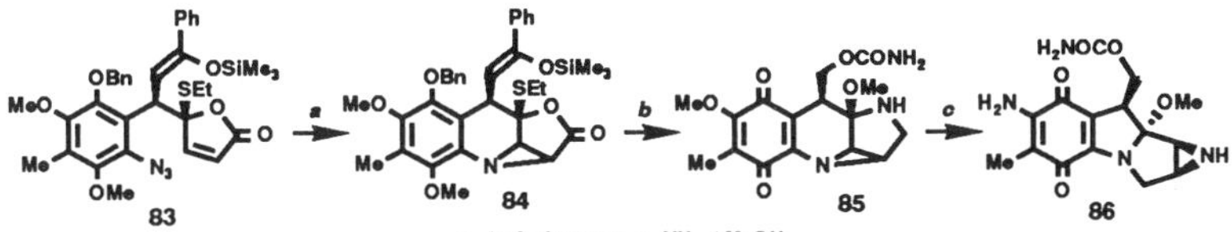

a, Δ; *b*, 9 steps; *c*, NH_3 / MeOH.

The functionalization of olefins using alkoximidoylnitrenes has been reported by the Lwowski group. (*N*-Cyano- or *N*-(methylsulphonyl)alkoxycarbimidoyl)nitrenes, generated *in situ* from the corresponding azides by 300 nm UV light, convert a variety of olefins cleanly and sterospecifically to the corresponding aziridines. These can readily be hydrolyzed to *N*-unsubstituted aziridines or ring-opened to allylic isoureas [89JOC3945].

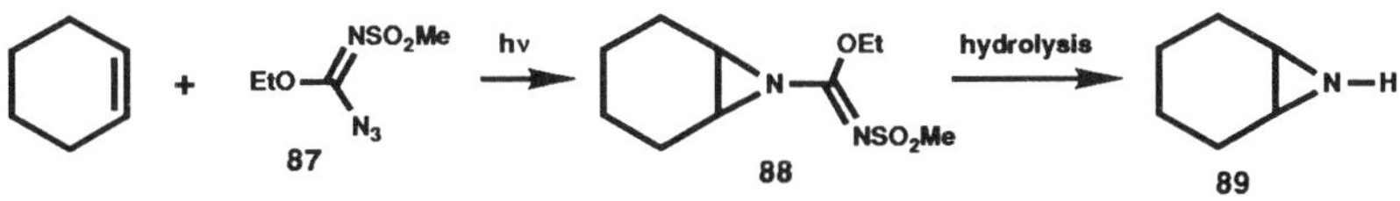

2.4.2 REACTION OF AZIRIDINES

A number of reports appeared during 1989 describing the ring opening reactions of aziridines as a method for preparing acyclic and cyclic amino compounds. Baldwin and coworkers have described a new entry to amino acids employing the ring opening reaction of aziridine carboxylate esters with organocuprate reagents [89CC1852]. The potential of the method was exemplified by the conversion of aziridine **90** with methyl cuprate to the N-protected β-amino acid **91**.

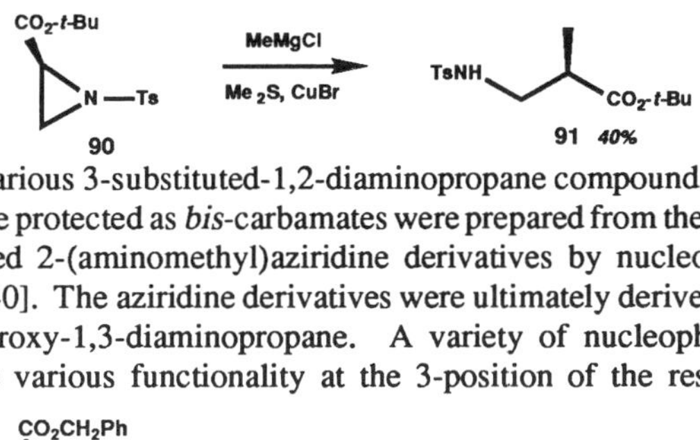

Various 3-substituted-1,2-diaminopropane compounds in which the amino groups were protected as *bis*-carbamates were prepared from the corresponding *N,N*-*bis*-protected 2-(aminomethyl)aziridine derivatives by nucleophilic ring opening [89JOC1940]. The aziridine derivatives were ultimately derived from readily available 2-hydroxy-1,3-diaminopropane. A variety of nucleophiles were added to incorporate various functionality at the 3-position of the resulting 1,2-diaminopropanes.

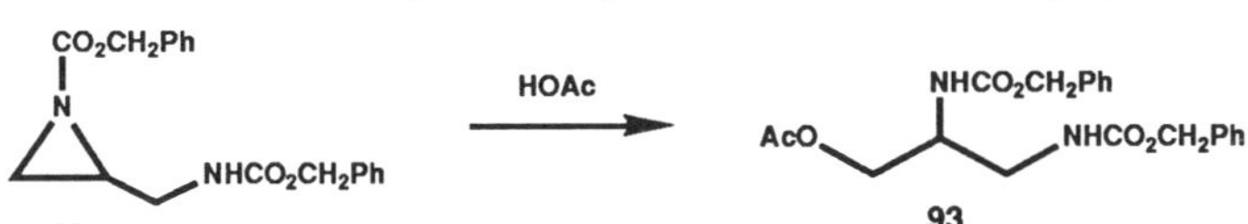

A series of optically active β-lactams were prepared by the enantio and stereospecific rhodium (I)-catalyzed carbonylation and ring expansion of racemic aziridines in the presence of *d* or *l*-menthol as an added chiral agent [89JA931]. For example, when *N*-*tert*-butyl-2-phenylaziridine **94** was exposed to carbon monoxide and a catalytic quantity of chlorodicarbonylrhodium (I) dimer, monocyclic β-lactam **95** was formed in quantitative yield. The process is both stereo and enantiospecific, occurring with retention of configuration.

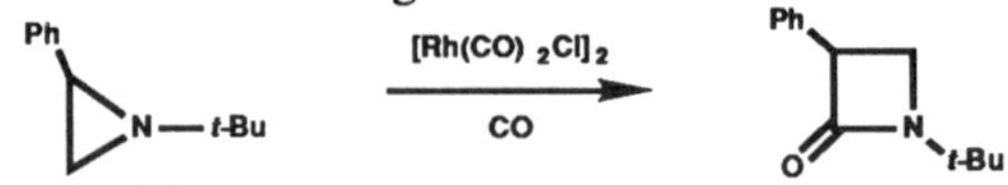

94 95

The reactivity of various *N*-protected *bis*-aziridines toward several carbon and heteronucleophiles has been the focus of a recent investigation [89JOC5324]. One interesting example involves the initial opening of one of the aziridine rings of the optically active *bis*-aziridine **96** with ammonium fluoride. The resulting amine reacts with the second aziridine ring in an intramolecular fashion to give piperidine **97**. Reaction orientation toward *bis*-opening is influenced by the nature of the N-protecting group, the nucleophile, and by Lewis acids.

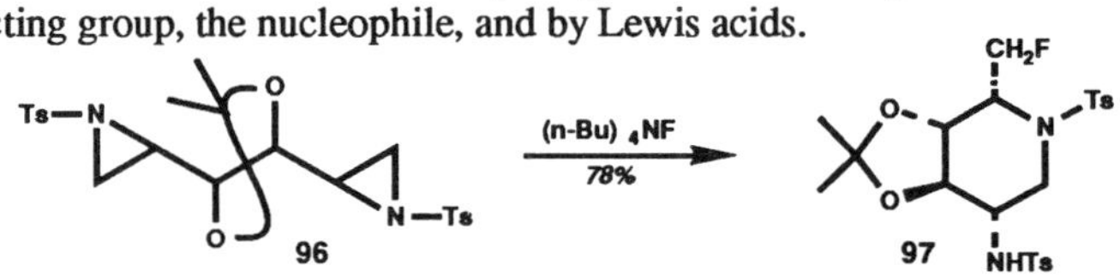

96 97

2.5 AZIRINES

A new method for the conversion of aziridines to azirines has been described. Desilylation of aziridine **99** by treatment with cesium fluoride gave azirine **100** in excellent yield [89TL2703]. The aziridinyl carbanion is long-lived enough to be captured by benzaldehyde, but in the absence of an electrophile, forms the azirine by expulsion of the quinazolone anion. Similar work with stannyl-substituted aziridines has been reported [89CC836].

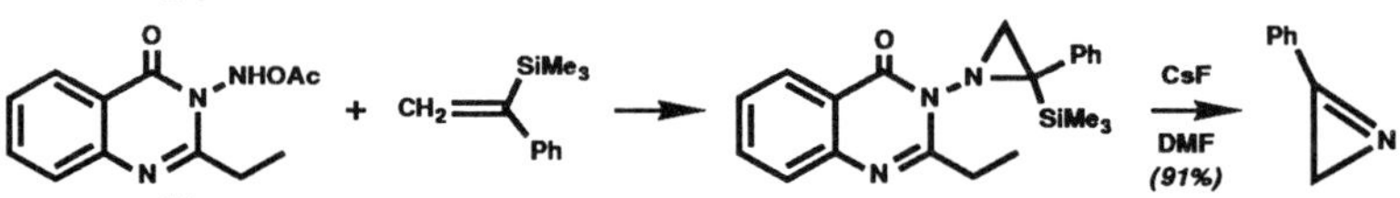

71 98 99 100

2.6 DIAZIRINES

2.6.1 Preparation

Aryltrifluoromethyldiazirines were prepared via Swern oxidation of the corresponding diaziridine precursors. The usual oxidation methods for converting diaziridines to diazirines (i.e. silver oxide or positive halogens) failed for these systems [89JCS(P1)1172]. The use of DMSO-oxalyl chloride represents a useful, mild, inexpensive procedure for the synthesis of diazirines and one that should find widespread use in photoaffinity labelling.

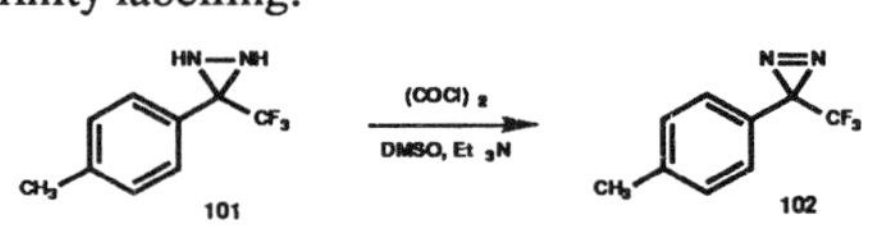

101 102

2.6.2 Reactions

Doyle and coworkers have reported the use of diazirines as stable diazoalkane precursors which, when heated in the presence of a rhodium (II) catalyst, undergo carbenoid-type reactions (i.e. C-H insertions and 1,2-hydride shifts) [89TL3049]. Minimal competition was observed from processes that were dominant when the diazo compound was used directly or when the diazirine was decomposed in the absence of catalyst.

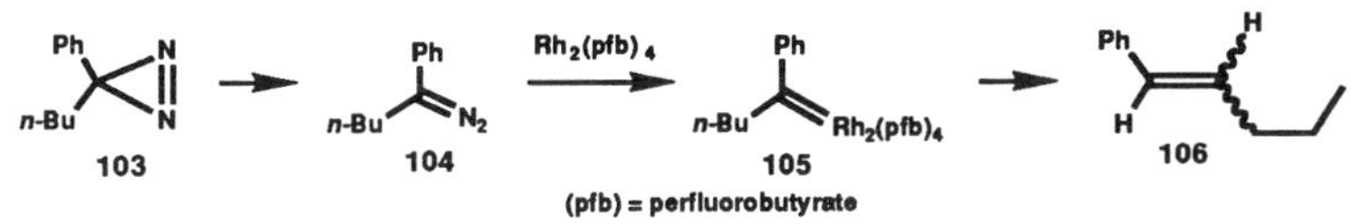

A general and convenient method for the preparation of vinyl anions in a flowing afterglow apparatus was reported. Vinyl anions were generated from diazirines using hydroxide ion in the gas phase. The reactivity of the vinyl anion produced was studied via mass spectrometry in the presence of nitrous oxide [89TL3045].

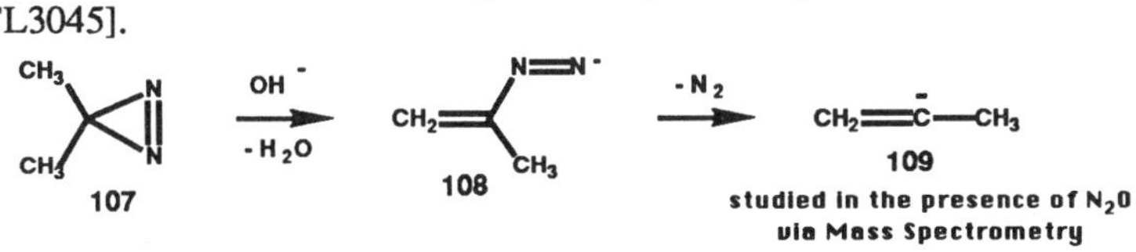

The mechanism of the diazirine exchange reaction with nucleophiles has been the focus of a recent study by Dailey and Bainbridge [89TL4901]. Treatment of isotopically labeled bromophenyldiazirine containing one ^{15}N atom with azide ion produces benzonitrile which contains 50% ^{15}N. This suggests that N-azido-diazirines are intermediates in the reaction. The transformation of **110** into **112** was suggested to proceed via a double SN2' pathway. The N-azidodiazirine was calculated to decompose directly to the nitrile and two molecules of nitrogen with a very small energy barrier.

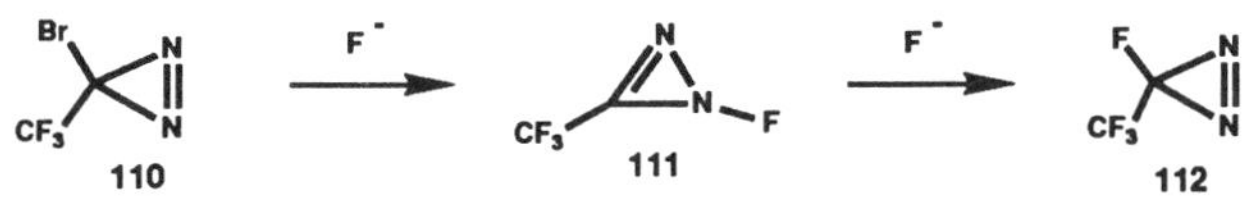

2.7 OXAZIRIDINES

The Davis oxaziridine continues to be used as an oxygen transfer reagent for the enantiospecific α-hydroxylation of carbonyl compounds [84JA3243]. Several natural product syntheses appeared in 1989 that relied upon this oxidation method. For example, during the course of the enantioselective synthesis of the antibiotic Macbecin I, Baker and Castro found that the enantioselective hydroxylation of the enolate anion of optically active oxazolidone **113** with the Davis reagent gave α-hydroxyamide **114** in high yield with excellent enantiomeric selectivity [89JCS(P1)190].

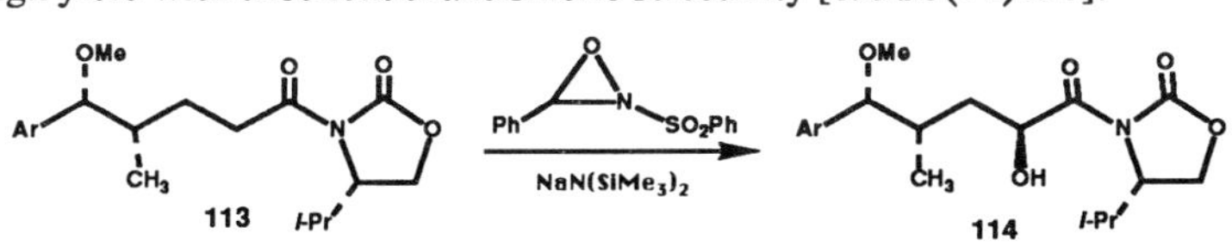

An additional example using oxaziridines to generate α-hydroxy carbonyls was reported by Smith [89JA8039] in the synthesis of the antibiotic (-)-Echinosporin. A related oxidation was used by Taschner [89TL1029] in connection with the total synthesis of Tirandamycin.

2.8 EPISULPHIDES

Huisgen and Langhals have recently reported on the use of thiones as super dipolarophiles [89TL5369]. According to kinetic measurements, the 1,3-dipolar cycloadditions of diphenyldiazomethane to thioketones proceed much faster than those to α,β-unsaturated carbonyl compounds and nitriles previously regarded as extremely reactive dipolarophiles. The primary cycloadducts are tetrasubstituted 1,3,4-thiadiazolines which extrude nitrogen and furnish episulphides via thiocarbonyl ylide intermediates.

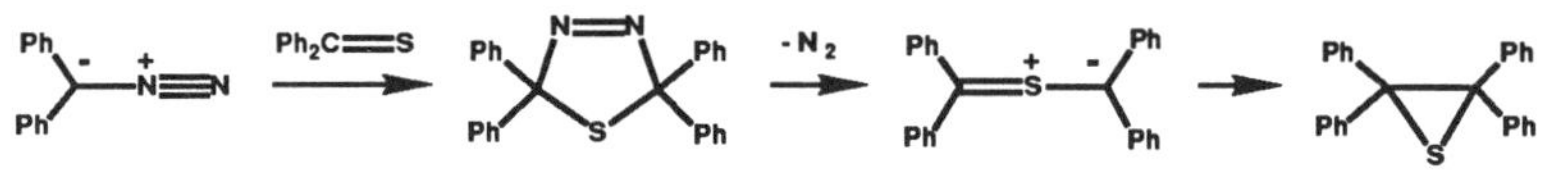

115 116 117 118

The first reported example of a photochemical [3+2]-cycloaddition which utilized 2,2,3,3-tetraarylthiiranes and tetracyanoethylene (TCNE) was described in 1989 [89CC557]. Episulphide **120**, when irradiated in the presence of TCNE, gave rise to tetrahydrothiophene **119**. This reaction was proposed to proceed via 1,3-dipolar cycloaddition of a thiocarbonyl ylide intermediate. Similar treatment of **120** with fumaronitrile resulted only in the formation of olefin **121** .

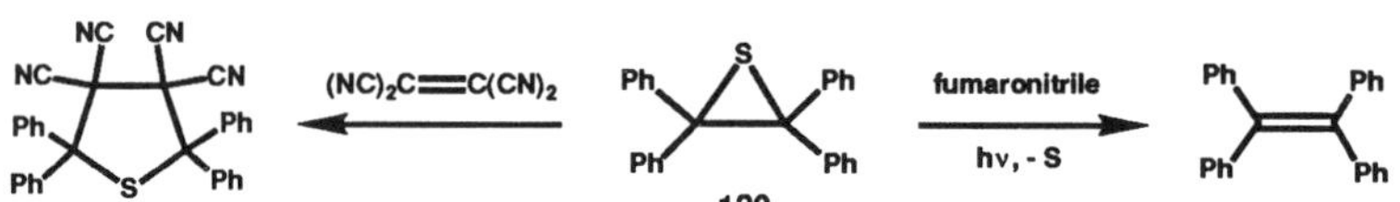

119 120 121

2.9 References

60JA563 M. C. Henry and G. Wittig; *J. Am. Chem. Soc.*, 1960, **82**, 563.

65CI(L)1726 A. W. Johnson and J. O. Martin; *Chem. Ind. (London)*, 1965, 1726.

75MI1 B. M. Trost and L. S. Melvin in "Sulfur Ylides", Academic Press, New York, 1975

80JA5976 T. Katsuki and K. B. Sharpless; *J. Am. Chem. Soc.*, 1980, **102**, 5976.

81JA1283 W. C. Still and V. J. Novack; *J. Am. Chem. Soc.*, 1981, **103**, 1283.

83OR345 J. K. Crandall and M. Apparu; *Org. React.*, 1983, **29**, 345.

84JA3243 F. A. Davis, L. C. Vishwakarma, J. M. Billmers and J. Finn; *J. Org. Chem.*, 1984, **49**,3243.

86JA2470 R. W. Murray, R. Jeyaraman and L. Mohan, *J. Am. Chem. Soc.*, 1986, **108**, 2470.

87CC1362 R. S. Atkinson, R. S. and B. J. Kelly; *J. Chem. Soc., Chem. Commun.*, 1987,

	1362.
87T3123	J. T. Welch; *Tetrahedron*, 1987, **43**, 3123.
88CC624	R. S. Atkinson and B. J. Kelly; *J. Chem. Soc., Chem. Commun.*, 1988, 624.
88JA7933	B. M. Trost and A. R. Sudhakar; *J. Am. Chem. Soc.*, 1988, **110**, 7933.
89CC557	M. Kamata and T. Miyashi; *J. Chem. Soc., Chem. Commun*, 1989, 557.
89CC562	R. S. Atkinson, B. J. Kelly and M. McNicolas; *J. Chem. Soc.*, 1989, 562.
89CC836	R. S. Atkinson and B. J. Kelly; *J. Chem. Soc., Chem.Commun*, 1989, 836.
89CC889	C. Querci and M. Ricci; *J. Chem. Soc., Chem. Commun.*, 1989, 889.
89CC1344	S. Takano, K. Samizu, T. Sugihara and K. Ogasawara; *J. Chem. Soc., Chem. Commun.*, 1989, 1344.
89CC1848	J. Ichihara and T.Hanafusa; *J. Chem. Soc., Chem. Commun.*,1989, 1848.
89CC1852	J. E. Baldwin, R. M. Adlington, I. A. O'Neil, C. Schofield, A. C. Spivey and J. B. Sweeney, *J. Chem. Soc. Chem. Commun.*, 1989, 1852.
89CL1149	K. Yamamoto and N. Yamamoto; *Chem. Lett.*, 1989, 1149.
89CPB184	T. Satoh, A. Sugimoto and K. Yamakawa; *Chem. Pharm. Bull.*, 1989, **37**, 184.
89JA203	W. Adam, M. Braun, A. Griesheck, V. Lucchini, E. Staab and B. Will; *J. Am. Chem. Soc.*, 1989, **111**, 203.
89JA931	S. Calet, F. Urso and H. Alper; *J. Am. Chem. Soc.*, 1989, **111**, 931.
89JA1123	J. S. M. Wai, I.Marko, J. S. Svendsen, M. G. Finn, E. N. Jacobsen and K. B. Sharpless; *J. Am. Chem. Soc.*, 1989, **111**, 1123.
89JA4525	T. V. Rajan Babu and W. A. Nugent; *J. Org. Chem. Soc.*, 1989, **111**, 4525.
89JA5330	K. C. Nicolaou, C. V. C. Prasad, P.K. Somers and C. K. Hwang; *J. Am. Chem. Soc.*, 1989, **111**, 5330.
89JA5335	K. C. Nicolaou, C. V. C. Prasad, P.K. Somers and C. K. Hwang; *J. Am. Chem. Soc.*, 1989, **111**, 5335.
89JA6280	M. Oshima, H. Yamazaki, I. Shimizu, M. Nisar and J. Tsugi; *J. Am. Chem. Soc.*, 1989, **111**, 6280.
89JA6431	K. Maruoka, T. Ooi and H.Yamamoto; *J. Am. Chem. Soc.*, 1989, **111**, 6431.
89JA6661	R. L. Halcomb and S. J. Danishefsky; *J. Am. Chem. Soc.*, 1989, **111**, 6661.
89JA7628	J. Wu and R. G. Bergman; *J. Am. Chem. Soc.*, 1989, **111**, 7628.
89JA8039	A. B. Smith III, G. A. Sulikowski and K. Fujimoto; *J. Am. Chem. Soc.*, 1989, **111**, 8039.
89JA8303	T. Fukuyama and L. Yang, *J. Am. Chem. Soc.*, 1989, **111**, 8303.
89JA9116	S. O'Malley and T. Kodadek; *J. Am. Chem. Soc.*, 1989, **111**, 9116.
89JCS(P1)190	R. Baker and J. L. Castro; *J. Chem. Soc., Perkin Trans I*, 1989, 190.
89JCS(P1)835	M. Ashwell and R. F. W. Jackson; *J. Chem. Soc., Perkin Trans. I*, 1989, 835.
89JCS(P1)1172	S. K. Richardson and R. J. Ife; *J. Chem. Soc., Perkin Trans I*, 1989, 1172.
89JOC1491	C. Najera and M. Yus; *J. Org. Chem.*, 1989, **54**, 1491.
89JOC1940	D. S. Jones, A. Srinivasan, S. Kasina, A. R. Fritzberg and D. W. Wilkening; *J. Org. Chem.*, 1989, **54**, 1940.
89JOC2826	C. J. Burns, C. A. Martin and K. B. Sharpless; *J. Org. Chem.*, 1989, **54**, 2826.
89JOC3114	G. A. Molander and S. W. Andrews; *J. Org. Chem.*, 1989, **54**, 3114.
89JOC3130	T. Satoh, T. Oohara, Y. Ueda and Y. Yamakawa; *J. Org. Chem.*, 1989, **54**, 3130.
89JOC3231	J. D. Hsi and M. Koreeda; *J. Org. Chem.*, 1989, **54**, 3231.
89JOC3588	L. A. Flippin, P. A. Brown and J. Jalali-Araghi; *J. Org. Chem.*,1989, **54**,

3588.

89JOC3945 A. Subbaraj, O. S. Rao and W. Lwowski, *J. Org. Chem.*, **54**, 3945.

89JOC3973 T. Satoh, T. Sato, T. Oohara and K. Yamakawa; *J. Org. Chem.*, 1989, **54**, 3973.

89JOC4016 P. R. Carlier and K. B. Sharpless; *J. Org. Chem.* 1989, **54**, 4016.

89JOC4042 G. A. Molander and K. Mautner; *J. Org. Chem.*, 1989, **54**, 4042.

89JOC4222 N. Furukawa, Y. Sugihara and H. Fujihara; *J. Org. Chem.*, 1989, **54**, 4222.

89JOC4249 M.K. Chenault and S. J. Danishefsky; *J. Org. Chem.*, 1989, **54**, 4249.

89JOC5324 A. Dureaut, I. Tranchepain and J. C. Depezay; *J. Org. Chem.*, 1989, **54**, 5324.

89TL1029 M. J. Taschner and A. S. Aminbhavi; *Tetrahedron Lett.*, 1989, **30**, 1029.

89TL2545 P. Chabert, C. Mioskowski and J. R. Falck; *Tetrahedron Lett.*, 1989, **30**, 2545.

89TL2623 B. B. Lohray, Y. Gao and K. B. Sharpless; *Tetrahedron Lett.*, 1989, **30**, 2623.

89TL2703 R. S. Atkinson and B. J. Kelly; *Tetrahedron Lett.*, 1989, **30**, 2703.

89TL3045 K. K. Anderson and S. R. Kass; *Tetrahedron Lett.*, 1989, **30**, 3045.

89TL3049 M. P. Doyle, K. G. High, S. M. Oon and A. K. Osborn, *Tetrahedron Lett.*, 1989, **30**, 3049.

89TL3893 B. M. Trost and R. Hurnaus; *Tetrahedron Lett.*, 1989, **30**, 3893.

89TL4717 H. N. Rao, A. G. Holkar and N. R. Ayyangar; *Tetrahedron Lett.*, 1989, **30**, 4717.

89TL4863 J. Lown, G. J. J. Out, J. L. Baan, F. J. de Kanter, F. Bickelhaupt and G. W. Klumpp *Tetrahedron Lett.*, 1989, **30**, 4863.

89TL4881 J. Letgers, L. Thijs and B. Zwanenburg; *Tetrahedron Lett.*, 1989, **30**, 4881.

89TL4901 K. E. Bainbridge and W. P. Dailey; *Tetrahedron Lett.*, 1989, **30**, 4901.

89TL5369 R. Huisgen and E. Langhals; *Tetrahedron Lett.*, 1989, **30**, 5369.

89TL5425 M. T. Reetz and J. Binder; *Tetrahedron Lett.*, 1989, **30**, 5425.

89TL5489 R. Grigg and J. Markandu; *Tetrahedron Lett.*, 1989, **30**, 5489.

89TL6023 B. Boubia, C. Mioskowski, S. Mana and J. R. Falck; *Tetrahedron Lett.*, 1989, **30**, 6023.

89TL6233 S. Colonna and N. Gaggero; *Tetrahedron Lett.*, 1989, **30**, 6233.

89TL6387 T. Itoh, T. Nagano, M. Sato and M. Hirobe; *Tetrahedron Lett.*, 1989, **30**, 387.

89TL6497 W. Adam, L. Hadjiarapoglou and X. Wang; *Tetrahedron Lett.*, 1989, **30**, 6497.

89TL6545 T. Higuchi, H. Ohtake and M. Hirobi; *Tetrahedron Lett.*, 1989, **30**, 6545.

89TL6899 B. M. Fraga, J. R. Hanson, M. G. Hernandez and F. G. Tellado; *Tetrahedron Lett.*, 1989, **30**, 6899.

CHAPTER 3

Four-Membered Ring Systems

JOHN PARRICK
Brunel University, Uxbridge, Middlesex, UK

3.1 AZETIDINES, AZETINES AND AZETES

The sterically stabilised azete (2), obtained by pyrolysis of the cyclopropene azide (1), forms the organo cobalt derivative (3), and photoelectron studies are described for both (2) and (3) (88AG1616). A one step synthesis of certain 3-azetidinols (4, R=1-adamantyl or cyclodecyl) is achieved in fair yield by treatment of epichlorohydrin with the appropriate hindered primary amine. Both X-ray crystallography and NMR studies show the substituents to be on the same side of the ring (89CJC1514).

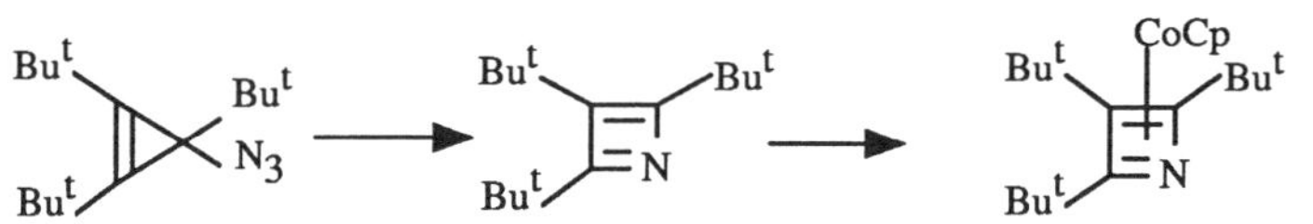

(1) (2) (3)

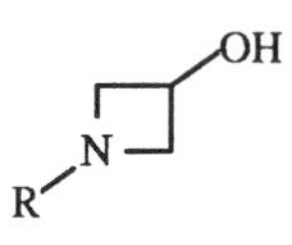

(4)

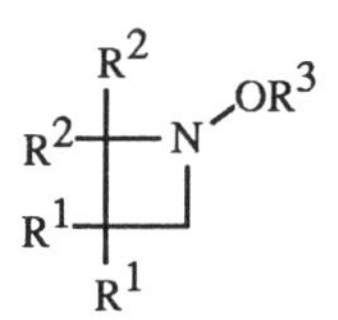

(5)

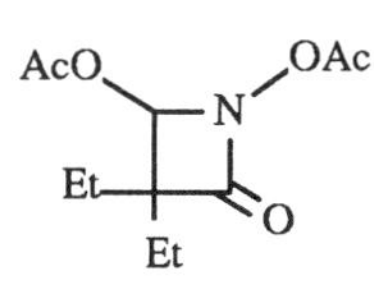

(6)

Reductive cyclisation of *o*-benzyl derivatives of 3-bromo aldehyde oximes yield 1-benzyloxyazetidines (5, e.g. R^1=Et, R^2=H, R^3=$CH_2C_6H_5$) which on treatment with lead tetraacetate undergo debenzylation, ring oxidation and substitution to give 1,4-diacetoxyazetidin-2-one (6). A similar reaction occurs with the spirocyclohexyl derivative (7) but this time both the acetoxyazetidinone (8, R=OAc, R^1=H) and the diacetoxy derivative (8, R=R^1=OAc) are formed (88RTC381). Oxidation of the 1-hydroxy derivatives gives nitrones such as (9) which on treatment with KCN give 1-hydroxyazetidine (10, R^3=CN) by addition at the less hindered side as also occurs in the formation of (10) by reaction of (9) with a Grignard reagent, R^3MgX (88RTC63). The nitrone (9, R=R^1=Me) on treatment with acetyl chloride in the absence of water yields (11), whereas (12) is obtained in a similar manner from (9, R^1=Ph, R^2=Me) (88RTC52).

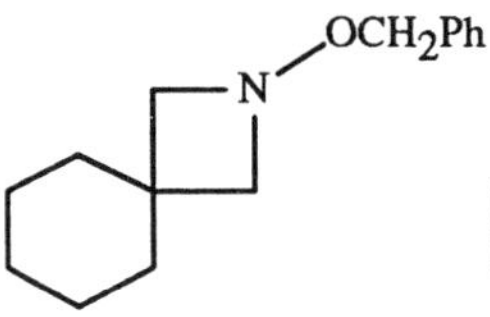

(7)

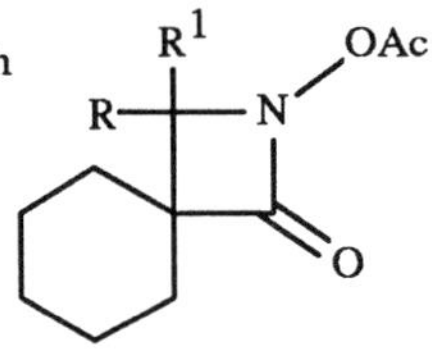

(8)

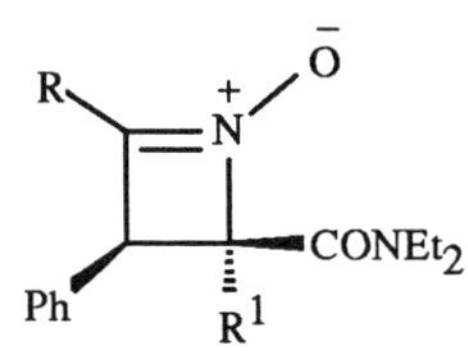

(9)

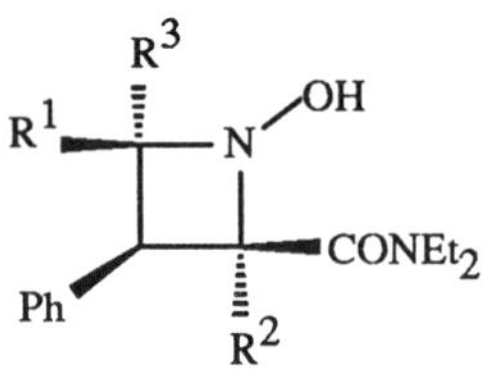

(10)

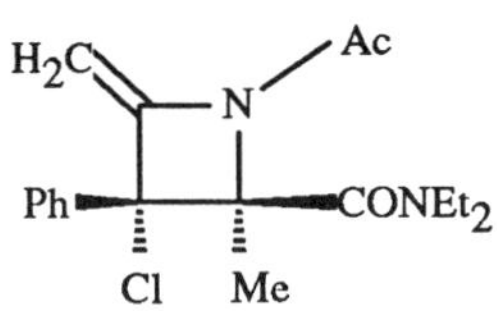

(11)

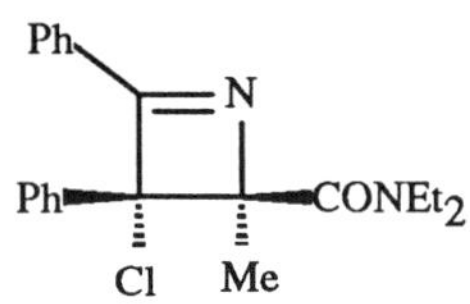

(12)

3.2 OXETANES AND THIETANES

1,1-Diethoxyethane undergoes thermal or photochemical [2 + 2]cycloaddition reactions with carbonyl compounds, e.g. ethyl pyruvate yields (13), whereas other electron rich olefins usually require the presence of Lewis acid catalysts. Generally the thermal and photochemical reactions are complementary in both the ease of reaction and selectivity. The $[1s^D + 1s^A]$ geometry of the transition state is favoured rather than $[2s^D + 1s^A]$ geometry adopted by symmetrical olefins (88H2153). Photochemical [2 + 2]cycloaddition of cycloalkenes or furan with α-ketoalkynes produces,stereoselectively, the oxetanes (e.g. 14) with an endo acetylenic substituent. Reduction of the oxetane (14) in the presence of a Lindlar catalyst gives the vinyl compound (88H125).

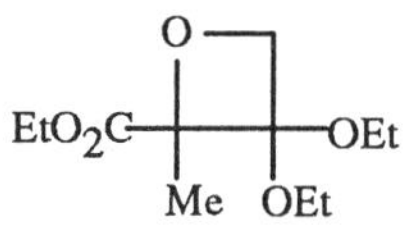

(13)

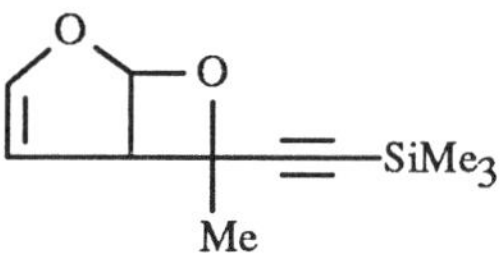

(14)

Treatment of a 2-substituted but-3-enoic acid with bromine in aqueous bicarbonate causes lactonisation to give exclusively the 3-substituted-4-bromomethyloxetane (e.g. 15) in good yield (89MI241). 4-Bromomethyloxetanes are readily obtained in moderate yield by treatment of an unsaturated carboxylic acid with thallium(I) carbonate and bromine (88S1009). The action of magnesium bromide on spiro-β-lactones unexpectedly gave 1-cyclohexenylacetic acids in high yield and isomeric purity (89TL531).

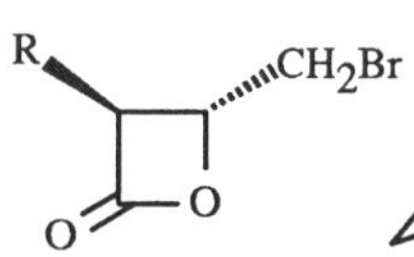

(15)

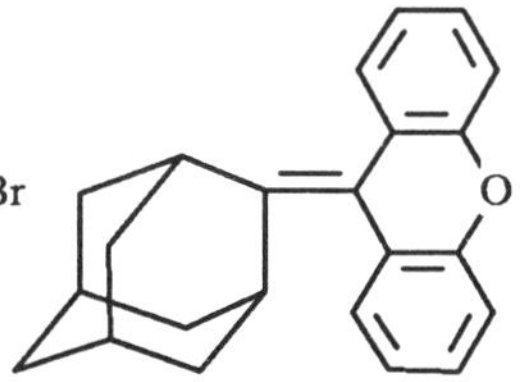

(16)

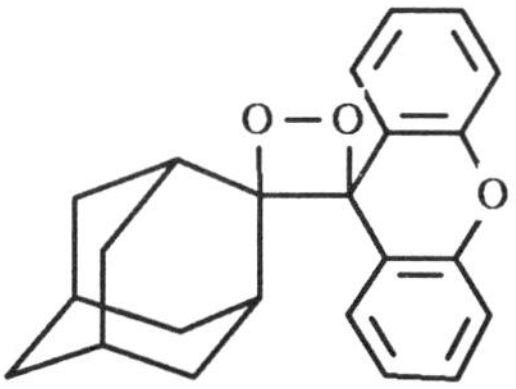

(17)

3.3 DIOXETANES, DITHIATANES AND SULTONES.

The xanthenylideneadamantane (16) (obtained from adamantinone and xanthene by Peterson olefination) undergoes photochemical cycloaddition of oxygen (88TL3137) to yield the 1,2-dioxetane (17) which has been described as a potential thermochemiluminescent label.

The 1,3-dithiatane (18) has been prepared in low yield (88ZOR1773) and the sultone (19) has been obtained by the action of chlorine and sulfur trioxide on norbornadiene (88MI55). Novel fluorinated sultones have been reported (89MI349).

3.4 RINGS CONTAINING BORON, PHOSPHORUS, SILICON OR TITANIUM.

9-Fluorenylidene (tetramethylpiperidino)boranes undergo

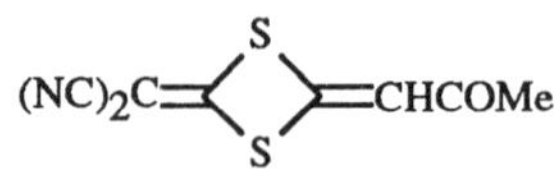

(18)

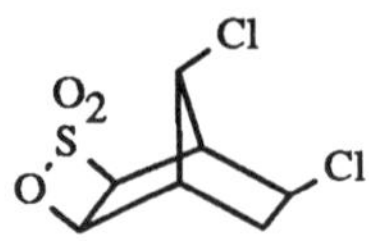

(19)

[2 + 2]cycloaddition reactions with alkynes (e.g. diethylamino-ethyne) to give substituted 1,2-dihydroboretes (e.g. 20; R=tetra methylpiperidine, R^1=fluorenylidene) while tetramethylpiperidino iminoboranes and alkynes yield 1,2,4-azadiboretidines (21, R=tetramethylpiperidine) (88ZN(B)449).

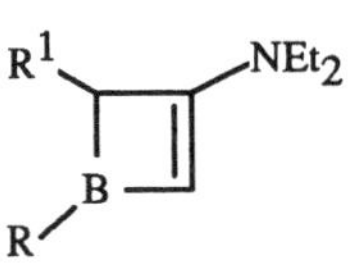

(20)

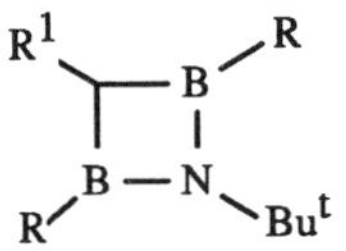

(21)

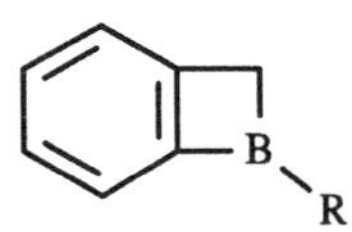

(22)

The first synthesis of a benzodihydroborete (22, R=NPr^i_2) has been obtained by intramolecular cyclisation of the o-disubstituted benzene, $BrCH_2C_6H_4B(Cl)NPr^i_2$, by the action of lithium metal. The amine (22, R=NPr^i_2) on treatment with HCl gave the dichlorodihydrobenzoborate (23). Treatment of the o-disubstituted benzene with potassium metal caused an intermolecular cyclisation to yield (24) (88JOM33). The lithium salt (25, R=aryl) with trimethyltin(IV) chloride or methyl iodide yielded the 1,3-diboretanes (26, R^1=$SnMe_3$ or Me, respectively) (89AG761) and aryldifluoroboranes ($ArBF_2$) and (25) gave (27) which have a strong Lewis acidity (89AG759).

The photolysis of 1,1-dimethyl-1-silacylobut-2-ene (28) in *t*-butanol gives a range of acyclic products and studies of the reaction of the deuterium labelled heterocycle have provided evidence that species containing a Si=C group, such as 1,1-dimethyl-1-sila-1,3-butadiene and silacyclopropylmethylene, act as intermediates in the reaction (89MI530). Silanes such as (28) react with two molar equivalents of isocyanides by an insertion process to give the 1-sila-3-azacyclobutane (29) (89MI850).

The alkylidenesilirane (30) undergoes a ring expansion reaction on oxidation with cyclohexyldimethylamine *N*-oxide to give two 1,2-oxasiletanes (31) and (32) as the *Z*-isomers. However, oxidation of (29) with diphenyl sulfoxide gives (32) in

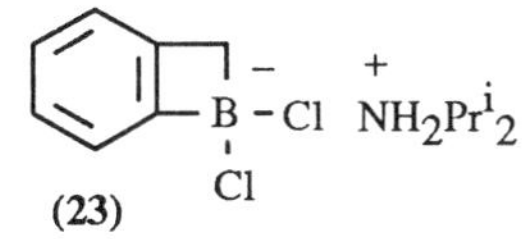

(23)

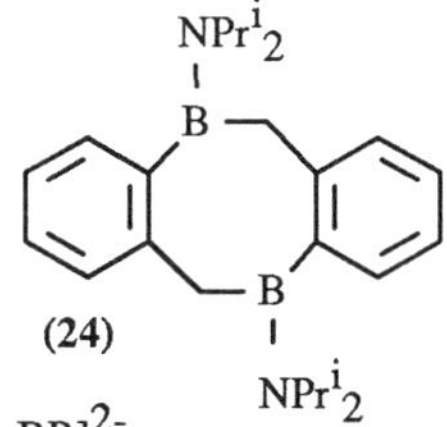

(24)

$2Li^+[(Me_3Si)_2C{=}B\text{-}CR{=}BR]^{2-}$

(25)

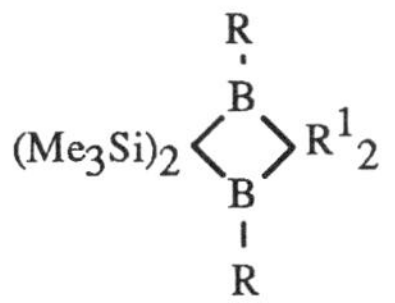

(26)

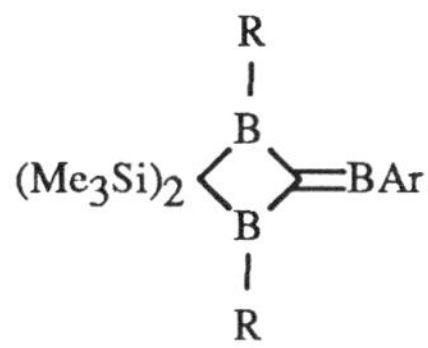

(27)

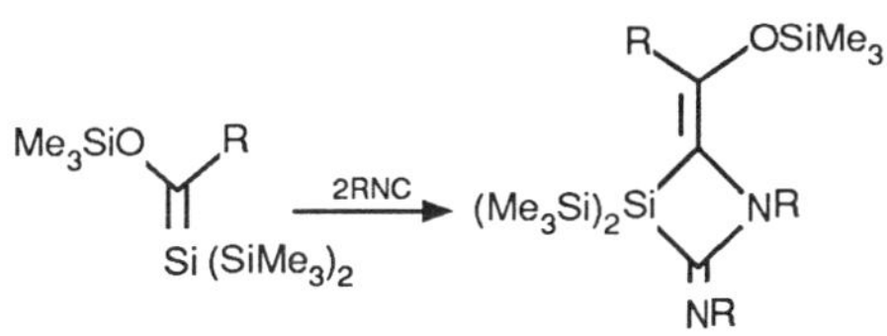

(28) **(29)**

70% yield (88TL4747) by insertion of an oxygen atom into the Si-sp^2 carbon bond which is thought to result from the interaction of the exocyclic π bond in (30) and the sulfur atom of the sulfoxide.

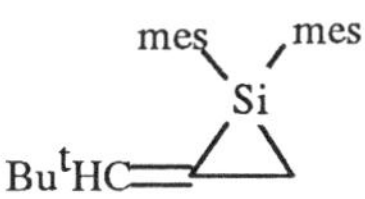

(30)

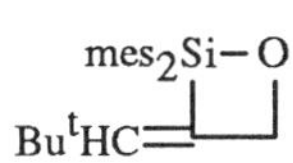

(31)

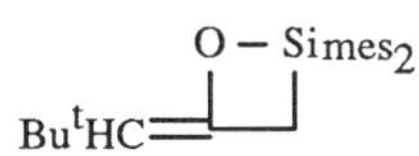

(32)

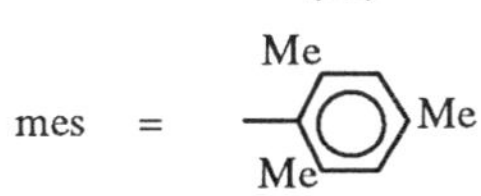

Photochemical decomposition of the α-diazo-α-silyl carbonyl compounds (33) gives 3-oxo-1-silaprop-1-enes (34), which cyclise intramolecularly to the heterocycle (35) if R is stereochemically demanding, otherwise the cyclic dimers (36) are formed (88CC72).

Titanocyclobutenes (e.g. 37), readily obtainable from an

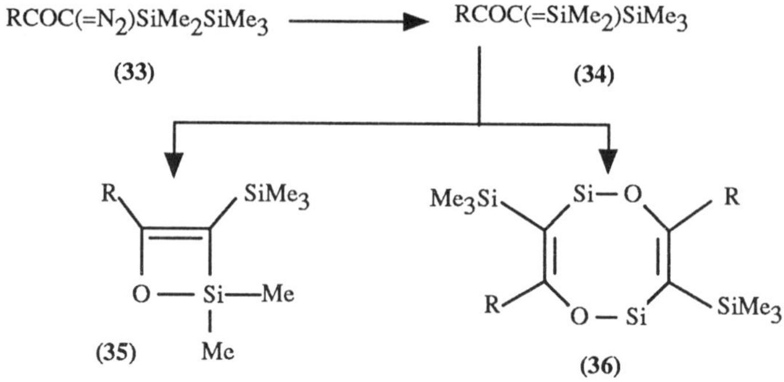

alkyne and Tebbe's reagent ($Cp_2TiCH_2AlClMe_2$), readily react with a variety of multiple bond systems containing a heteroatom (88BCJ171). Thus acetone reacts with (37) to give the titanaoxa cyclohexene (39). Treatment alkyl (or aryl) dichlorophosphine has produced the first phosphacyclobutene (38) (89JA9129).

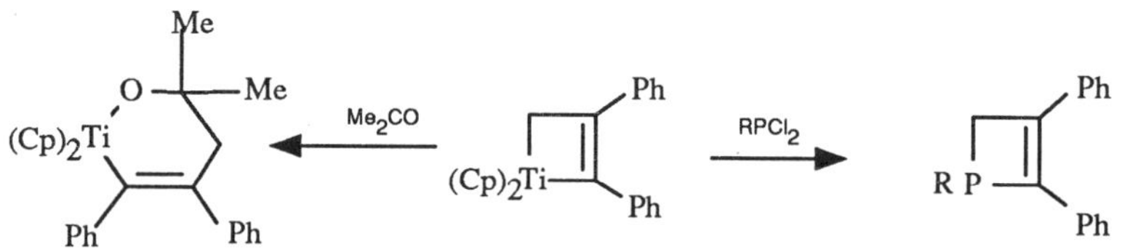

(39) (37) (38)

The synthesis of 2-oxo-1,2-λ^5-oxaphosphetanes (43) from the azo compound (40) and p-toluenesulfinyl chloride has provided evidence of an α-phosphanyl-α'-sulfinylcarbene (41) to α-phosphanyl-α'-sulfanylcarbene (42) rearrangement (89JOC4426). The rate of pseudorotation of some oxyphosphatanes has been studied by NMR spectroscopy and the findings discussed in relation to their significance for the stereochemistry and mechanistic detail of the Wittig reaction (89JA1519).

3.5 AZETIDIN-2-ONES

The activity in this area has continued and about half the references this year fall in this section, even though the area covered is limited to developments in the chemistry of simple monocyclic β-lactams.

Lactam formation to give (44) without epimerisation of activated esters of β-aminopropionic acid has been caused using

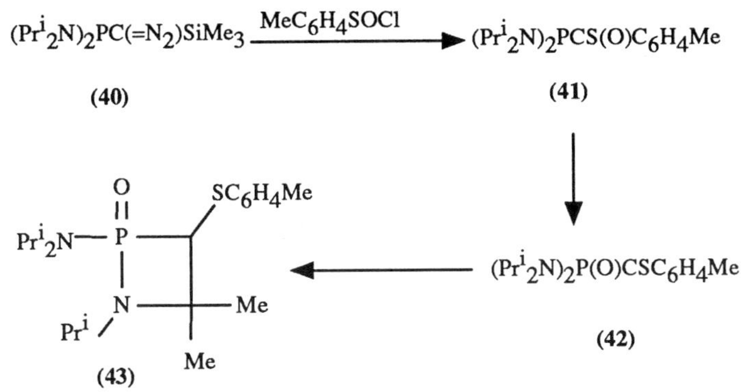

copper(I) trifluoromethane sulfonate (89JOC3511). Acetoacetates have been used as starting materials for the synthesis of the *cis*-4 substituted monobactam (45) (89JCS(P1)67) and for the preparation of 3-unsubstituted β-lactams (e.g. 46) (89JA1073).

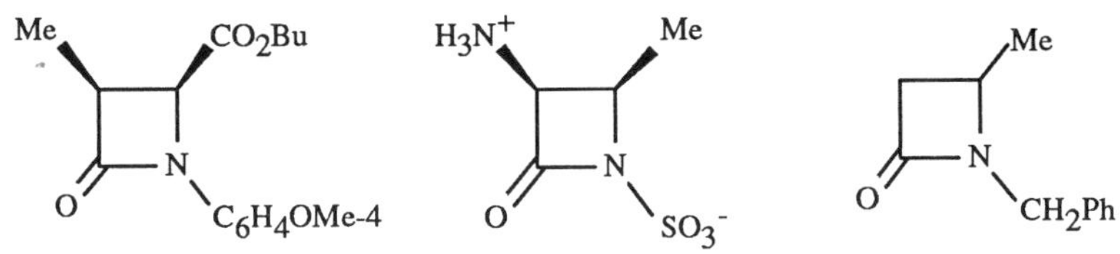

(44) (45) (46)

Glycine derivatives have been converted into β-lactams (47) by the reaction of imines in the presence of zinc chloride in a process which affords a considerable degree of stereochemical control. The choice of solvent and the substituents on the imine are major factors affecting the stereochemical outcome of the reaction (89TL765).

Benzaldehyde imines have been converted into β-lactams (48) in a Reformatsky type of reaction with 4-bromocrotonates; pyrimidones were identified amongst the by-products (88MI1805).

The ring expansion of 2-phenylaziridines (49) has been achieved in an enantiospecific reaction in the presence of carbon monoxide, the dimer of chloro(1,5)-cyclooctadiene)rhodium (I) in catalytic quantities, and (+) or (-) menthol as the chiral reagent (89JA931). 4-Unsubstituted β-lactams were also obtained from the readily prepared dithioketal (50) by treatment with Raney

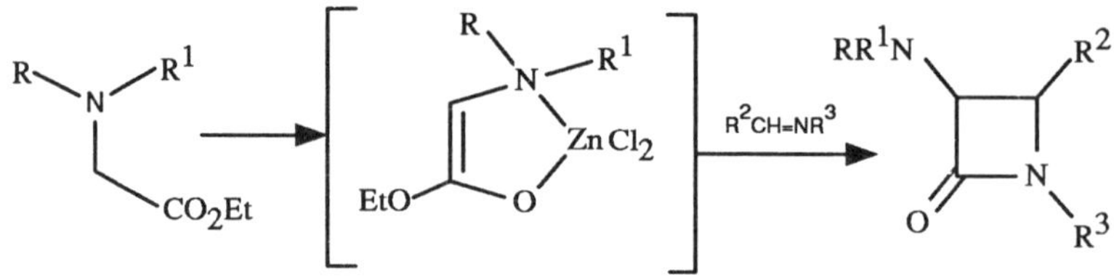

(47)

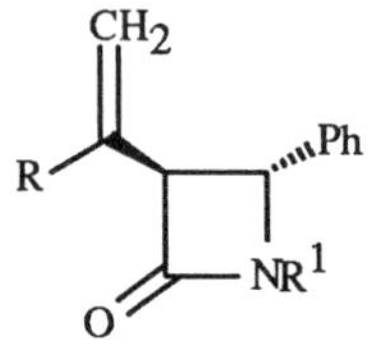

(48)

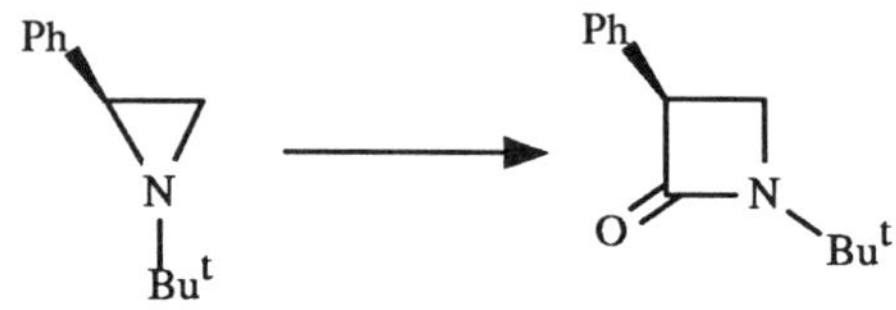

(49)

nickel(87S990). The reaction of an imine with an acid chloride in the presence of base to give β-lactams has been investigated by FT-IR spectroscopy and shown to occur exclusively through the ketene intermediate (89JOC3792).

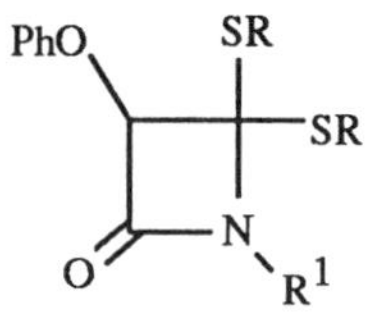

(50)

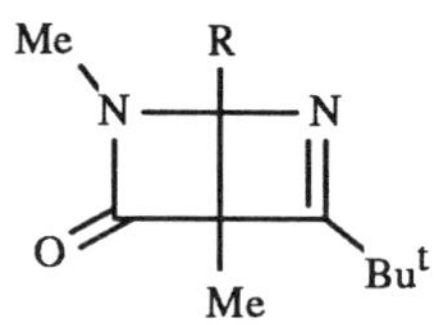

(51)

Evidence has been presented (88JCS(P1)2653) to show that (51) is a likely intermediate in the formation of (53) and (54) upon photolysis of the pyridone (52) in acetic acid - acetonitrile at 0°C.

A Japanese group (89TL113) has described the stereoselective synthesis of the *trans*-β-lactam (56) from the chiral monoester (55); this synthesis is complementary to the preparation of the *cis*-β-lactam (57) mentioned last year (88TL1057).

The two carboxylic acid groups of tartaric acid have been differentiated by use of other derivatives (58) and this allowed the selective *t*-butyldimethylsilylation of each of the hydroxyl groups by the simple procedure of using different molar ratios of imidazole in the ether formation step (89T3071). A correction of an earlier (87TL1861) structural assignment was made. The silyl

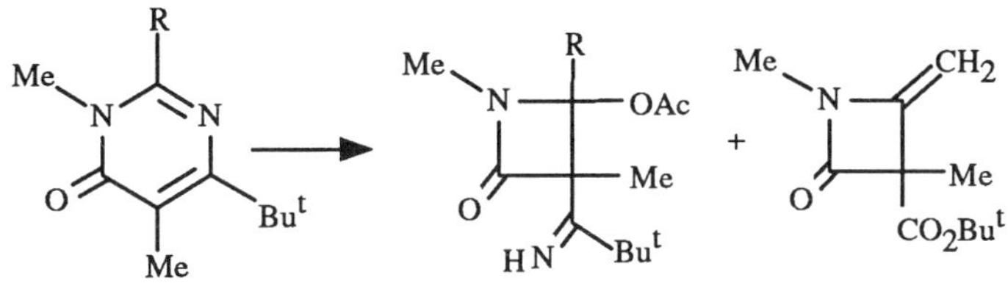

(52) (53) (54)

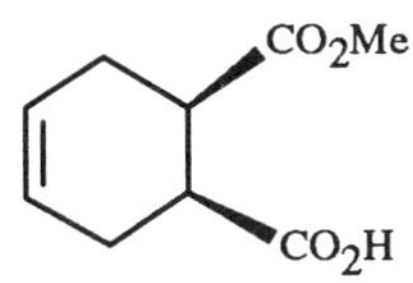

(55)

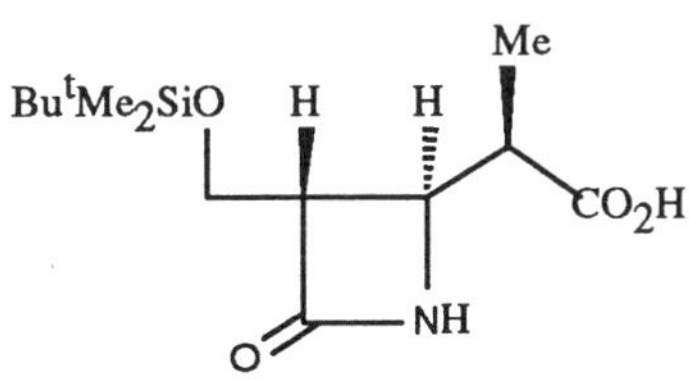

(56)

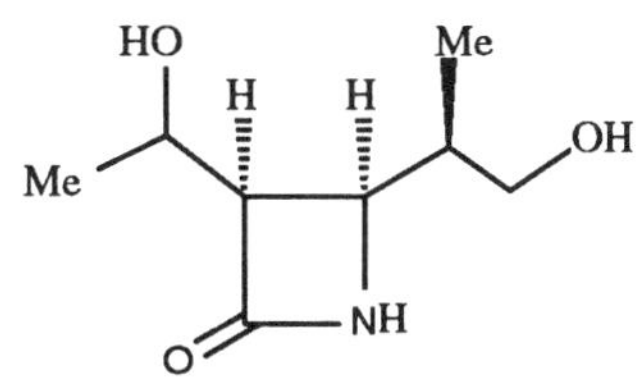

(57)

ethers were cyclised to β-lactams with triphenylphosphine and dimethyl azodicarboxylate to give the *trans*-β-lactams, e.g. (59) from (58, R=$SiMe_2Bu$-t, R^1=H) in 40% yield.

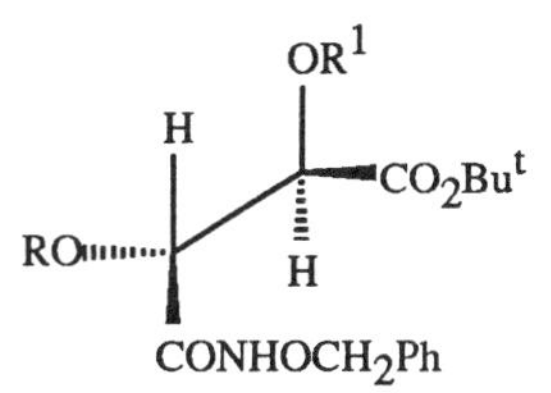

(58)

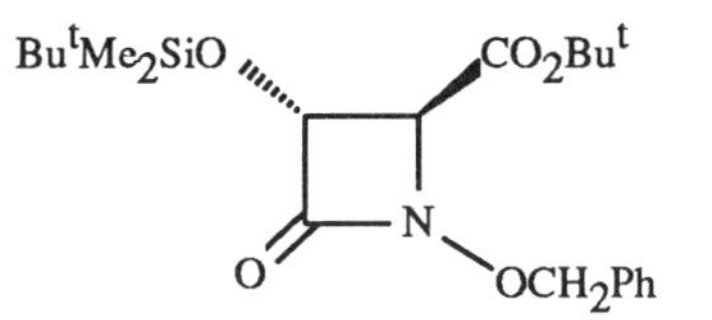

(59)

Acyclic amides have been converted to β-lactams by stereoselective oxidative coupling of their dianions, e.g. the dianion of (60) yielded (61) by oxidation with *N*-iodosuccinimide (NIS). The addition of (R)-$PhCHMeNH_2$ produced a higher degree of asymmetric induction (89JA6843).

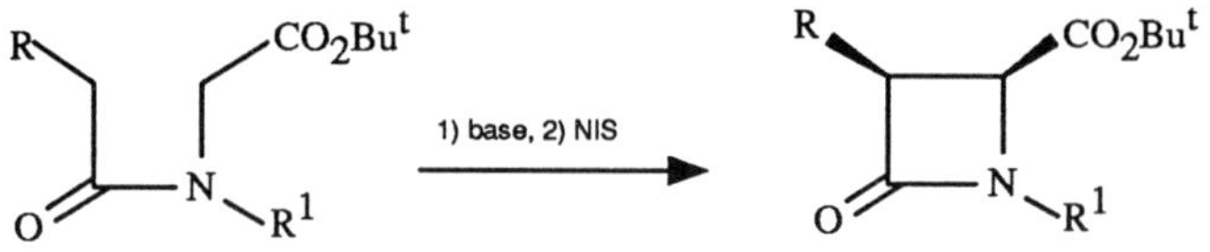

(60) **(61)**

The Arnstein tripeptide analogue (62) has been cyclised in the presence of a silicon enolate (CH_2=CMeOSiMe$_2$Bu-t) and zinc iodide in a Pummerer type of rearrangement to the *cis*-β-lactam (63) mainly (40%), with (15%) of the *trans* isomer (89TL729).

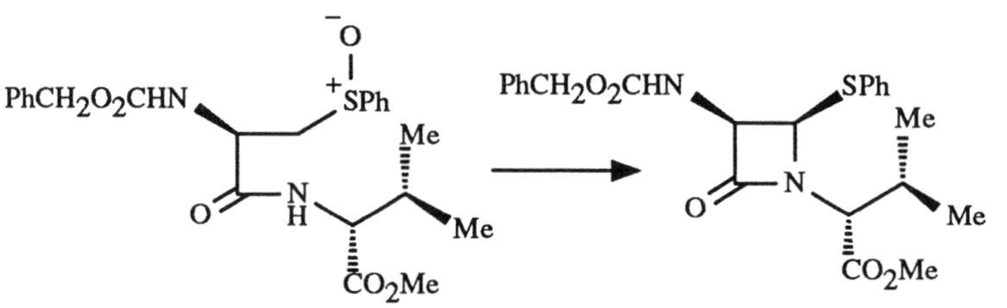

(62) **(63)**

Commercially available *cis*-but-2-ene-1,4-diol has been converted to the epoxide of an unsaturated acid amide and thence to a *trans*-3,4-disubstituted β-lactam (88TL815) of the type often obtained from naturally occurring precursors.

A reliable large scale synthesis of 4-acetoxy-2-azetidinone by [2 + 2]cycloaddition from vinyl acetate has been reported (87OS135) and a review of the use of β-lactams as synthons by molecular rearrangement or lactam cleavage has appeared (88H1755). The splitting of the CO stretching band in the infrared spectrum of β-lactams is due to the effects of the solvent and ring substituents and not to H-bonded association (88MI123).

Compounds of the type (ArRC=C(OR^1)$_2$ undergo [2 + 2]cycloaddition reactions with β-lactams whether or not R and R^1 are H, whereas earlier work had indicated that neither R nor R^1 could be H (89S54). Certain 1-azabutadienes (Ar^{1}CH=CMeCH=NAr^2) in a Reformatsky type of reaction with ethyl bromoacetate in the presence of trimethylsilyl chloride gave the 3-unsubstituted β-lactams (64, R=H) which undergo substitution at the 3-position to give *trans*-3,4-disubstituted β-lactams (64, R $\neq$ H) (89CC74). Addition of a diethylzinc as a complexing agent was the first process in a three-step, one-pot conversion of methyl (R)-3-hydroxybutyrate into the cis-β-lactam (65) (88CC1376). An innovative synthesis is the photoconversion in the solid state of either enantiomer of $PhCOCONPr^i_2$ into the β-lactam (66) in high optical and chemical yield (89JA697).

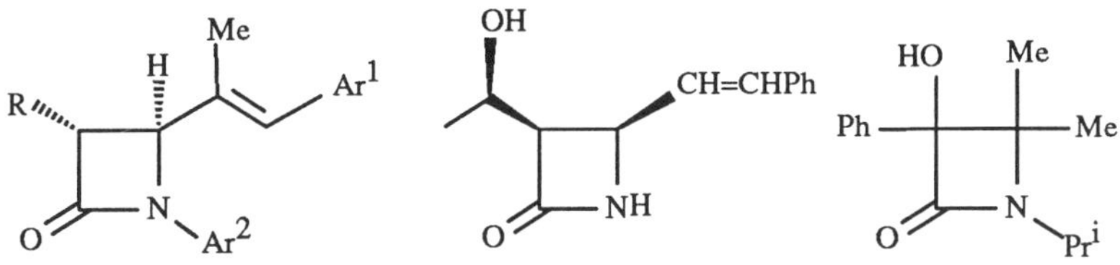

(64) (65) (66)

Peterson alkenylation of bis(trimethylsilyl)methylimines in the presence of fluoride ions has been used both for the preparation of 2-aza-1,3-dienes and to convert *N*-bis (trimethylsilyl)methyl β-lactams into *N*-vinyl-β-lactams (89CC72). Organozinc reagents have been used to replace a 4-phenylthio group by an alkenyl substituent (88TL5391). The synthesis and crystal structure of the anti-Bredt 1,3-bridged β-lactams (68) has been reported (89JA1073). The key synthetic step is the formation of the carbene from the diazo compound (67) and its Rh(II) catalysed insertion into the NH bond of the lactam.

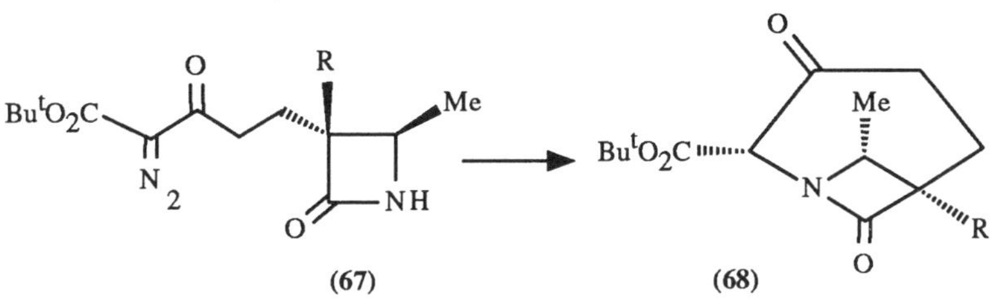

(67) (68)

REFERENCES

87OS135 S.J. Mickel, C.N. Hsiao, and M.J. Miller; Org. Synth., 1987, 65, 135.

87S990 S.D. Sharma, U. Mehra, J.P.S. Khurana, and S.B. Pandhi; Synthesis, 1987, 990

87TL1861 T. Kolasca and M.J. Miller; Tetrahedron Lett., 1987, 28, 1861.

88AG1616 M. Ledermann, M. Regitz, K. Angermund, P. Binger, C, Krueger, R. Mynott, R. Gleiter, and I. Hyla-Kryspin;

Angew. Chem., 1988, 100, 1616.

88BCJ171 J.D. Meinhart and R.H. Grubbs; Bull. Chem. Soc. Jpn., 1988, 61, 171.

88CC72 G. Mass, K. Schneider, and W. Ando; J. Chem. Soc., Chem. Commun., 1988, 72.

88CC1376 N. Oguni and Y. Ohkawa; J. Chem. Soc., Chem. Commun., 1988, 1376.

88H125 A. Feigenbaum, J.P. Pete, and A.S. Poquet-Dhimane; Heterocycles, 1988, 27, 125.

88H1755 M.S. Manhas, D.R. Wagle, J. Chiang, and A.K. Bose; Heterocycles, 1988, 27, 1755.

88H2153 J. Mattay and K. Buchkremer; Heterocycles, 1988, 27, 2153.

88JCS(P1)2653 T. Takahashi, S. Hirokami, M. Nagata, and T. Yamazaki; J. Chem. Soc., Perkin. Trans. 1, 1988, 2653.

88JOM33 W. Schacht and D. Kaufmann; J. Organomet. Chem., 1988, 339, 33.

88MI55 K.A. Potekhin, Yu. T. Struchkov, A.S. Koz'min, V.V. Zhadkin, V.D. Sorokin, and N.S. Zefirov; Sulfur Lett., 1988, 8, 55.

88MI123 M. Gil, J. Plumet, N. Iza, and J. Morcillo; J. Mol. Struct.,1988, 174, 123.

88MI1805 J.L. Moreau; C.R. Acad. Sci., Ser. 2, 1988, 307, 1805.

88RTC52 P.J.S.S. Van Eijk, W.P. Trompenaars, D.N. Reinhoudt and S. Harkema; Recl. Trav. Chim. Pays-Bas, 1988, 107, 52.

88RTC63 P.A. Van Elburg, D.N. Reinhoudt, and S. Harkema; Recl. Trav. Chim. Pays-Bas, 1988, 107, 63.

88RTC381 P.A. Van Elburg and D.N. Reinhoudt; Recl. Trav. Chim. Pays-Bas, 1988, 107, 381.

88S1009 R.C. Cambie, P.S. Rutledge, R.F. Somerville, and P. Woodgate; Synthesis, 1988, 1009.

88TL815 C. Bonini and R. Di Fabio; Tetrahedron Lett., 1988, 29, 815.

88TL1057 H. Kaga, S. Kobayashi, and M. Ohno; Tetrahedron Lett.,1988, 29, 1057.

88TL3137 A.C. Brouwer, J.C. Hummelen, T.M. Luider, F. Van Bolhuis, and H. Wynberg; Tetrahedron Lett., 1988, 29, 3137.

88TL4747 H. Saso, H. Yoshida, and W. Ando; Tetrahedron Lett., 1988, 29, 4747.

88TL5391 S. Mori, H. Iwakura, and S. Takechi; Tetrahedron Lett., 1988, 29, 5391.

88ZN(B)449 B. Glaser, E.P. Mayer, H. Noeth, W. Rattay, and U. Wietelmann; Z. Naturforsch., Teil B, 1988, 43, 449.

88ZOR1773 V.N. Drozd and E.N. Komarova; Zh. Org. Khim., 1988, 24,

1773.

89AG759	R. Hunold, M. Pilz, J. Allwohn, M. Stadler, W. Massa, P.v.R. Schleyer, and A. Berndt; Angew. Chem., 1989, 101, 759.
89AG761	M. Pilz, M. Stadler, R. Hunold, J. Allwhon, W. Massa, and A. Berndt; Angew. Chem., 1989, 101, 761.
89CC72	J. Lasarte, C. Palomo, J.P. Picard, J. Dunogues, and J.M. Aizpurua; J. Chem. Soc., Chem. Commun., 1989, 72.
89CC74	F.P.Cossio, J.M. Odriozola, M. Oiarbide, and C. Paloma; J. Chem. Soc., Chem. Commun., 1989, 74.
89CJC1514	M. Laguerre, C.Boyer, J.M. Leger, and A. Carpy; Can. J. Chem., 1989, 67, 1514.
89MI241	I. Shibata, M. Toyota, A. Baba, and H. Matsuda; Chem. Express, 1989, 4, 241.
89MI349	J. Mohtasham, M. Brennen, Z. Yu, J.L. Adcock, and G.L. Gard; J. Fluorine Chem., 1989, 43, 349.
89MI530	M.G. Steinmetz, B.S. Udayakumar, and M.S. Gordon; Organometallics, 1989, 8, 530.
89MI850	A.G. Brook, A.K. Saxena, and J.F. Sawyer; Organometallics, 1989, 8, 850.
89JA697	A. Sekine, K. Hori, Y. Ohashi, M. Yagi, and F. Toda; J. Am. Chem. Soc., 1989, 111, 697.
89JA931	S. Calet, F. Urso, and H. Alper; J. Am. Chem. Soc., 1989, 111, 931.
89JA1073	R.M. Williams, B.H. Lee, M.M. Miller, and O.P. Anderson; J. Am. Chem. Soc., 1989, 111, 1073.
89JA1519	E. Vedejs and C.F. Marth; J. Am. Chem. Soc., 1989, 111, 1519.
89JA6843	T. Kawabata, K. Sumi, and T. Hiyama; J. Am. Chem. Soc., 1989, 111, 6843.
89JA9129	K.M. Doxsee, G.S. Shen, and C.B. Knobler: J. Am. Chem. Soc., 1989, 111, 9129.
89JCS(P1)67	P. Fernandez-Resa, R. Herranz, S. Conde, and E. Arribas; J. Chem. Soc., Perkin. Trans. 1, 1989, 67.
89JOC3511	N. Miyachi, F. Kanda, and M. Shibasaki; J. Org. Chem., 1989, 54, 3511.
89JOC3792	J.E. Lynch, S.M. Riseman, W.L. Laswell, R.P. Volante, G.B. Smith, I. Shinkai, and D.M. Tschaen; J. Org. Chem., 1989, 54, 3792.
89JOC4426	G. Sicard, H. Gruetzmacher, A. Baceiredo, J. Fischer, and G. Bertrand; J. Org. Chem., 1989, 54, 4426.
89S54	M.L. Graziano and G. Cimminiello; Synthesis, 1989, 54.
89T3071	T. Kolasa and M.J. Miller; Tetrahedron,1989, 45, 3071.
89TL113	H. Kaga, S Kobayashi, and M. Ohno; Tetrahedron Lett., 1989, 30, 113.
89TL531	T.H. Black and S.L. Maluleka; Tetrahedron Lett., 1989, 30, 531.
89TL729	Y. Kita, O. Tamura, T. Miki, H. Tono, N. Shibata, and Y. Tamura; Tetrahedron Lett., 1989, 30, 729.
89TL765	F.H. Van der Steen, H. Kleijn, J.T.B.H. Jastrzebski, and G. Van Koten; Tetrahedron Lett., 1989, 30, 765.

CHAPTER 4.1

Five-Membered Ring Systems: Thiophenes & Se & Te Analogs

JEFFERY B PRESS & RONALD K RUSSELL
R W Johnson Pharmaceutical Research Institute, Raritan, N J, USA

4.1.1 INTRODUCTION

Thiophene research continued to result in theoretical and preparative advances during 1988 and 1989. As a result of the unforseen omission of 1988 coverage in **Volume 1**, we will attempt to highlight the major chemical advances in both 1988 and 1989 in this review.

This paper is organized in sections which discuss the electronic character, ring substitution, ring formation, and synthesis of new thiophene derivatives as well as ring opening of thiophene to other compounds of interest. Brief mention is made of thiophene oligomers and polymers and their electroconductive properties. Finally, the few literature reports on selenophene and tellurophene syntheses are summarized. Only papers describing chemically interesting work are reviewed. The large number of biologically active thiophene derivatives that are prepared by routine chemical transformations are not discussed but a major review of this area has appeared in 1985 <85HC353> and an update is in press. A brief review of the synthesis of thiophene-containing heterocycles has been published <88MI49> during this time.

While no further discussion follows in this chapter, it should be noted that the unique coordination and stabilization properties of thiophene has made it essential as a non-transferable ligand in the chemistry of "higher order" organocuprates developed by Lipshutz <84TL5005, 87TL945, 87S325>. An excellent method of preparation of these reagents is reported <88JOC5366>.

4.1.2 AROMATICITY AND RADICAL STABILIZATION

Thiophene has always played a role as the most aromatic of the 5-membered ring heterocycles in terms of its electronic character and in chemical reactivity. New theoretical concepts to predict aromaticity have been proposed including a prediction of aromatic energies based upon heats of formation <89H1135>. A measure of aromaticity based upon absolute and relative hardness has recently been proposed <89JA7371>. A method to determine above-plane ring currents based upon a model for aromatic solvent-induced chemical shifts in the NMR is in good agreement with diamagnetic exaltation measurements <89JA6544>. The aromatic stabilities of thiophene analogues of helicenes are predicted by a conjugated circuits model; parent helicenes are always more aromatic than the thiophene analogues <88JST111>. For halogenated selenophene and tellurophene derivatives, a GVFF procedure was developed for the determination of substituent effects on heteroaromatic ring force fields <88MI243>.

The properties of organic diradicaloid species has been studied using semiempirical M.O. theory. While **1** is a ground state singlet with non-delocalized geometry, the second heteroatom completes the (4n+2) π delocalized electron shell in **2** <89JOC958>. The intensely purple **1** is also predicted to be a ground state singlet in another study <89JA3671>. The solid state NMR of **1** as a "captive intermediate" at 77°K allows the accurate measurement of bond lengths and angles <89JA1533>. 2-Thienyl and 2-furyl are second only to 2-pyridyl *N*-oxide for the stabilization of radicals as determined by $\gamma\cdot$ values <89JOC2904>. A study of the solution phase chemistry of **1** shows stereoselective reaction with maleonitrile to give **3** <89JA3659>. Interestingly, the furan isostere of **1** reacts to give predominantly the bicyclic isomer resulting from reaction of the furan ring.

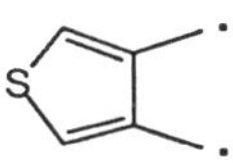

1

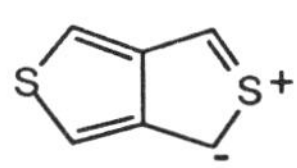

2

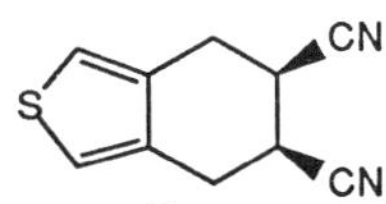

3

Flash vacuum pyrolysis of 2-chloromethyl-5-methylthiophene gives **4** which is characterized by variable temperature photoelectron spectroscopy (VTPES) <88TL6239>. In a similar way, **5** is generated and characterized only as its dimer **6** <88TL117>.

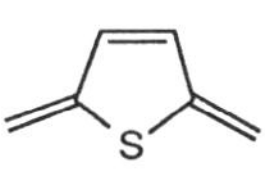

4

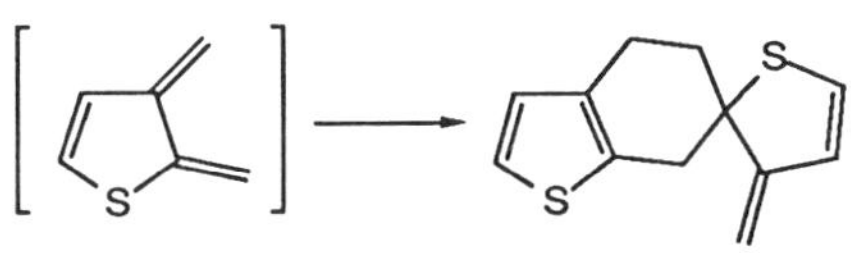

5 **6**

The issue of tautomerism of hydroxythiophenes has received much study over the years. Studies of the monohydroxy derivatives of thiophene, furan, pyrrole and their benzo analogues show the furan and thiophene compounds to be the strongest acids <89JA5346>. In a similar manner, the diene-diol tautomerism of thiosuccinic anhydride has been studied <89JOC1211>. The tautomerism of 2-mercaptothiophene results in the formation of an unusual dimer as a consequence of a Michael-type addition of the mercaptan to the corresponding α,β-unsatutated thione <89JOC3223>.

Hydrodesulfurization of thiophene has tremendous theoretical and commercial importance. While a complete summary of this area is beyond the scope of this review, several results are noteworthy. In particular, η^4-thiophene bound ligand **7** was prepared as part of mechanistic studies of hydrodesulfurization and the unusually basic sulfur strongly coordinates with BH_3 <89MI2277>.

Another study shows that sulfur-bound metal complexes (**8**) have reactive dienes which further coordinate with iron <89JA8753>. Several mechanistic studies of hydrodesulfurization show deuterium incorporation into benzo[*b*]thiophene using a molybdenum catalyst <88MI36> and that 2,3-dihydrothiophene inserts into platinum hydride catalysts <89MI1856>. Finally, several easily prepared nickel-containing complex reducing agents (NiCRAs and NiCRALs) are non-pyrophoric and are alternatives to Raney nickel and nickel boride for desulfurization reactions <89JOC4848>.

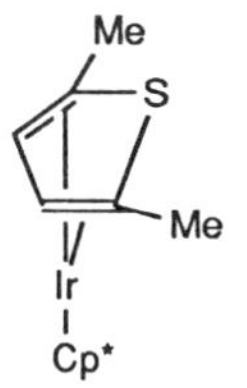

7

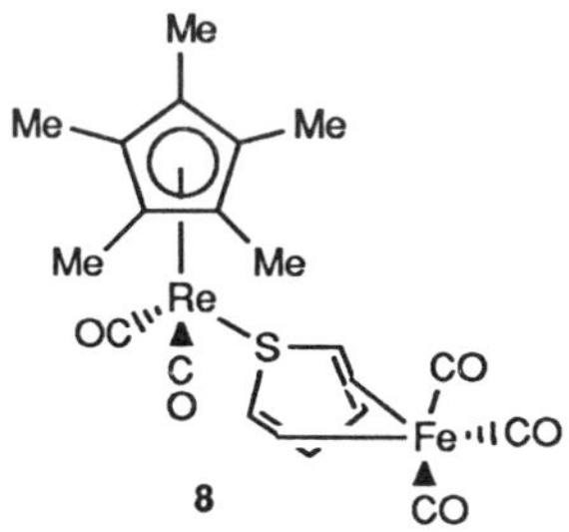

8

4.1.3 RING SUBSTITUTION

Direct substitution reactions on the thiophene ring may be accomplished by essentially classical electrophilic reactions, aromatic nucleophilic substitution as well as by use of thienyl organometallics including 2-thienyl lithium. The latter reaction type is too well exemplified for complete coverage, but several such reactions are particularly illustrative. In a study of lithiated heterocycles, addition of 2-thienyl lithium to 2-aminoaldehydes occurs with 80-95% non-chelation control which is consistent with the Felkin-Ahn model of addition <89H707>. 2,5-Dilithiothiophene may be prepared from 2,5-dibromothiophene by lithium-halogen exchange <89B1984> or by direct dimetallation of thiophene <88S316> and reacts to give disubstituted thiophenes in either case. Bithienyl lithium may also be prepared by lithium halogen exchange <89JHC533>. Use of ultrasound activation of a bromo dihydrothiophene derivative results in reaction with magnesium to form a useful Grignard reagent <89JOC4138>.

Electrophilic reactions also functionalize thiophene. A convenient selective bromination to prepare 2-bromo- and 2,5-dibromothiophene is reported <88S890>. Friedel-Crafts acylation of thiophene produces α-sulfenylated derivatives <89SC443>. In the case of 2,5-dichlorothiophene, aluminum chloride catalyzed reaction with aromatic compounds leads to a simple preparation of 3-arylthiophene; hydrogenation removes the residual chlorine to produce **9** <88BCJ3779>. Whereas Vilsmeier formylation of thiophene is difficult, stannic chloride catalyzed addition of dichloromethyl butyl ether to thiophene gives formyl derivatives which are precursors to nitrile oxides <88MI221>. Stannic tetrachloride also catalyzes the addition of sugars to thiophene to produce **10** <88TL3537>. Formyl derivatives of thieno[3,4-*c*]thiophenes (**11**) may be prepared from thioethers under Vilsmeier conditions <89CC223>.

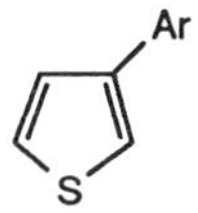
9

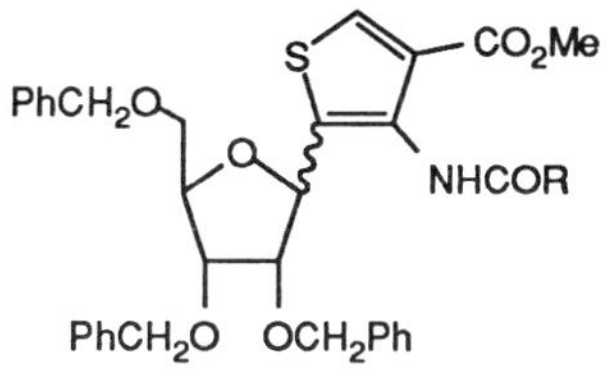
10

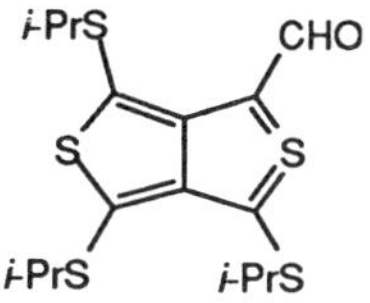
11

Palladium catalyzed coupling reactions also lead to thiophene derivatives. Pd(0) coupling of organotin derivatives with bromothiophenes results in bithiophene precursors to all isomers of dithienopyridines, for example, Equation 1 <88CS281, 89S130>.

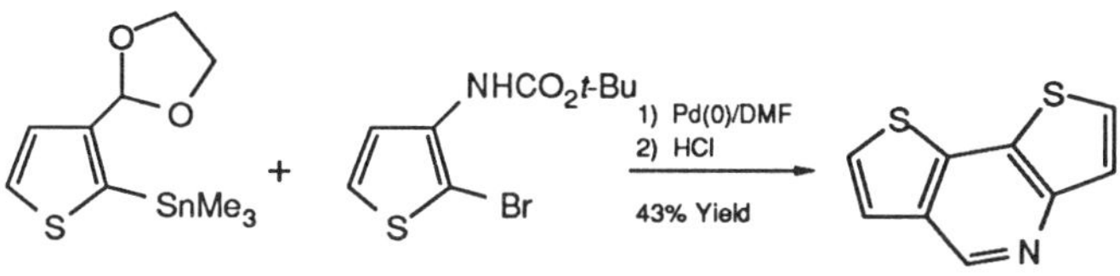
(Eq. 1)

Other bithiophene derivatives may also be prepared <89SC307>. Pd(0) catalyzed coupling of phenyl boric acid with 3-bromo-4-nitrothiophene leads to the precursor of thieno[*c*]quinolines <89JHC865>. In addition to aryl-aryl coupling, palladium catalyzed Heck alkylation of bromothiophene derivatives leads to aromatic retinoid precursors <88TL4409>. Coupling of acetylenes with 3-iodobenzo[*b*]thiophene produces 3-ethynyl derivatives in good yields <88CPB2248>. Tetrakistriphenylphosphine nickel catalyzes the regioselective perfluoroalkylation of thiophene <89MI385>.

Radical reactions also lead to substituted thiophenes. Irradiation of bis(heptafluorobutyryl) peroxide presumably leads to a radical intermediate which reacts on thiophene. While a 98% yield of 2-substitution on thiophene occurs, product ratios depend on the initial substitution pattern of the precursor thiophene <88BCJ3549>. Irradiation of 2-iodo-5-carbomethoxythiophene in aromatic solvents leads to 2-aryl substituted thiophene derivatives <89MI191>.

Ylide rearrangements result in some novel thiophene substitution reactions. Thiophenium ylide **12** rearranges cleanly to give the 2-thienylmalonate derivative (Eq. 2) <89CS97>. Photochemical decomposition of the bis(phenylsulfonyl)methylide of thiophene gives **13** cleanly in moderate yields <88JHC1599>. Sommelet-Hauser rearrangement of the ylide formed from 3-bromomethylthiophene (or benzo[*b*]thiophene) and dimethylsulfide gives **14** in good yields <89BCJ958>.

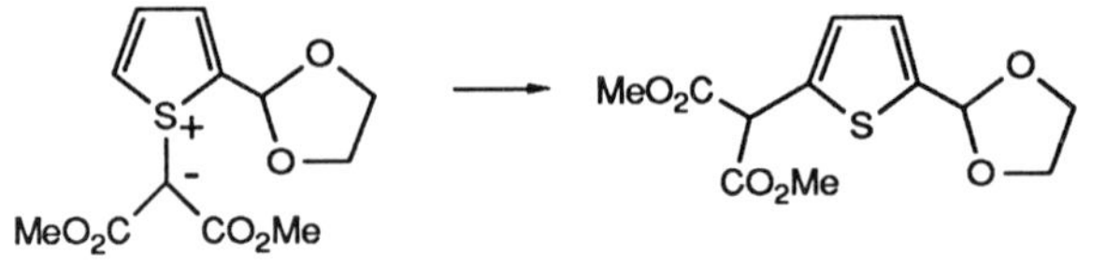

12

(Eq. 2)

Nucleophilic addition to activated thiophenes has been reported. Meisenheimer-type adduct **15** forms from 3-nitrothiophene and sodium methoxide <88CS349>. Oxidation of benzothiophene to the 1,1-dioxide derivative also activates the system to nucleophilic attack <89JOC3311>. 3-Amino derivatives of thiophene (**16**) are formed by nucleophilic displacement of the 3-bromo precursor <88MI553>.

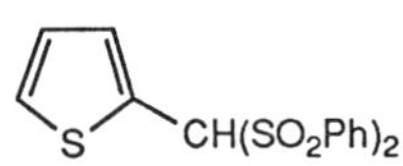

13

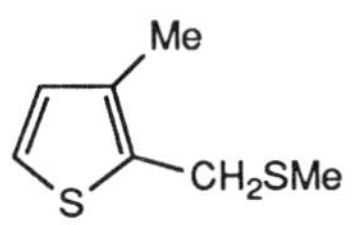

14

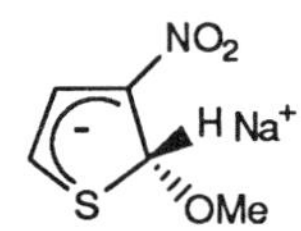

15

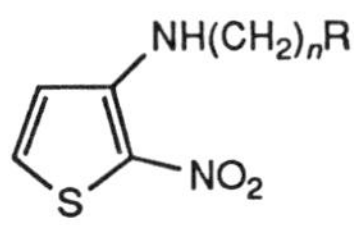

16

4.1.4 THIOPHENE RING FORMATION

Preparation of the thiophene ring has been reported using both classical and non-classical approaches. There are several examples of Gewald-type syntheses wherein a 2-aminothiophene derivative forms from condensation of a mercaptide anion with an unsaturated nitrile as in Equation 3 <89ZN(B)488, 88JPR585>.

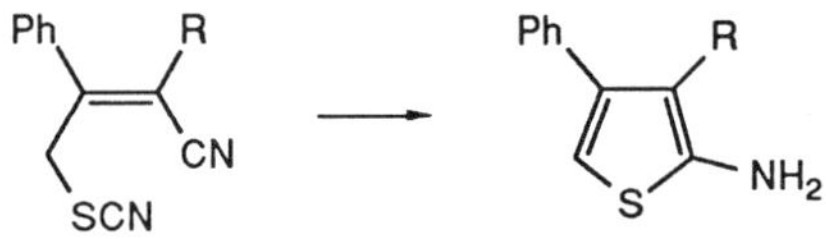

(Eq. 3)

Perkin condensation of mercaptoacetate derivatives with aldehydes also produces thiophene compounds in preparative yields (Equation 4) <89TL3315>. Other examples of this synthetic approach have also been reported <89ZC166, 88MI537, 88JIC695>.

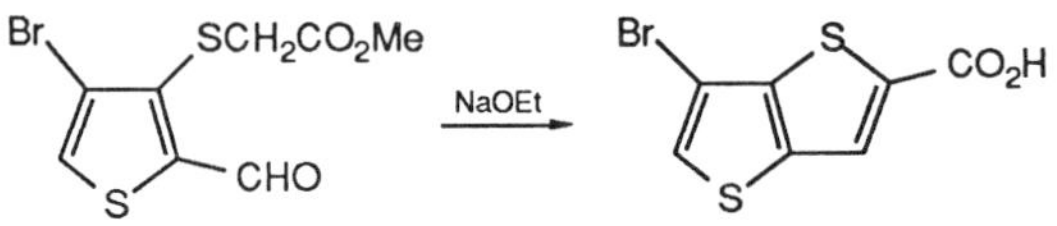

(Eq. 4)

An interesting application of this technology is the synthesis of thieno[2,3-*c*]quinolin-2-one derivatives wherein a Michael-type addition of a mercaptide leads to a thiophene which further lactamizes to product (Equation 5) <89MI377>. 2-Mercapto substituted thiophenes may also be prepared with this method

<88S556>. A novel synthesis of thieno[2,3-*c*]isothiazoles uses this approach <88FES457>.

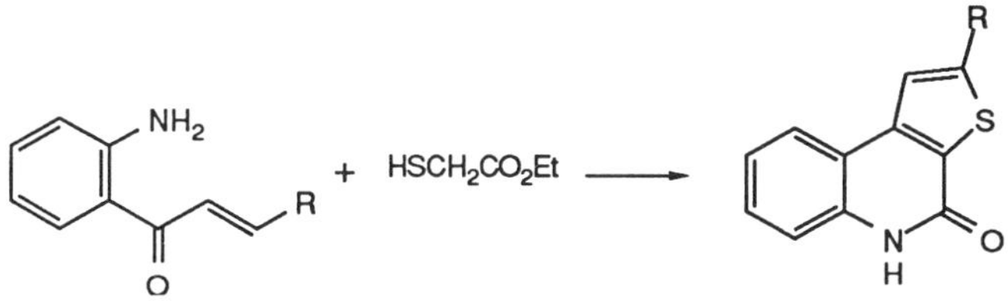

(Eq. 5)

A similar condensation on nitrile derivatives leads to 3-aminothiophene derivatives (Equation 6) <88LA633>. Other examples include <88BCJ4431, 88MI358, 88JCR(S)50, 89JA285>; the latter provides a pathway to the synthesis of (±)-urothione (molybdenum cofactor). Thieno[3,2-*a*]indolizine derivatives also may be prepared in this manner <89BCJ119>.

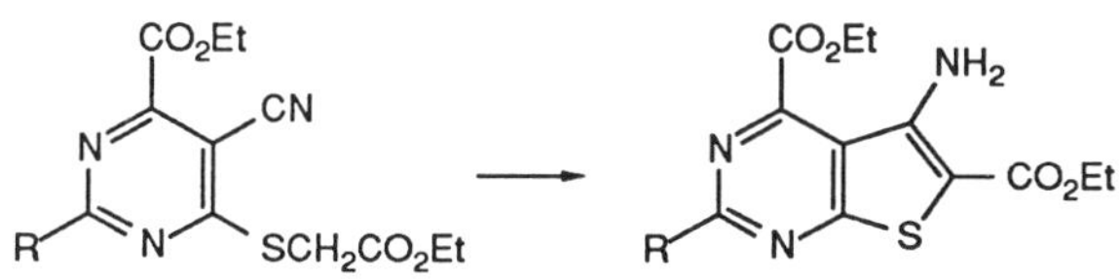

(Eq. 6)

Benzo[*b*]thiophenes form by acid catalyzed ring closure of the reaction product between benzenethiol derivatives and bromoacetaldehyde diethyl acetal <89JMC2548, 88JHC1271>. A one-pot method to prepare polysubstituted benzo[*b*]thiophenes involves aryl zirconium coupling with acetylenes followed by incorporation of sulfur using sulfur dichloride (Equation 7) <89JOC2793>.

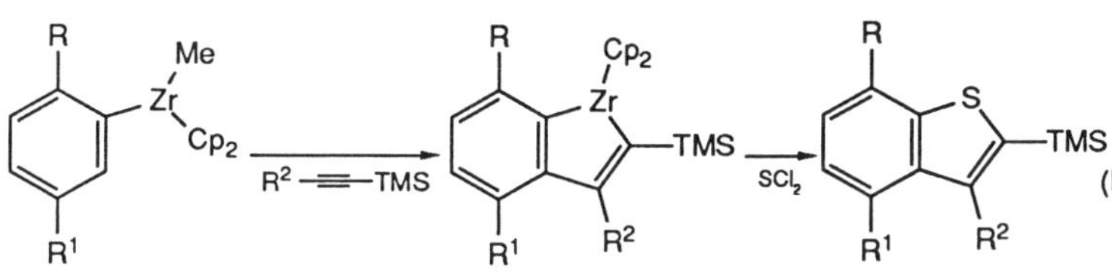

(Eq. 7)

Electrocyclic reactions are a powerful methodology to produce thiophene derivatives. Thus, 1,2,4-triazines undergo inverse electron demand Diels-Alder reactions to give dihydrothienopyridines (Equation 8) <89JOC4984>.

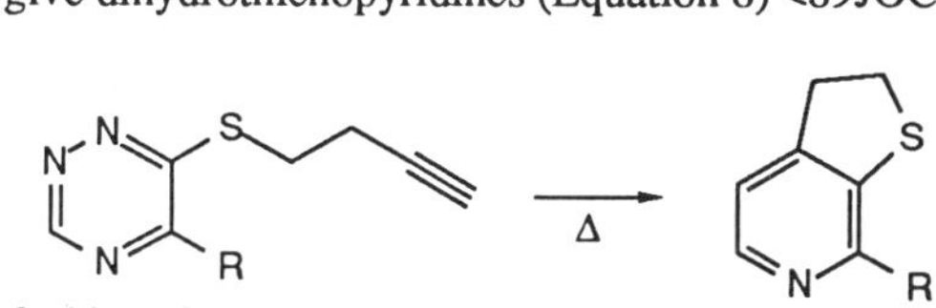

(Eq. 8)

In a similar fashion, furans undergo reaction to give dihydrobenzo[*c*]thiophenes (Equation 9) <88AP403>.

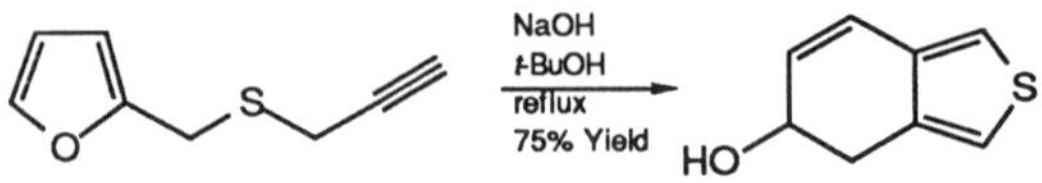

(Eq. 9)

An extremely interesting example of electrocyclic thiophene formation is the intramolecular closure of a tropolone to form a thieno-fused bicyclic derivative which may be induced to undergo photochemical isomerization to produce a novel thienopyridine derivative (Equation 10) <89CB657>.

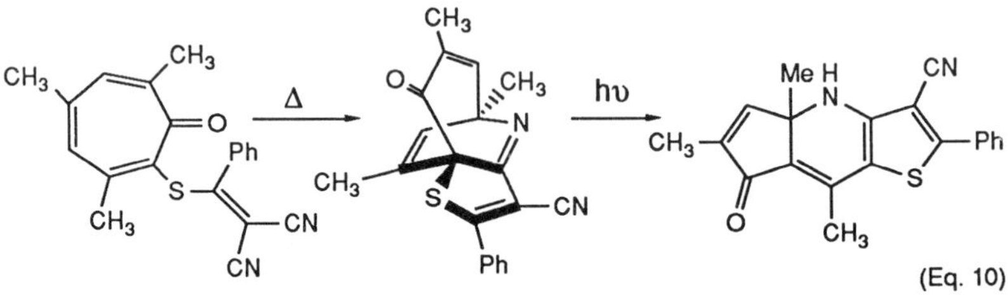

(Eq. 10)

Thiocarbonyl ylides undergo electrocyclic addition reaction with DMAD to give the 2,5-bridged bis(thiophenes) (Equation 11) <88CB313>.

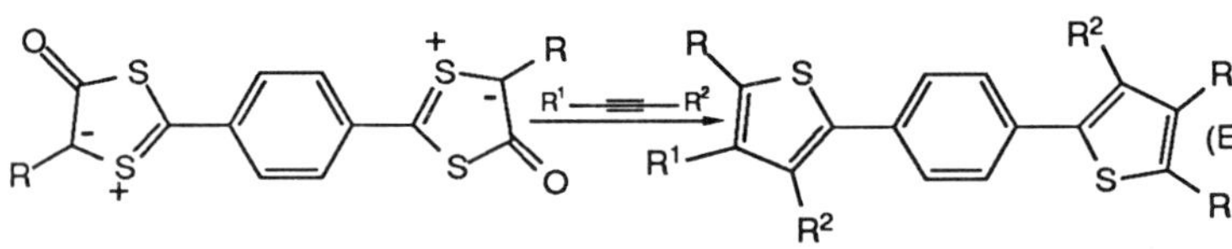

(Eq. 11)

Tetrahydrothiophene derivatives may form by reaction of thiocarbonyl compounds with diazomethane and subsequent electrocyclic reaction with a dienophile (Equation 12) <89JOC3664>.

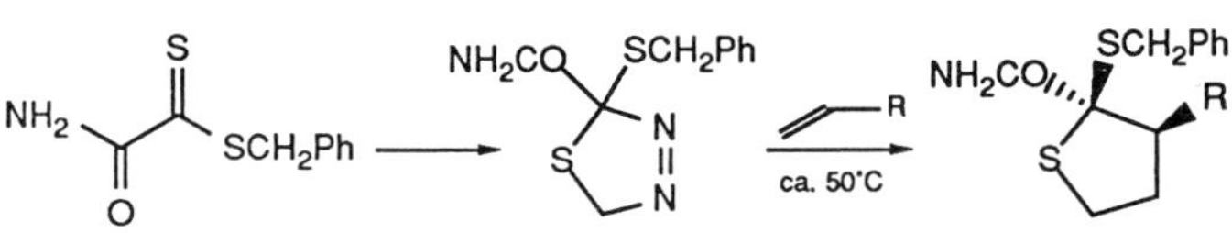

(Eq. 12)

Ring expansion may also be utilized to prepare thiophenes. Thus, treatment of the anion of an appropriate cyclopropane with carbon disulfide leads to good yields of 2-methylthiothiophenes <88LA465>. Conversion of β,γ-episulfido ketones with TFA give excellent yields of thiophenes <88S467>.

Other rings may form thiophene derivatives in surprising ways. Treatment of a dibenzothiepin with hydroiodic acid leads to a spirobenzo[*b*]thiophene presumably by cleavage of the benzylic thiomethyl bridge (Equation 13) <89CCC235>.

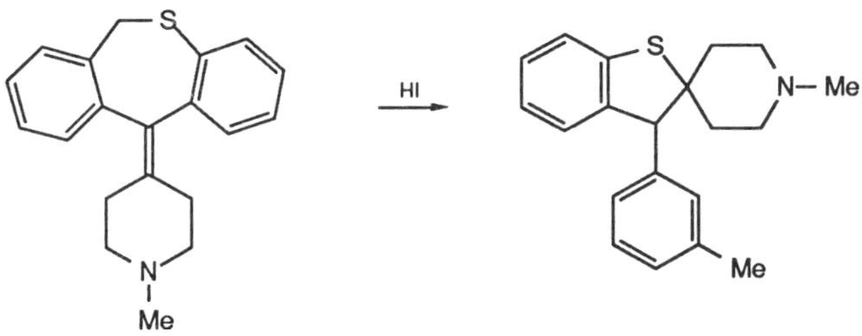

(Eq. 13)

Titanium tetrachloride induced pinacol condensation of dione **17** gives **18** after acid catalyzed dehydration <88JCS(C)959>. Alternatively, the theoretically interesting thieno[3,4-*c*]thiophene system may also be conveniently prepared by phosphine-catalyzed dimerization of **19** <88JHC559>.

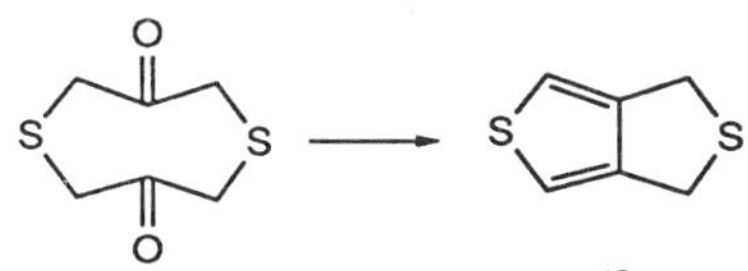

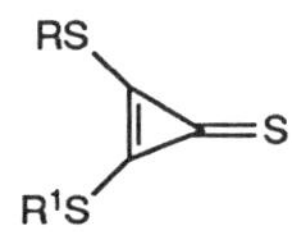

17 **18** **19**

Ring contraction of thiachromone leads to a substituted dihydrobenzo[*b*]thiophene (Equation 14) <88T2397>.

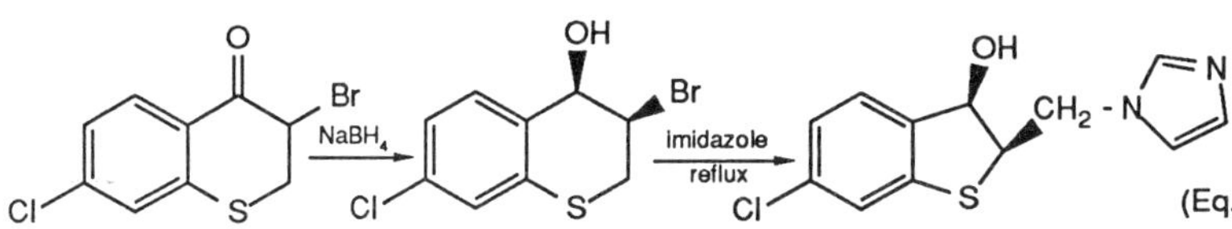

(Eq. 14)

Photochemical reactions also lead to the formation of thiophenes. Irradiation of **20** gives the corresponding R, R^1 substituted benzo[*b*]thiophene <88CL1637>. Heteroatom directed intramolecular photoarylation of **21** produces a naphthothiophene <89TL795>.

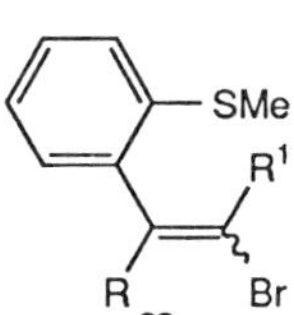

20

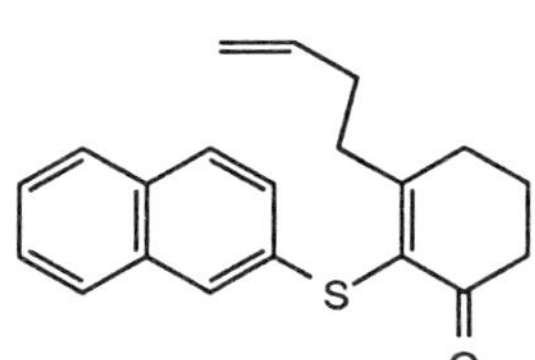

21

Several additional thiophene forming reactions need to be mentioned. α-Haloketones may be converted to α-*t*-butylthioketones which in turn react with acetylenic magnesium bromides to give 3,5- and 2,3,5-alkylthiophenes <89S210>. Addition of DMSO anion to quinoxaline produces thieno[3,4-*b*]quinoxalines <89H141>. 1,4-Dichlorides react with sodium sulfide to give tetrahydrothiophenes <89JMC875>. In an interesting 1-step conversion of methylthiobenzene, dilithiation and quenching with an acid chloride gives 3-substituted benzo[*b*]thiophene derivatives in moderate overall yields (Equation 15) <88S888>.

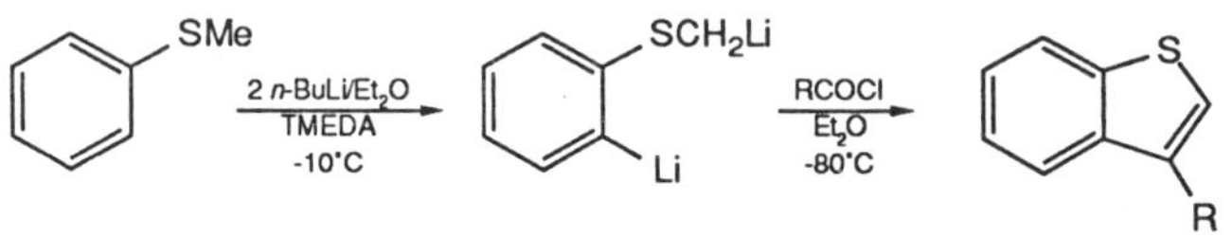

(Eq. 15)

In a different approach, reaction of α-bromoketones with Lawesson's reagent also produces 2,5-diarylthiophenes as well as thieno-fused polycycles depending on choice of the methylene ketone precursor <88PS127>. Introduction of sulfur to form a thiophene ring may also be accomplished with either hydrogen sulfide at 500°C over a heterogeneous catalyst <89JHC345> or by the reaction of thionyl chloride under normal laboratory conditions (Equation 16) <88JHC1363>.

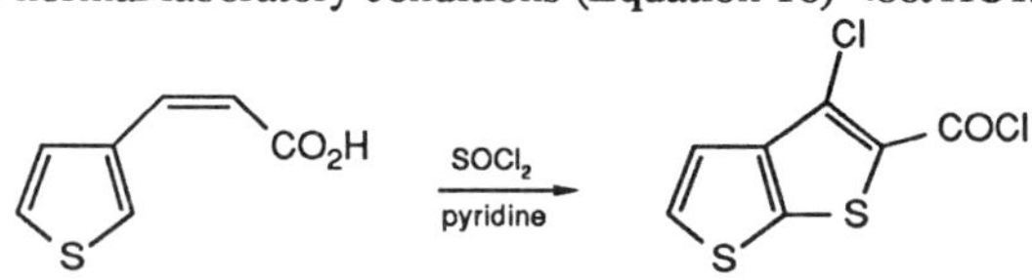

(Eq. 16)

4.1.5 RING ANNELATION ON THIOPHENE

Some very interesting electrocyclic reactions lead to novel thiophene derivatives. Intramolecular Diels-Alder reaction of thienofuran **22** provides a route to thiamarmelerin <89LA405>.

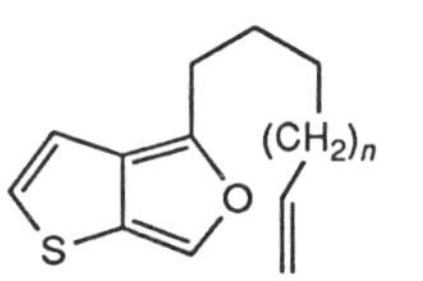

22

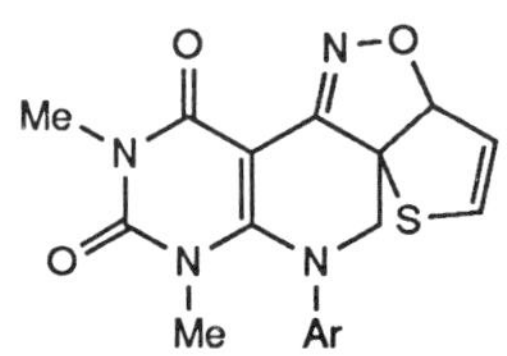

23

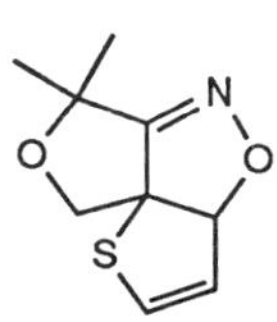

24

1,3-Dipolar intramolecular cycloaddition of a nitrile oxide onto the thiophene moiety gives **23** <88S342>. In a similar fashion, one double bond of thiophene acts as a dipolarophile to produce derivative **24** <89JOC5277>. Iminium ion-mediated cycloaddition of bithiophene with 1,4-dihydropyridine gives **25**; thiophene is not reactive enough for this condensation <89JOC4134>. Photochemically induced intramolecular reaction of thiophenones with olefins in methanol produces novel tricyclic thiolane derivatives such as **26** <88HCA502>.

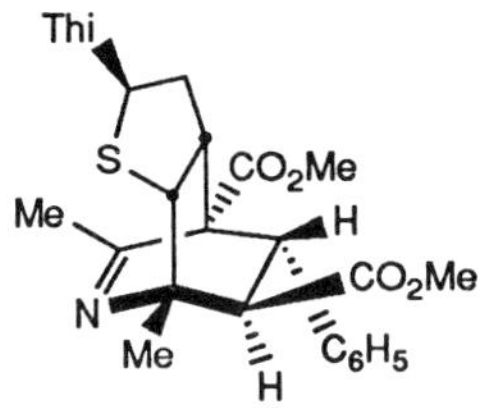

25

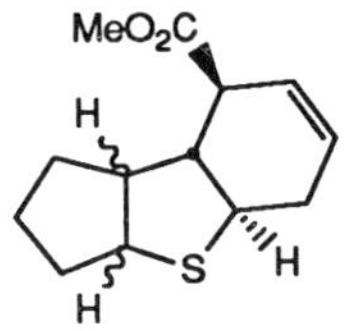

26

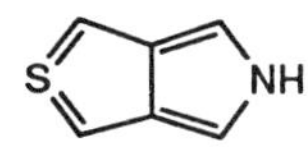

27

The theoretically interesting 4*H*-thieno[3,4-*c*]pyrrole (**27**) is formed by an intramolecular azide cycloaddition on an olefin and subsequent loss of diazomalonate. The compound undergoes trapping reaction with *N*-phenylmaleimide <88CC320>. *N*-Aryl derivatives of **27** as well as the [2,3-*c*]-isomers may be formed by condensations of anilines with thiophenedialdehydes <88JHC1531>. Intramolecular azide cycloaddition on a thiophene double bond leads to 4*H*-thieno[3,2-*b*]pyrrole derivatives <89TL1655>. Thieno[3,2-*c*]isothiazole **28** forms by oxidative closure of 2-amino-1-thioamidothiophene; **28** acts as a Diels-Alder diene and forms interesting substitution products upon diazotization <89JCR(S)29>. Thienoisoxazoles may be formed by intramolecular substitution of the oxime <88AP391>. 2-Azido-3-nitrothiophene reacts to form thieno[3,2-*c*]furazan oxide **29** which rapidly equilibrates with the [2,3-*c*] isomer <89JCS(PII)127>. Thieno[2,3-*c*]furan derivative **30** forms from an oxazolium iodide by reaction with phenyl Grignard reagent and subsequent ring closure and dehydration <89CB1119>.

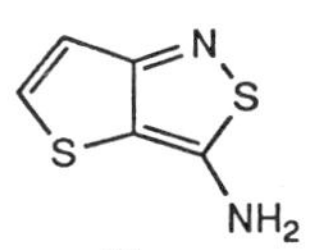

28

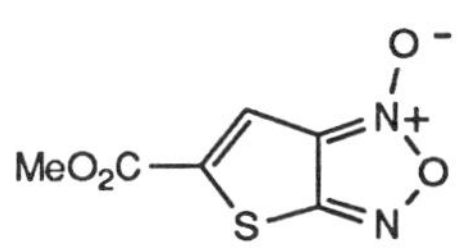

29

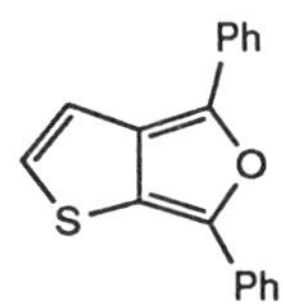

30

Benzothiophenes may be formed by fusion of a 6-membered ring onto the thiophene moiety. Acetylenes react with 2-vinylthiophenes via Diels-Alder methodology to give substituted benzo[*b*]thiophenes <89JHC397>. Cyclocarbonation of 2-allylthiophene derivatives with carbon monoxide/palladium catalysis leads to 4-oxybenzo[*b*]thiophenes <89TL95>. Benzo[*b*]thiophenes also form by rhodium (II) catalyzed intramolecular diazoketone insertion into a double bond of thiophene and silica gel catalyzed rearrangement of the cyclopropylketone intermediate <89JOC299>. Diels-Alder reaction of pyrano[3,4-*b*]thiophene-5-one (**31**) with acetylenes gives ester derivatives <88TL5817>. Nitrile **32** arises from a 4-keto precusor and provides a route to the preparation of the thieno[2,3-*g*]indole system <89JCS(PI)603>. The methyl carboxylate analogue of **32** is also reported <89JMC1265>. Oxidative photocyclization of **33** provides a route to 6-

substituted naphtho[2,1-*b*]thiophenes <88JHC1729>.

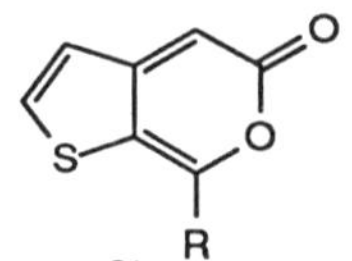
31

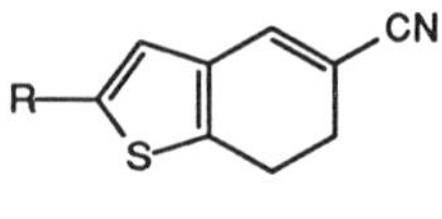
32

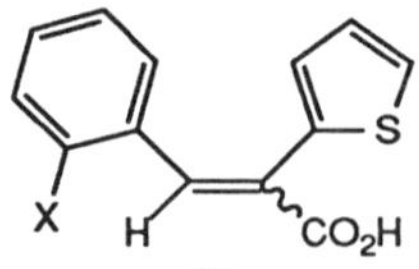
33

Thieno-fused 6-membered ring heterocycles form in a variety of ways. Thieno[2,3-*c*]pyridin-5-ol (**34**) and its isomers may be prepared from 3-substituted thiophenes using classical approaches <88JHC1571>. Compound **34** reacts with electrophiles on the pyridine ring α- to the hydroxyl moiety (X). Thienopyridine **35** is produced by intramolecular cyclization of an isocyanate and may be converted to amidine derivatives <89JMC1147>. The related thieno[2,3-*c*]pyridin-7-one system **36** forms from a Schmidt ring expansion or Beckmann rearrangement <88H1637>.

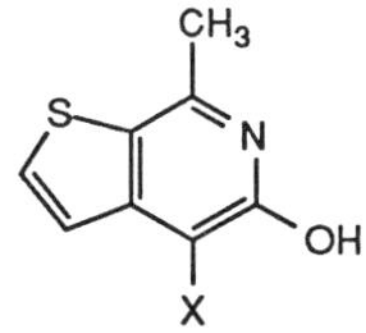
34

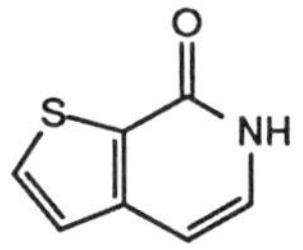
35

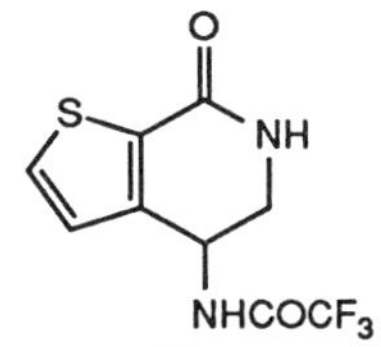
36

Thiophenic NADH models **37** may be prepared by dithionite reduction of the aromatic thienopyridine precursors <88T1079>. Some thieno analogues of protoberberine <88AJC111, 88AJC575> as well as of 1-benzylisoquinolines <88JCR(S)240> are reported. 1,3-Dipolar cycloaddition of a thienopyridine with DMAD produces a novel tricyclic system (Equation 17); similar reactions occur with thienopyridazine and thienopyrimidine analogues <89JOC990>.

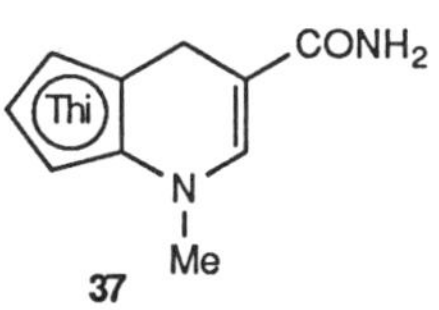
37

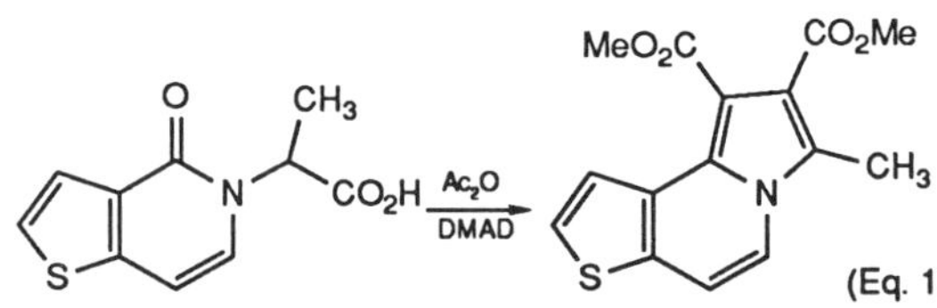
(Eq. 17)

Dinitrogen thieno-fused six-membered ring heterocyclic systems also are reported. All isomers of thienopyrimidinedione are prepared from appropriate aminothiophene carboxylates <88JMC1786> and intermediates in the synthesis also form novel tricyclic derivatives such as **38** <89CPB1197>. 2-Amino-3-cyanothiophene thiourea derivatives form 4-amino-2-thiothienopyrimidines **39** <89AP227, 89MI153>. 2-Aminothiophene-3-carboxylates also may be converted to thieno[2,3-*d*]pyrimidin-4-ones <88MI453>. Transformation of selected thienopyrimidin-4-one derivatives leads to imidazo[1,2-*c*]thieno[3,2-*e*]pyrimidines **40** <88MI756>. Diels-Alder reaction of 2,3-bismethylenebenzo[*b*]thiophene with diethyl azodicarboxylate gives the corresponding pyridazine <88CB1203>. Condensation of thiophene-2,3-diones with *o*-phenylenediamine produces either the thienopyrazine or furanopyrazine depending on reaction conditions <88M367>.

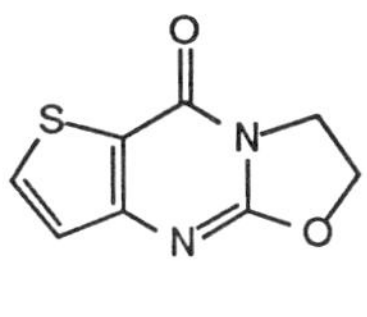

38

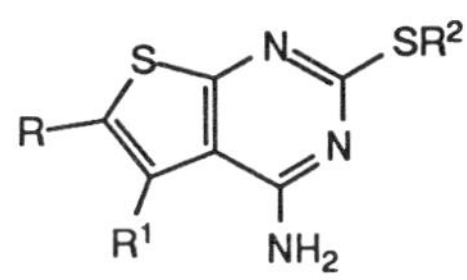

39

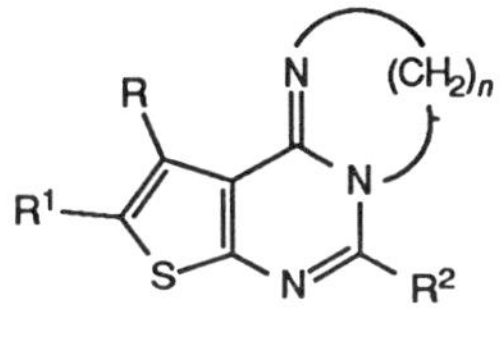

40

Thienothiopyrans **41** are formed by Michael-type addition of 2-mercaptothiophene to acrylates and subsequent Friedel-Crafts ring closure <88JOC9>. This ring system may also be prepared by Diels-Alder reaction technology <88CS311>. Aminothiophene carboxylates react with chlorosulfonamide and subsequent ring closure to form thieno[1,2,6]thiadiazine 2,2-dioxides such as **42** <89S280>. Dehydrocyclization over chromia-alumina catalyst at elevated temperatures form the novel silicon-containing benzo[*b*]thiophene heterocycle **43** <89KGS419, 88MI163>.

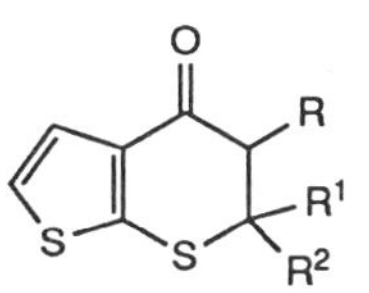

41

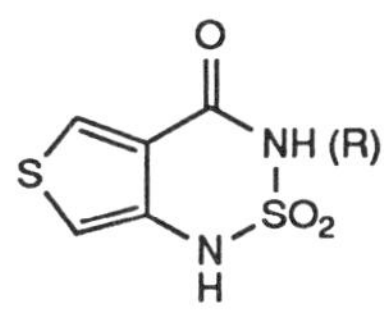

42

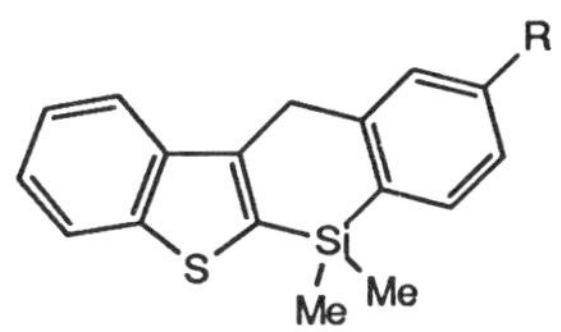

43

Other thieno-fused heterocycles have been reported. Thieno[2,3-*d*]azepine **44** forms by photolysis of benzo[*b*]thiophene azide in the presence of amines <88JCS(PI)1933>. Triazolo- and tetrazolo[1,5-*d*]thieno[3,2-*f*]-1,4-diazepin-6-ones **45** are prepared in two steps from an aminothiophene precursor <88JHC615>. Thieno[2,3-*c*][1,2,4]triazepines also may be prepared from aminothiophenes <88MI753>. Thermal ring expansion of *N*-oxide derivatives produce thieno[3,2-*e*][1,2]oxazepine **46** <88AJC1815>.

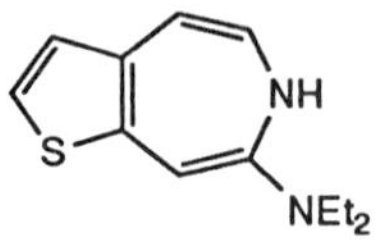

44

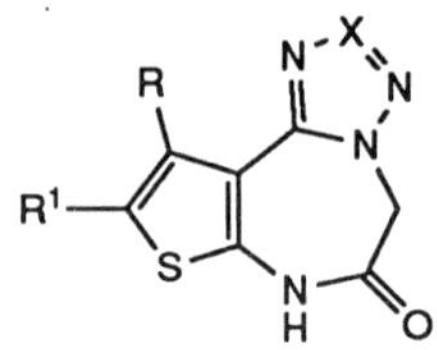

45

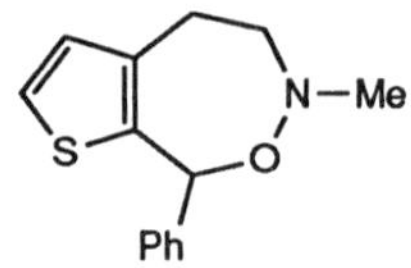

46

Some miscellaneous fused thiophene derivatives include the dimer **47** of 3,4-dibromothieno[2,3-*b*]thiophene <89AG1254>. A Stevens rearrangement produces a one carbon ring expansion to form amine **48** <89S327>. Several macrocyclic thiophene derivatives have selective metal ion coordination properties <89BCJ838, 89BCJ1346, 89CJC639>.

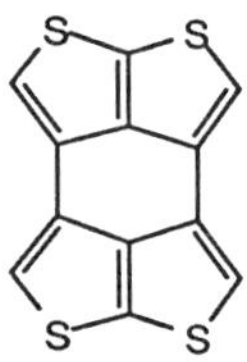

47

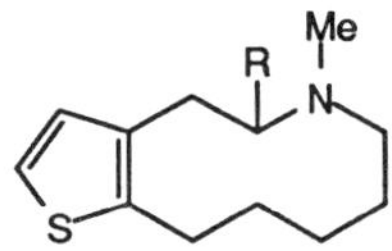

48

4.1.6 THIOPHENES AS INTERMEDIATES

Thiophene destruction is a useful method to produce novel non-thiophene ring systems. For example, 2,5-dichlorothiophene derivatives react with diazo compounds under rhodium catalysis to produce S,C-thiophenium methylides which undergo smooth thermal rearrangement to 1,4-oxathiocines **49** (Equation 18) <88CC138, 89JOC4822>. The oxathiocines are heteroanalogues of eight-membered ring 10π systems but have no aromatic character.

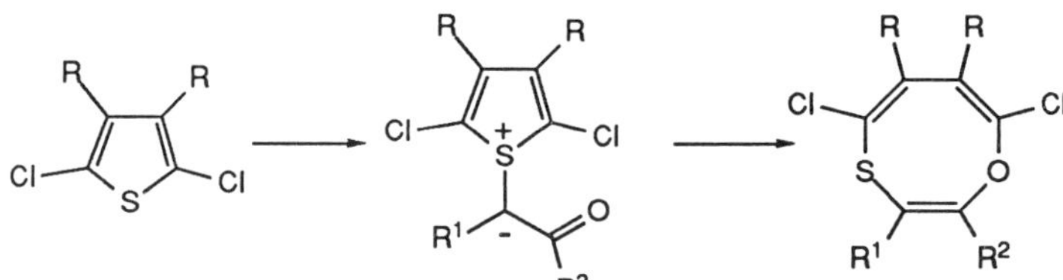

(Eq. 18)

49

Thermolysis of 2-azidobenzo[*b*]thiophene in the presence of olefins leads to the regioselective formation of thiochroman **50** presumably by the intermediacy of a ring-opened cyano "enethione" <88JCS(PI)3375>. 2-Thiophenamines react with hydrazine to form ring-opened cyanothione intermediates which further react with hydrazine to form novel heterocycles (Equation 19) <88CB573>.

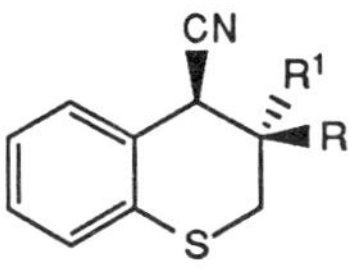

50

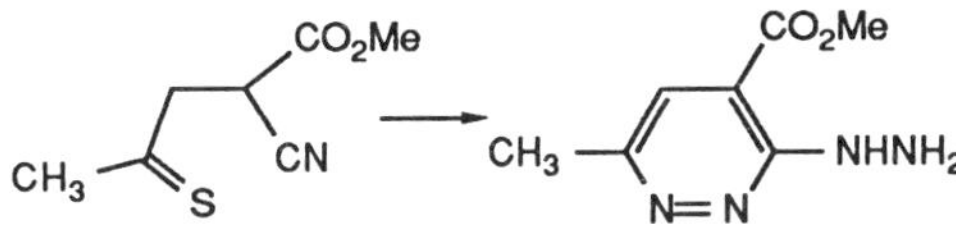

(Eq. 19)

Reaction of 2,5-dihydrothiophenes with benzyne produces sulfonium ylide intermediates which rearrange to 1-phenylthio-1,3-butadienes in good yields <89TL847>. Reaction of thiophene derivatives with benzyne results in predominant [4+2] cycloaddition to produce naphthalenes; however, 2-halogen substituted thiophenes react in a [2+2] fashion presumably as a consequence of the enhanced 2π enamine-like character of these thiophene derivatives <88JOC5799, 89JOC2142>. Rhodium catalyzed carbenoid insertion of a diazo precursor produces 1,3-dihydrothiophene 2,2-dioxide **51** in a regiospecific manner. Compound **51** serves as a starting point for diene preparation as well as for dianion based chemistry <89CJC1071>. Menthyltrihalothiophene S,S-dioxides are efficient sources of asymmetric dienes in Diels-Alder reactions which produce regio- and stereoisomerically pure adducts (Equation 20) <88T3127>. Photolysis of thiophen-2(5*H*)-ones in methanol results in the formation of 4-mercaptobutenoates <89HCA83>.

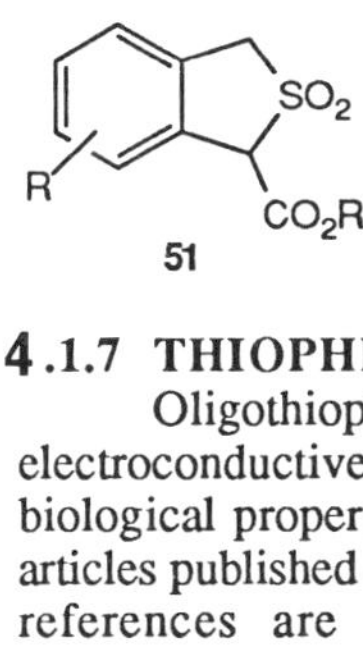

51

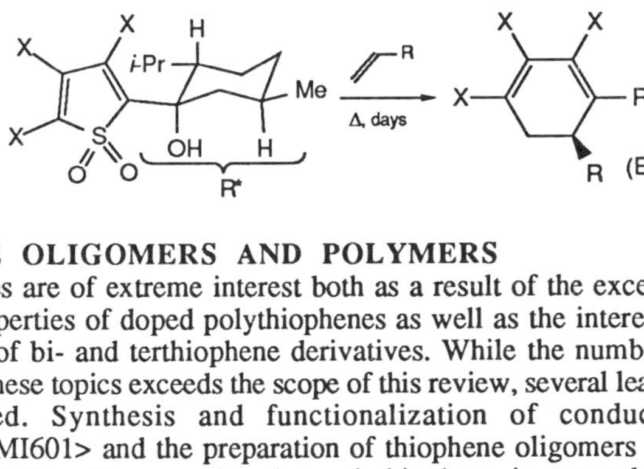

(Eq. 20)

4.1.7 THIOPHENE OLIGOMERS AND POLYMERS

Oligothiophenes are of extreme interest both as a result of the excellent electroconductive properties of doped polythiophenes as well as the interesting biological properties of bi- and terthiophene derivatives. While the number of articles published on these topics exceeds the scope of this review, several leading references are noted. Synthesis and functionalization of conducting poly(thiophenes) <88MI601> and the preparation of thiophene oligomers have been reviewed <88H1731>. The paradox that polythiophene is not stable at electrochemical potentials used for its synthesis has been discussed <89MI263>.

Polyether functionality on the structure of poly(thiophenes) produces highly conducting polymers <89CC679>. Electropolymerization of chirally substituted thiophene gives a "chiral metal" which might exhibit exotic properties <88CC658, 88CC917>. The electropolymerization of cyclopenta[*c*]thiophene and its isomers produces material intermediate in conductivity between poly(3-methylthiophene) and poly(3,4-dimethylthiophene) <88MI93>. Poly(dithieno-[3,4-*b*:3',4'-*d*]thiophene) is a new transparent conducting polymer <88CC246>.

Extended donors have utility in redox systems. Diphenoquinone analogue **52** is a novel amphoteric multistage redox system with electrical conductivity. The amphotericity is enhanced by heteroatom stabilization of radicals <89JA5483>. Compound **53** is a "Wertz-type" donor with a good coplanar configuration in its oxidized state <89TL2091>. A thieno[3,2-*b*]thiophene derivative prepared from a cyclopropenethione is a novel three color electrochromic system <88AG(E)560>.

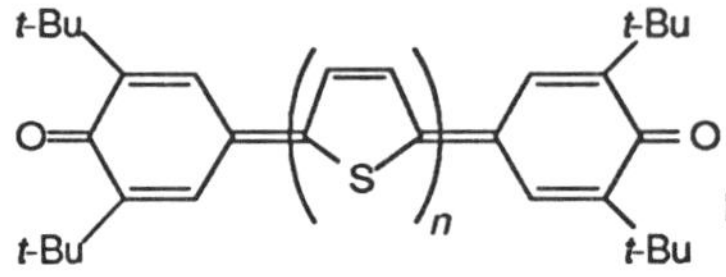

52

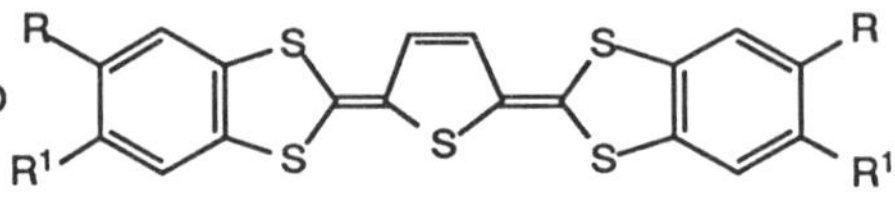

53

4.1.8 SELENOPHENES AND TELLUROPHENES

There are very few recently reported examples of these ring systems. The most synthetically useful synthesis of selenophenes utilizes titanium tetrachloride induced pinacol coupling to form a diol derivative which eliminates two moles of water on treatment with tosic acid (Equation 21) <88TL1399>. 3-Methylselenophene forms by reaction of 2-methylbutadiene with Se vapor at 450°C in a pyrex glass tube <89S181>.

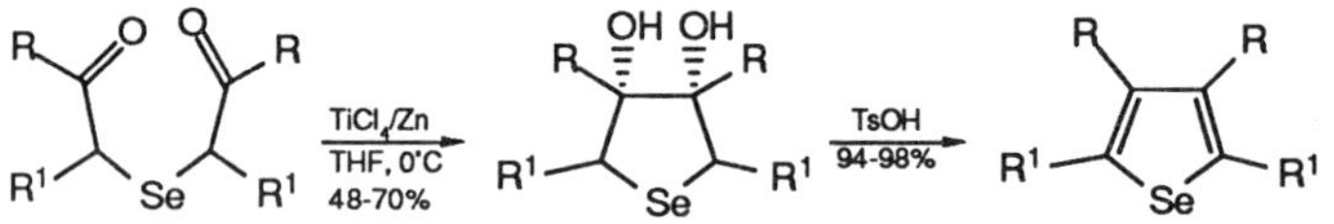

(Eq. 21)

Seleno[3,4-*d*]thiazole **54** forms by condensation of a bromoketone with the appropriate selenamide precursor <89JHC709>. Studies of the electrochemical properties of poly(selenophene) are hampered by poor availability of material but **55** has been studied <88CL1179>. Mixed oligomers containing thiophene and selenophene are prepared by nickel catalyzed cross-coupling reactions <88CC988>.

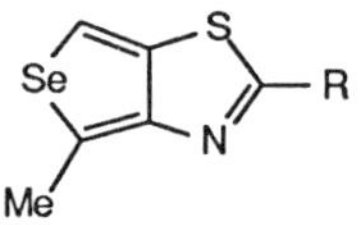

54

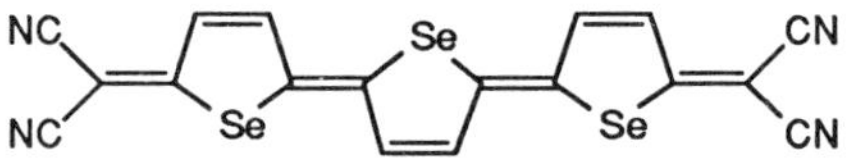

55

The sole example of tellurophene synthesis utilizes an acetylenic epichlorohydrin reaction with sodium telluride (Equation 22) <88TL4923>.

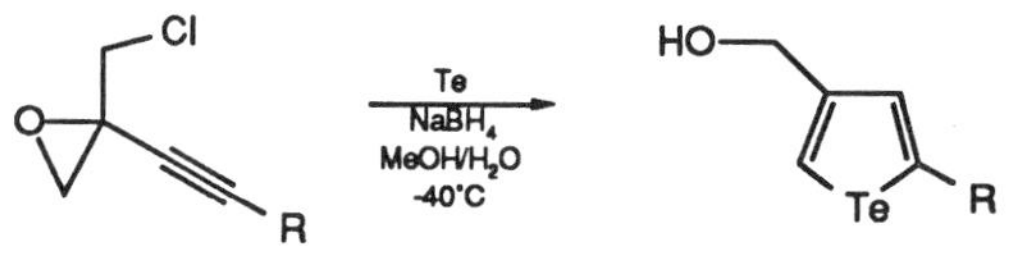

(Eq. 22)

4.1.9 REFERENCES

84TL5005	B.H. Lipshutz, R.S. Wilhelm and J.A. Kozlowski; *Tetrahedron Lett.*, 1984, **24**, 5005.
85HC353	J. B. Press; Thiophene and Its Derivatives, Part 1, Chemistry of Heterocyclic Compounds, 1985, **44**, 353-456.
87S325	B.H. Lipshutz; *Synthesis*, 1987, 325.
87TL945	B.H. Lipshutz, M. Koerner and D.A. Parker; *Tetrahedron Lett.*, 1987, **28**, 945.
88AG(E)560	T. Sugimoto, T. Nagatomi, H. Ando and Z.-ichi Yoshida; *Angew. Chem., Int. Ed. Engl.*, 1988, **27**, 560.
88AJC111	J.B. Bremner, E.J. Browne, L.M. Engelhardt, G.S. James and A.H. White; *Aust. J. Chem.*, 1988, **41**, 111.
88AJC575	J.B. Bremner, E.J. Browne, L.M. Engelhardt, B.A. Hyland, G.S. James and A.H. White; *Aust. J. Chem.*, 1988, **41**, 575.
88AJC1815	J.B. Bremner, E.J. Browne, L.M. Engelhardt, C.S. Greenwood and A.H. White; *Aust. J. Chem.*, 1988, **41**, 1815.
88AP391	D. Binder, C.R. Noe, K. Baumann and W. Holzer; *Arch. Pharm.* (Weinheim, Ger.), 1988, **321**, 391.
88AP403	H.F.G. Linde, N. Krämer and A. Flohr; *Arch. Pharm.* (Weinheim, Ger.), 1988, **321**, 403.
88BCJ3549	M. Yoshida, T. Yoshida, N. Kamigata and M. Kobayashi; *Bull. Chem. Soc. Jpn.*, 1988, **61**, 3549.
88BCJ3779	T. Sone, M. Inoue and K. Sato; *Bull. Chem. Soc. Jpn.*, 1988, **61**, 3779.
88BCJ4431	G.E.H. Elgemie, H.A. Elfahham and H.A. Nabey; *Bull. Chem. Soc. Jpn.*, 1988, **61**, 4431.
88CB313	H. Gotthardt, W. Pflaumbaum and P. Gutowski; *Chem. Ber.*, 1988, **121**, 313.
88CB573	K. Gwald, U. Hain and M. Gruner; *Chem. Ber.*, 1988, **121**, 573.
88CB1203	G. Dyker and R.P. Kreher; *Chem. Ber.*, 1988, 121, 1203.
88CC138	O. Meth-Cohn and E. Vuorinen; *J. Chem. Soc., Chem. Commun.*, 1988, 138.
88CC246	A. Bolognesi, M. Catellani, S. Destri, R. Zamboni and C. Taliani; *J. Chem. Soc., Chem. Commun.*, 1988, 246.
88CC320	C.-K. Sha, C.-P. Tsou and S.-L. Wang; *J. Chem. Soc., Chem. Commun.*, 1988, 320.
88CC658	M. Lemaire, D. Delabouglise, R. Garreau, A. Guy and J. Roncali; *J. Chem. Soc., Chem. Commun.*, 1988, 658.
88CC917	D. Kotkar, V. Joshi and P.K. Ghosh; *J. Chem. Soc., Chem. Commun.*, 1988, 917.
88CC988	R. Shabana, A. Galal, H.B. Mark, Jr., H. Zimmer, S. Gronowitz and A.B. Hörnfeldt; *J. Chem. Soc., Chem. Commun.*, 1988, 988.
88CL1179	K. Yui, Y. Aso, T. Otsubo and F. Ogura; *Chem. Lett.*, 1988, 1179.
88CL1637	T. Kitamura, S. Kobayashi and H. Taniguchi; *Chem. Lett.*, 1988, 1637.
88CPB2248	T. Sakamoto, T. Nagano, Y. Kondo and H. Yamanaka; *Chem. Pharm. Bull.*, 1988, **36**, 2248.
88CS281	S. Gronowitz, A.-B. Hörnfeldt and Y. Yang; *Chem. Scr.*, 1988, **28**, 281.
88CS311	M.A.Kalik, Ya L. Gol'dfarb and M.M. Karyushkin; *Chem. Scr.*, 1988, **28**, 311.
88CS349	F. Sancassan; *Chem. Scr.*, 1989, **28**, 349.

88FES457	G. Romeo, E. Bousquet, M.S. Pappalardo, G. Ronsisvalle, S. Oliveri and E. Cammarata; *Farmaco E. Sci.*, 1988, **43**, 457.
88H1637	P. Dallemagne, S. Rault, M. C. de Sévricourt and M. Robba; *Heterocycles*, 1988, **27**, 1637.
88H1731	J. Nakayama, T. Konishi and M. Hoshino; *Heterocycles*, 1988, **27**, 1731.
88HCA502	R. Kiesewetter, A. Graff and P. Margaretha; *Helv. Chim. Acta*; 1988, **71**, 502.
88JCR(S)50	T.T. Raj and S.Y. Ambekar; J. *Chem. Res. (S)*; 1988, 50.
88JCR(S)240	P.R. Huddleston, J.M. Barker, B. Stickland, M.L. Wood and L.H.M. Guindi; *J. Chem. Res.* (S), 1988, 240.
88JCS(C)959	J. Nakayama, A. Ishii, Y. Kobayshi and M. Hoshino; *J. Chem. Soc.* (C), 1988, 959.
88JCS(PI)1933	C.M. Daly, B. Iddon, H. Suschitzky, U. Jordis and F. Sauter; *J. Chem. Soc., Perkin Trans. I*, 1988, 1933.
88JCS(PI)3375	P. Spagnolo and P. Zanirato; *J. Chem. Soc., Perkin Trans. I*, 1988, 3375.
88JHC559	S. Yoneda, K. Ozaki, A. Tsubouchi and H. Kojima; *J. Heterocycl. Chem.*, 1988, **25**, 599.
88JHC615	C.J. Shishoo, M.B.Devani, G.V. Ullas, S. Ananthan and V.S. Ghadti; *J. Heterocycl. Chem.*, 1988, **25**, 615.
88JHC1271	P.A. Plé and L.J. Marnett; *J. Heterocycl. Chem.*, 1988, **25**, 1271.
88JHC1363	S.L. Castle, J.-K. Luo, H. Kudo and R.N. Castle; *J. Heterocycl. Chem.*, 1989, **25**, 1363.
88JHC1531	F. Benachenhou, M.A. Mesli, M. El Borai, B. Hanquet and R. Guilard; *J. Heterocycl. Chem.*, 1988, **25**, 1531.
88JHC1571	J.B. Press and J.J. McNally; *J. Heterocycl. Chem.*, 1988, **25**, 1571.
88JHC1599	L. Hatjiarapoglou and A. Varvoglis; *J. Heterocycl. Chem.*, 1988, **25**, 1599.
88JHC1729	Y. Ming and D.W. Boykin; *J. Heterocycl. Chem.*, 1988, **25**, 1729.
88JIC695	S. El-Bahaie, E.E. Bayoumy and Sh. Youssif; *J. Indian Chem. Soc.*, 1988, **65**, 695.
88JMC1786	R.K. Russell, J.B. Press, R.A. Rampulla, J.J. McNally, R. Falotico, J.A. Keiser, D.A. Bright and A. Tobia; *J. Med. Chem.*, 1988, **31**, 1786.
88JOC9	G. S. Ponticello, M.B. Freedman, C.N. Habecker, M.K. Holloway, J.S. Amato, R.S. Conn and J.J. Baldwin; *J. Org. Chem.*, 1988, **53**, 9.
88JOC5366	E. Piers and R.D. Tillyer; *J. Org. Chem.*, 1988, **53**, 5366.
88JOC5799	D.Del Mazza and M.G. Reinecke; *J. Org. Chem.*, 1988, **53**, 5799.
88JPR585	F.M. Abdelrazek and H.A. Ead; *J. Prakt. Chem.*, 1988, **330**, 585.
88JST111	M. Randic, B.M. Gimarc, S. Nikoli´c and N. Trinajstic; *J. Mol. Struct.*, 1988, **181**, 111.
88LA465	C. Brüekner and H.-U. Reißig; *Liebigs Ann. Chem.*, 1988, 465.
88LA633	W. Ried and G. Beller; *Liebigs Ann. Chem.*, 1988, 633.
88M367	E. Terpetschnig, W. Ott, G. Kollenz, K. Peters, E.-M. Peters and H.G. von Schnering; *Monatsh. Chem.*, 1988, **119**, 367.
88MI36	S.C. Huckett, R.J. Angelici, M.E. Ekman and G.L. Schrader; *J. Catal.*, 1988, **113**, 36.
88MI49	F. Sauter; *Mitteilungsbl.-Chem. Ges. D.D.R.* 1988, **35**, 49. [Chem. Abstr., 1988, **109**, 128722y].
88MI93	R. Garreau, J. Roncali, F. Garnier and M. Lemaire; *J. Chem. Phys. Phys.-Chim. Biol.*, 1988, **86**, 93; [Chem. Abstr., 1989, **111**, 58458g].
88MI163	V.M. Polosin, A.A. Astakhov, A.V. Ivashenko, M.A. Ryashentseve, E.P. Belanova and K.M. Minachev; *Sulfur Lett.* 1988, **8**, 163.
88MI221	L.I. Belenky, A.A. Loktionov and M.M. Krayushkin; *Stud. Org. Chem.* (Amsterdam), 1988, **35**, 221; [Chem. Abstr., 1988, **110**, 114604v].
88MI243	A. Santucci, S.M. Murgia, G. Paliani and R.S. Cataliotti; *Spectrosc. Lett.*, 1988, **21**, 243; [Chem. Abstr., 1988, **108**, 212670s].
88MI358	H. Vieweg, S. Leistner and G. Wagner; *Pharmazie*, 1988, **43**, 358.
88MI453	M. Perrissin, M. Favre, C.L. Duc, F. Huguet, C. Gaultier and G. Narcisse;

Eur. J. Med. Chem., 1988, **23**, 453.
88MI537 S. El-Bahaie, A.M. Kadry, M.G. Assy and Y.A. Ibrahim; *Pharmazie*, 1988, **43**, 537.
88MI553 G. Ronsisvalle, M.S. Pappalardo, F. Vittorio, L. Pasquinucci, A. Caruso, A. Felice and M. Amico-Roxas; *Eur. J. Med. Chem.*, 1988, **23**, 553.
88MI601 J. Roncali, R. Garreau, D. Delabouglise, F. Garnier and M. Lemaire; *Makromol. Chem. Macromol. Symp.*, 1988, **20/21**, 601.
88MI753 P. Richter, D. Oertel and F. Oertel; *Pharmazie*, 1988, **43**, 753.
88MI756 S. Leistner, M. Gütschow, H. Vieweg, G. Wagner, Th. Strohscheidt and R. Grupe; *Pharmazie*, 1988, **43**, 756.
88PS127 C.-D. Czogalla and F. Boberg; *Phosphorous Sulfur*, 1988, **35**, 127.
88S316 B.L. Feringa, R. Hulst, R. Rikers and L. Brandsma; *Synthesis*, 1988, 316.
88S342 D. Prajapati and J.S. Sandhu; *Synthesis*, 1988, 342.
88S467 M.S.F.L.K. Jie and Y.G. Zheng; *Synthesis*, 1988, 467.
88S556 A. Dalta, H. Ila and H. Junjappa; *Synthesis*, 1988, 556.
88S888 S. Cabiddu, D. Cancellu, C. Floris, G. Gelli and S. Melis; *Synthesis*, 1988, 888.
88S890 M.A. Keegstra and L. Brandsma; *Synthesis*, 1988, 890.
88T1079 J. Cazin, T. Trefouel, G. Dupas, J. Bourguignon and G. Quequiner; *Tetrahedron*, 1988, **44**, 1079.
88T2397 D.F Rane, R.E. Pike, M.S. Puar and J.J. Wright; *Tetrahedron*, 1988, **44**, 2397.
88T3127 J.L.M. Dillen, O. Meth-Cohn, C. Moore and P.H. van Rooyen; *Tetrahedron*, 1988, **44**, 3127.
88TL117 P.M.S. Chauhan, G. Jenkins, S.M. Walker and R.C. Storr; *Tetrahedron Lett.*, 1988, **29**, 117.
88TL1399 J. Nakayama, F. Murai, M. Hoshino and A. Ishii; *Tetrahedron Lett.*, 1988, **29**, 1399.
88TL3537 S.P. Rao, K.V.B. Rao, B.A. Otter, R.S. Klein and W.-Y. Ren; *Tetrahedron Lett.*, 1988, **29**, 3537.
88TL4409 J. Eustache, J.-M. Bemardon and B. Shroot; *Tetrahedron Lett.*, 1988, **29**, 4409.
88TL4923 R.P. Discordia and D.C. Dittmer; *Tetrahedron Lett.*, 1988, **29**, 4923.
88TL5817 P.M. Jackson, C.J. Moody and P. Shah; *Tetrahedron Lett.*, 1988, **29**, 5817.
88TL6239 N. Münzel, K. Kesper, A. Schweig and H. Specht; *Tetrahedron Lett.*, 1988, **29**, 6239.
89AG1254 Y. Kono, H. Miyamoto, Y. Aso, T. Otsubo, F. Ogura and T. Tanaka, M. Sawada; *Angew. Chem.*, 1989, **101**, 1254.
89AP227 S. Leistner, M. Gütschow and G. Wagner; *Arch. Pharm.* (Weinheim, Ger.), 1989, **322**, 227.
89B1984 W.D. Wilson, F.A. Tanious, R.A. Watson, H.J. Barton, A. Strekowska, D.B. Harden and L. Strekowski; *Biochemistry*, 1989, **28**, 1984.
89BCJ119 A. Kakehi, S. Ito, T. Fujii, Y. Morimoto, S.-i. Matsumoto and M. Shiohara; *Bull. Chem. Soc. Jpn.*, 1989, **62**, 119.
89BCJ838 T. Sone, K. Sato and Y. Ohba; *Bull. Chem. Soc. Jpn.*, 1989, **62**, 838.
89BCJ958 M. Yamamoto, M. Kakinuma, S. Kohmoto and K. Yainada; *Bull. Chem. Soc. Jpn*, 1989, **62**, 958.
89BCJ1346 T. Sone, Y. Ohba and R. Watanabe; *Bull. Chem. Soc. Jpn.*, 1989, **62**, 1346.
89CB1119 A. Schöning, T. Debaerdemaeker, M. Zander and W. Friedrichson; *Chem. Ber.*, 1989, **122**, 1119.
89CB657 K. Hartke, W.F. Richter, W. Massa, G. Baren; *Chem. Ber.*, 1989, **122**, 657.
89CC223 A. Tsubouchi, N. Matsumura, H. Inoue, N. Haimasaki, S. Yoneda and K. Yanagi; *J. Chem. Soc., Chem. Commun.*, 1989, 223.
89CC679 J. Roncali, R. Garreau, D. Delabouglise, F. Garnier and M. Lemaire; *J.*

	Chem. Soc., Chem. Commun., 1989, 679.
89CCC235	Z. Polívka, V. Valenta, K. Sindelár, J. Holubek, M. Budesínsky, M. Ryska, I. Koruna, J. Kohoutová, J. Metys, J. Metysová, M. Valchár and M. Protiva; *Collect. Czech. Chem. Commun.*, 1989, **54**, 235.
89CJC639	C.R. Lucas, S. Liu, M.J. Newlands, J.-P. Charland and E.J. Gabe; *Can. J. Chem.*, 1989, **67**, 639.
89CJC1071	S.D. Babu, M.D. Hrytsak and T. Durst; *Can. J. Chem.*, 1989, **67**, 1071.
89CPB1197	H. Fukumi, M. Sugiyama and T. Sakamoto; *Chem. Pharm. Bull.*, 1989, **37**, 1197.
89CS97	S. Bien, S. Gronowitz and A.B. Hörnfeldt; *Chem. Scr.*, 1989, **29**, 97.
89H141	J.-M. Vierfond, L. Legendre, J. Mahuteau and M. Miocque; *Heterocycles*, 1989, **29**, 141.
89H707	M.T. Reetz, W. Reif and X. Holdgrün; *Heterocycles*, 1989, **28**, 707.
89H1135	M.J.S. Dewar and A.J. Holder; *Heterocycles*, 1989, **28**, 1135.
89HCA83	R. Kiesewetter and P. Margaretha; *Helv. Chim. Acta*, 1989, **72**, 83.
89JA285	E.C. Taylor and L.A. Reiter; *J. Am. Chem. Soc.*, 1989, **111**, 285.
89JA1533	K.W. Zilm, R.A. Merrill, G.G. Webb, M.M. Greenberg and J.A. Berson; *J. Am. Chem. Soc.*, 1989, **111**, 1533.
89JA3659	K.J. Stone, M.M. Greenberg, S.C. Blackstock and J.A. Berson; *J. Am. Chem. Soc.*, 1989, **111**, 3659.
89JA3671	M.M. Greenberg, S.C. Blackstock, K.J. Stone and J.A. Berson; *J. Am. Chem. Soc.*, 1989, **111**, 3671.
89JA5346	B. Capon and F.-C. Kwok; *J. Am. Chem. Soc.*, 1989, **111**, 5346.
89JA5483	K. Takahashi and T. Suzuki; *J. Am. Chem. Soc.*, 1989, **111**, 5483.
89JA6544	H. Stamm and H. Jäckel; *J. Am. Chem. Soc.*, 1989, **111**, 6544.
89JA7371	Z. Zhou and R.G. Parr; *J. Am. Chem. Soc.*, 1989, **111**, 7371.
89JA8753	M.G. Choi and R.J. Angelici; *J. Am. Chem. Soc.*, 1989, **111**, 8753.
89JCR(S)29	J.M. Barker, P.R. Huddleston and P.W. Needs; *J. Chem. Res.* (S), 1989, 29.
89JCS(PI)603	S. Datta and A. De; *J. Chem. Soc., Perkin Trans. I*, 1989, 603.
89JCS(PII)127	R. Noto, R. Rainieri and C. Arnone; *J. Chem. Soc., Perkin Trans. II*, 1989, 127.
89JHC345	L.H. Klemm, E. Hall, L. Cousins and C.E. Klopfenstein; *J. Heterocycl. Chem.*, 1989, **26**, 345.
89JHC397	C.K. Lee and Y.M. Ahn; *J. Heterocycl. Chem.*, 1989, **26**, 397.
89JHC533	Y.M. Kao and W.-S. Hwang; *J. Heterocycl. Chem.*, 1989, **26**, 533.
89JHC709	A. Shafiee, A. Shafaati and B. Habibi-Khamench; *J. Heterocycl. Chem.*, 1989, **26**, 709.
89JHC865	Y. Yang, A.-B. Hörnfeldt and S. Gronowitz; *J. Heterocycl. Chem.*, 1989, **26**, 865.
89JMC875	M.W. Harrold, R.A. Wallace, T. Farooqui, L.J. Wallace, N. Uretsky and D.D. Miller; *J. Med. Chem.*, 1989, **32**, 875.
89JMC1147	J.S. New, W.L. Christopher, J.P. Yevich, R. Butler, R.F. Schlemmer, Jr., C.P. VanderMaelen and J.A. Cipollina; *J. Med. Chem.*, 1989, **32**, 1147.
89JMC1265	Y. Amemiya, A. Terada, K. Wachi, H. Miyazawa, N. Hatakeyama, K. Matsuda and T. Oshima; *J. Med. Chem.*, 1989, **32**, 1265.
89JMC2548	S.L. Graham, K.L. Shepard, P.S. Anderson, J.J. Baldwin, D.B. Best, M.E. Christy, M.B. Freedman, P. Gautheron, C.N. Habecker, J.M.Hoffman, P.A. Lyle, S.R. Michelson, G.S. Ponticello, C.M. Robb, H. Schwam, A.M. Smith, R.L. Smith, J.M. Sondey, K.M. Strohmaier, M.F. Sugrue and S.L. Varga; *J. Med. Chem.*, 1989, **32**, 2548.
89JOC299	A. Padawa, T.J. Wisnieff and E.J. Walsh; *J. Org. Chem.*, 1989, **54**, 299.
89JOC958	P.M. Lahti, A.S. Ichimura and J.A. Berson; *J. Org. Chem.*, 1989, **54**, 958.
89JOC990	J.S. New, W.L. Christopher and P.A. Jass; *J. Org. Chem.*, 1989, **54**, 990.
89JOC1211	B. Capon and Z.-P. Wu; *J. Org. Chem.*, 1989, **54**, 1211.
89JOC2142	M.G. Reincke and D.Del Mazza; *J. Org. Chem.*, 1989, **54**, 2142.
89JOC2793	S.L. Buchwald and Q. Fang; *J. Org. Chem.*, 1989, **54**, 2793.

89JOC2904	X. Creary, M.E. Mehrsheikh-Mohammadi and S, McDonald; *J. Org. Chem.*, 1989, **54**, 2904.
89JOC3223	G.S. Ponticello, C.N. Habecker, S.L. Varga and S.M. Pitzenberger; J. *Org. Chem.* 1989, **54**, 3223.
89JOC3311	R.P. Gajewski, J.L. Jackson, N.D. Jones, J.K. Swartzendruber and J.B. Deeter; *J. Org. Chem.*, 1989, **54**, 3311.
89JOC3664	V.A. Montero, I.T. Hernandez, J.dePasaual Teresa, J.R. Moran and R. Olabarrieta; *J. Org. Chem.*, 1989, **54**, 3664.
89JOC4134	G.D. Hartman, B.T. Phillips, W. Halczenko and S.M. Pitzenberger; *J. Org. Chem.*, 1989, **54**, 4134.
89JOC4138	H.-H. Tso, T. Chow and Y.-L. Lai; *J. Org. Chem.*, 1989, **54**, 4138.
89JOC4822	O. Meth-Cohn, E. Vuorinen and T.A. Modro; *J. Org. Chem.*, 1989, **54**, 4822.
89JOC4848	S. Becker, Y. Fort, R. Yanderesse and P. Caubère; *J. Org. Chem.*, 1989, **54**, 4848.
89JOC4984	E.C. Taylor and J.E. Macor; *J. Org. Chem.*, 1989, **54**, 4984.
89JOC5277	A. Hassner, K.S.K. Murthy, A. Padwa, U. Chiacchio, D.C. Dean and A.M. Schoffstall; *J. Org. Chem.*, 1989, **54**, 5277.
89KGS419	V.M. Polosin, A.A. Astakhov, A.V. Irashchenko, M.A. Ryashentseva, E.P. Belanova, A.S. Shashkov and Kh.M. Minachev; *Khim. Geterotsikl. Soedin*, 1989, 419; [Chem. Abstr. 1990, **112**, 7567e].
89LA405	S. Schöning and W. Friedrichsen; *Liebigs Ann.* Chem., 1989, 405.
89MI153	S. Leistner, M. Gütschow and G. Wagner; *Pharmazie*, 1989, **44**, 153.
89MI191	M. D'Auria, A. DeMico, F. D'Onofrio, D. Mendola and G. Piancatelli; *J. Photochem. Photobiol.* A, 1989, **47**, 191.
89MI377	J.E. Olnbetta, S. Lyet, A. Xichina, J.F. Robert and J.J. Panouse; *Ann. Pharm. Fr.*, 1989, **46**, 377.
89MI385	Q. Zhow and Y. Huang; *J. Fluorine Chem.*, 1989, **43**, 385.
89MI1865	G.N. Glaudee, L.M. Daniels and R.J. Angelici; *Organometallics*, 1989, **8**, 1856.
89MI2277	J. Chen and R.J. Argelici; *Organometallics*, 1989, **8**, 2277.
89MI263	B. Krische and M. Zagorska; *Synth. Meth.*, 1989, **28**, C263.
89S130	Y. Yang, A.-B. Hörnfeldt and S. Gronowitz; *Synthesis*, 1989, 130.
89S181	G. Barbey, G. Dian, N. Merlet, F. Outurquin and C. Paulmier; *Synthesis*, 1989, 181.
89S210	K. Eichinger, P. Mayr and P. Nussbaumer; *Synthesis*, 1989, 210.
89S280	P. Goya, J. Lissavetzky and I. Rozas; *Synthesis*, 1989, 280.
89S327	A. Elmasmodi, P. Cotelle, D. Barbry, B.Hasiak and D. Couturier; *Synthesis*, 1989, 327.
89SC307	G.T. Crisp; *Synth. Commun.*, 1989, **19**, 307.
89SC443	H. Ishibashi, I. Takamuro, Y.-i. Mizukami, M. Irie and M. Ikeda; *Synth. Commun.*, 1989, **19**, 443.
89TL95	M. Iwasaki, J.P. Li, Y. Kobayashi, H. Matsuzaka, Y. Ishii and M. Hidai; *Tetrahedron Lett.*, 1989, **30**, 95.
89TL795	J.P. Dittami, H. Ramauathau and S. Breining; *Tetrahedron Lett.*, 1989, **30**, 795.
89TL847	J. Nakayama, Y. Kumano and M. Hoshino; *Tetrahedron Lett.*, 1989, **30**, 847.
89TL1655	T. Aratani, H. Yoshihara and G. Suzakamo; *Tetrahedron Lett.*, 1989, **30**, 1655.
89TL2091	K. Takahashi, T. Nihira, K. Takase and K. Shibata; *Tetrahedron Lett.*, 1989, **30**, 2091.
89TL3315	Y. Mazaki and K. Kobayashi; *Tetrahedron Lett.*, 1989, **30**, 3315.
89ZC166	M. Weissenfels, A. Hantschmann, T. Steinfuehrer and E. Birkner; Z. *Chem.*, 1989, **29**, 166.
89ZN(B)488	F.M. Abdelrazek; Z. *Naturforsch.*, Teil B, 1989, **44**, 488.

CHAPTER 4.2

Five-Membered Ring Systems: Pyrroles and Benzo Derivatives

RICHARD J SUNDBERG
University of Virginia, Charlottesville, VA, USA

One of the most active areas of research in pyrrole and indole chemistry over the past year has involved application of organometallic intermediates in both synthesis and reactions of the heterocycles. A review of the role of transition metals in synthesis and functionalization of indoles appeared in 1988. The reactions discussed include palladium-mediated substitution and cyclizations, copper-catalyzed cyclizations and nucleophilic addition to η^6-chromium tricarbonyl complexes.<88AG(E)1113> Methods for cyclizative formation of indoles are also included in a review of Pd-catalyzed reactions for the synthesis of condensed heteroaromatic compounds.<88H2225> Another active area of research has been the extension of cycloaddition chemistry to pyrrole and indole derivatives. A review of the cycloaddition reactions of indolequinodimethane intermediates emphasizes the factors controlling the regioselectivity and stereoselectivity of these intermediates.<89CR1681>

4.2.1 Syntheses Involving Ring Closure

Several new palladium-mediated processes for cyclization of indole rings have been developed. In each instance well-established reactivity patterns of organopalladium intermediates are applied to aryl halides and reactants which can lead to formation of an indole ring. Palladium-catalyzed cross coupling of 2-bromoacetanilide or 2-(trifluoromethylsulfonyloxy)acetanilides with alkynyltrialkyl stannanes provides good yields of 2-alkynylacetanilides. These compounds are cyclized to indoles in the presence of Pd(II) salts. The yields are good for systems with R being alkyl or aryl. The reaction can tolerate ester, methoxy, chloro and trifluoromethylsulfonyloxy group as ring substituents.<89JOC5856>

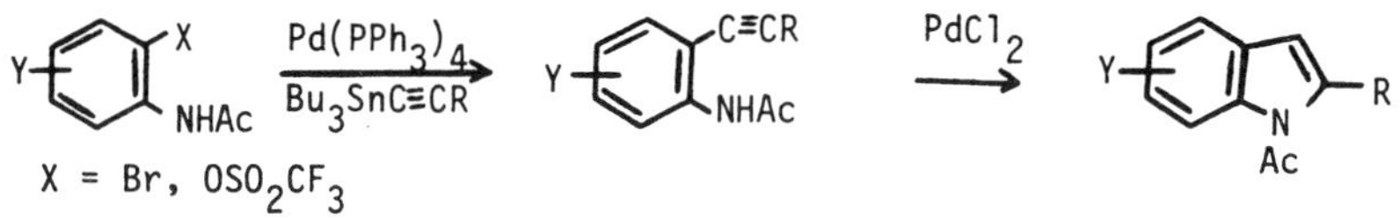

2-Ethynylaniline, prepared by Pd-Cu mediated coupling of 2-bromoaniline with trimethylsilylacetylene, is another starting material for palladium-mediated indole synthesis. The aniline undergoes cross-coupling with either aryl or vinyl triflates which, in turn, can be cyclized to 2-substituted indoles.<89TL2581>

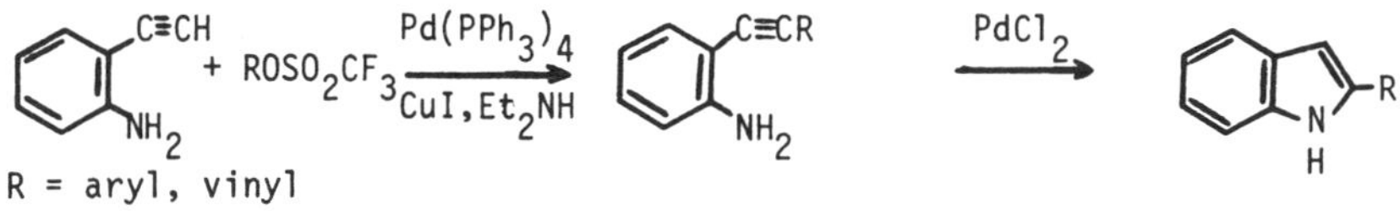

A variety of R groups including substituted aromatic, heteroaromatic, and cycloalkenyl gave yields mostly in the 50-80% range.

Two other procedures form indoles from 2-bromonitrobenzenes or 2-bromoacetanilides. The reaction of ethylene with several 2-bromonitrobenzenes was investigated under the conditions of the Heck reaction (catalytic amount of $Pd(OAc)_2$, Ph_3P, triethylamine). Some reactants gave 2-nitrostyrenes, the products of a normal Heck reaction, but others produced the corresponding indole as a by-product. The 2-nitrostyrenes could be converted to indoles by a two-step sequence involving Pd-catalyzed addition of 1,3-dihydroxypropane, followed by reduction and acid-catalyzed cyclization. Alternatively, starting with 2-bromoacetanilides, a Heck reaction, followed by Pd-catalyzed oxidative cyclization afforded indoles. This latter variant gave good yields for methyl, methoxy, and methoxycarbonyl-substituted reactants.<89JHC1405>

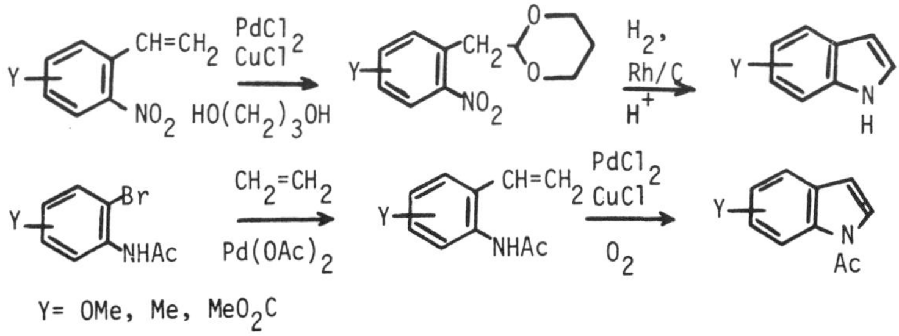

The vinyl palladium intermediates arising from cyclization of N-propargylic-2-iodoacetanilides can undergo tandem coupling with phenylzinc chloride. The Reformatsky reagent $BrZnCH_2CO_2Et$ effects an analogous coupling. The palladium species can also be trapped with B-vinylborolanes.<89TL1135>

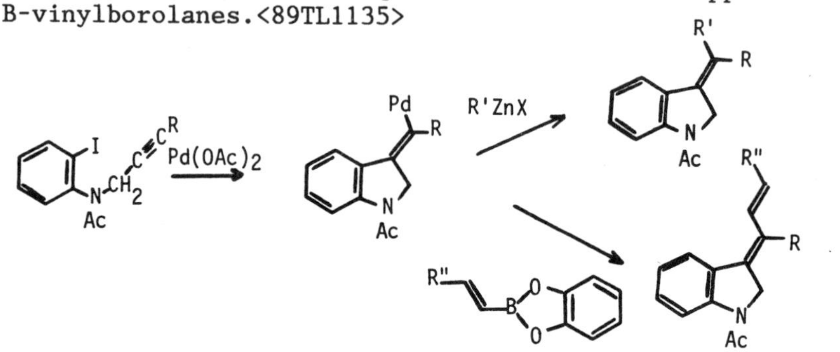

Electrophilic cyclization of 3-butenylamine derivatives by Pd(II) can lead to pyrroles. A new procedure for pyrrole synthesis based on this concept starts with methyl cyanomethyl sulfone which can be prepared by successive allylation and alkylation of α-methanesulfonylacetonitrile. The amines resulting from subsequent $LiAlH_4$ reduction are converted to 2,4-disubstituted pyrroles by $PdCl_2$ in the presence of $CuCl_2$. The reaction is a Pd-induced exo-5-cyclization. The $CuCl_2$ functions to reoxidize the palladium catalyst.<89CL737>

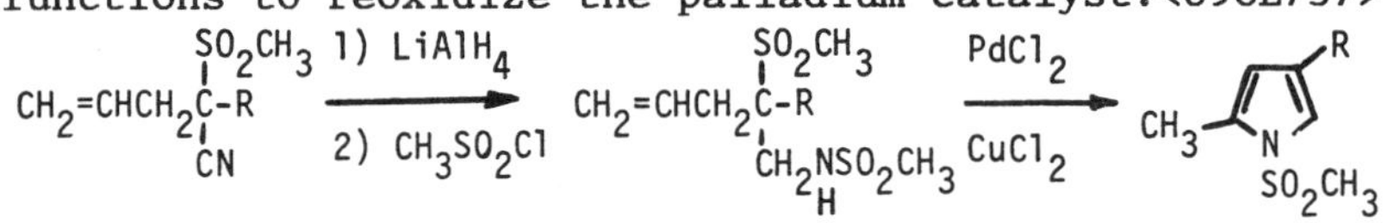

Benzylzinc intermediates play a key role in a new procedure for indole synthesis *via* cyclization of o-aminobenzyl ketones. *Ortho*-methylanilines are protected as *bis*-trimethylsilyl derivatives and then brominated on the methyl group. The resulting benzyl bromides are converted to mixed zinc-copper organometallics. These reagents can be acylated in good yield by acyl halides. After hydrolysis the final products are 2-substituted indoles.<89TL4795>

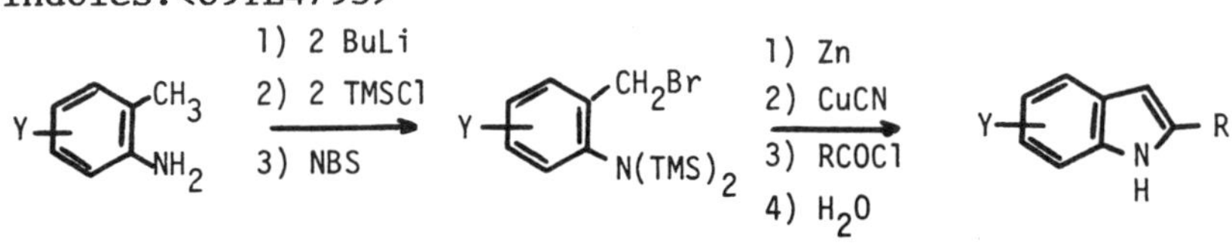

A novel synthesis adaptable to 2,3, 2,4, 3,4 and 2,3,4-substituted pyrroles is based on reaction of acetylenes with zirconocene complexes of imines. These complexes are formed by addition of the lithium N-trimethylsilylamides to the zirconium reagent. The orientation of the acetylene is evidently governed by steric factors. In the cases studied, the smaller acetylene substituent adopts the R^3 position.<89JA776>

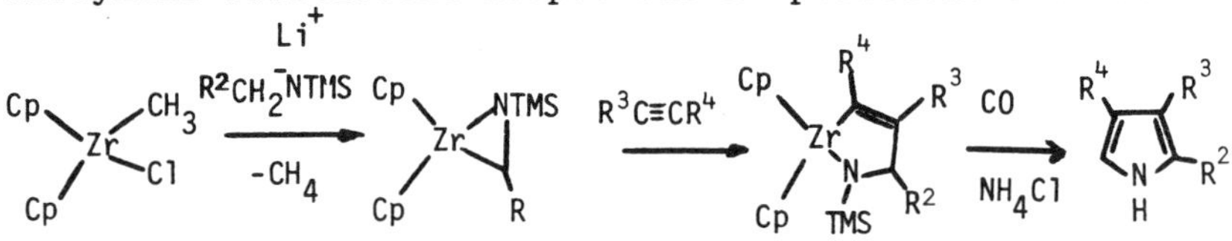

Niobium chloride promotes reaction of esters or dimethylformamide with α,β-unsaturated imines leading to pyrrole formation. The reaction is formulated as an initial nucleophilic addition of the azadiene unit, induced by net oxidation of Nb(III)→Nb(I).

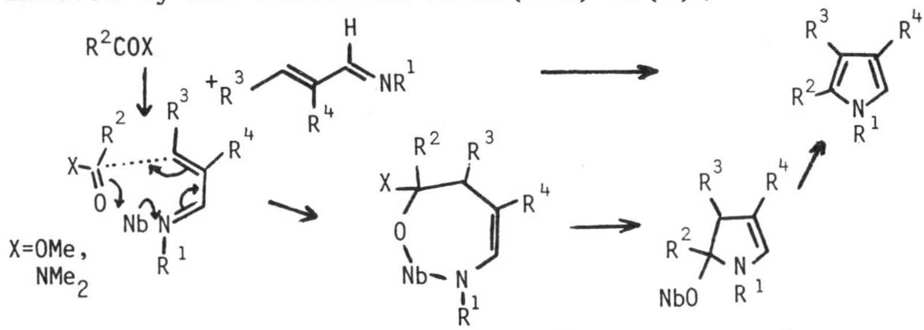

In most of the examples studies $R^3 = CH_3$ and R^4 = Ph but other cases were also successful, with most yields in the 40-60% range.<89JOC4736>

Formation of o-styrylnitrenes or o-biphenylnitrenes generally leads to indoles and carbazoles, respectively. Two new examples of nitrene-based synthesis of fused ring indoles were reported. The feasibility of a nitrene-based route for the formation of benz[c,d]indoles has been demonstrated for the first time. Tetralone was converted to the vinyl azide **1**. It gave the insertion product **2** in 67% yield in boiling mesitylene. An analogous reaction starting with benzosuberone furnished the homolog in 72% yield.<89TL4017>

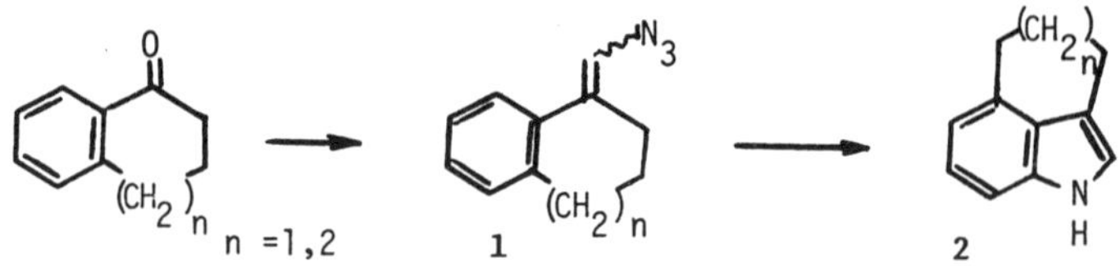

The continuing interest in the antitumor activity of ellipticine and its analogs has spurred development of syntheses of the pyrido[4,3-b]carbazole derivatives. A nitrene cyclization was successfully adapted to this objective. The 2-bromoisoquinoline **3** was phenylated in 99% yield using $Pd(PPh_3)_4$-catalyzed coupling with phenylboronic acid. Diazotization, azide displacement and thermolysis afforded ellipticine.<89TL297>

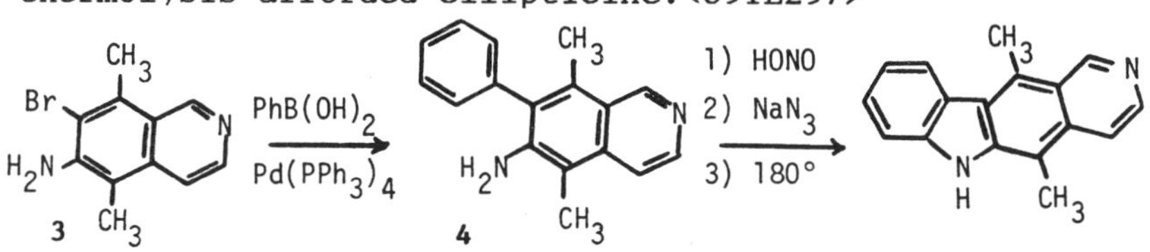

A new route to N-methoxy oxindoles is believed to proceed by electrophilic attack by an N-acylnitrenium ion intermediate.

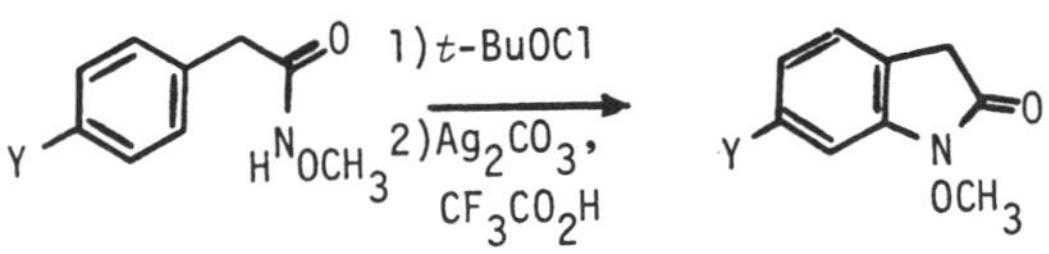

Excellent yields were obtained for most <u>para</u>-substituents examined. Only the <u>p</u>-nitro group sharply decreased the yield. With <u>ortho</u>-substituted reactants the yields tended to be lower because of competing reactions.<89JOC3394>

A novel intramolecular Diels-Alder reaction involving an allene was used to construct the 6,7-fused intermediate **5**. After aromatization with chloranil, the intermediate indole was subsequently converted to the <u>E</u> and <u>Z</u> isomers of <u>cis</u>-trikentin B, an alkaloid isolated from a marine sponge.<89TL6559>

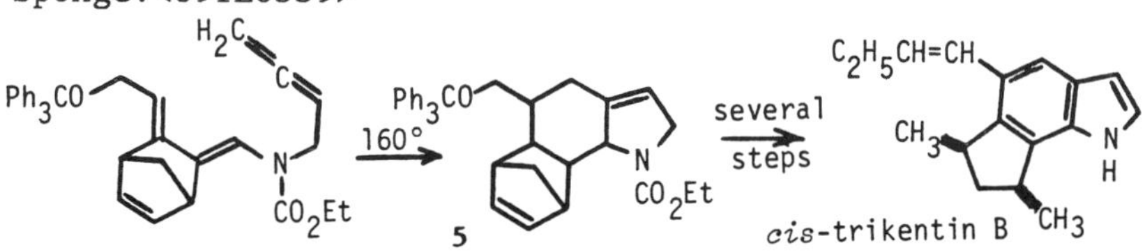

A useful methodological improvement was introduced in the conversion of β,2-dinitrostyrenes to indoles. Use of ammonium formate-formic acid in refluxing methanol in the presence of Pd/C catalyst gave better yields than previously used reductants.<89H415>

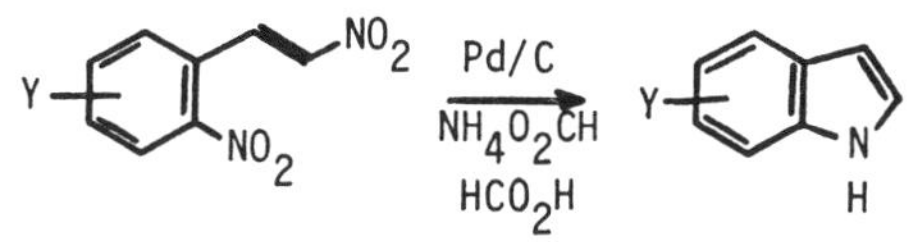

A number of 4-alkylpyrrole-2-carboxylate esters and 4-alkylpyrrole-2-carbonitriles have been prepared by an expedient procedure which involves 3-ethoxyacrolein as a key synthon. Most of the yields are in the range 30-50% but are lower (10-20%) for the 4-unsubstituted pyrrole.<89S337>

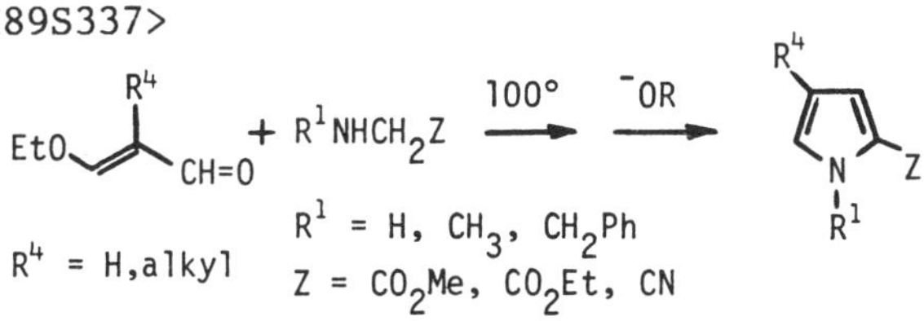

The reaction is carried out in two steps; thermal replacement of the ethoxy group followed by an intramolecular aldol condensation under basic conditions.

4.2.2. Ring Substitution and Substitutent Group Interconversions

Organometallic intermediates have also played a prominent role in the development of new methods for introduction and/or modification of pyrrole and indole ring substituents. The facile mercuration of pyrroles provides an alternative to oxidative addition as a means of access to the arylpalladium intermediates required for Heck vinylation. 2- and 3-(Chloromercuri)pyrroles prepared from N-protected (benzyloxymethyl, methyl, t-butoxycarbonyl) pyrroles react with methyl acrylate in the presence of $LiPdCl_3$ to give good yields of the corresponding pyrroleacrylates.<89JOC4801>

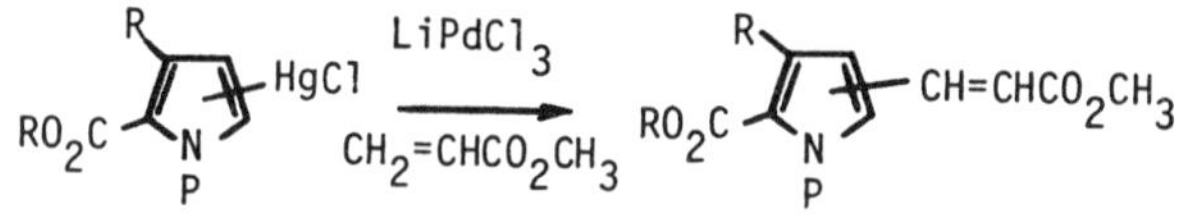

N-unsubstituted pyrroles give insoluble materials which are presumably mercury pyrrolates.

Hegedus and co-workers explored a range of organometallic intermediates which can effect fusion of a six-membered ring to the (c,d)-side of indole and thus create the ABC rings of the ergot alkaloid skeleton. The 2-(ethoxycarbonyl)-propenyl group can be introduced at the 4-position by use of a nickel reagent.

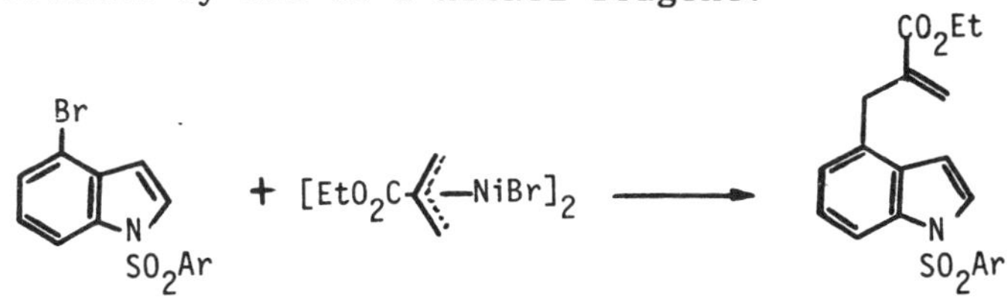

Both the allyl and 2-(ethoxycarbonyl)propenyl groups could be introduced on the 3-position via palladium-catalyzed cross-coupling with a chloromercuri derivative.

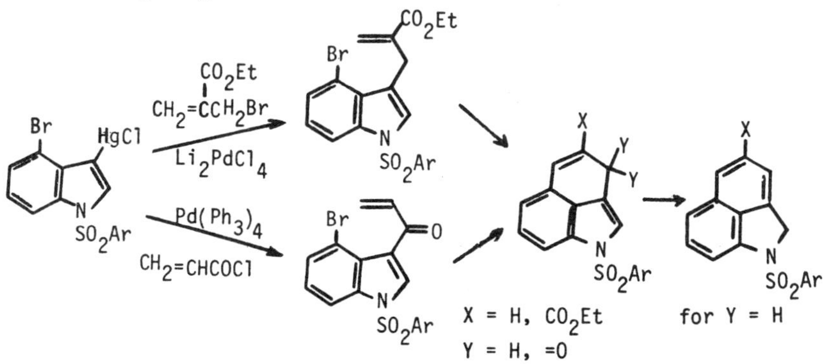

The same intermediate afforded a 3-acryloyl indole on reaction with acryloyl chloride in the presence of a catalylic amount of $[Ph_3P]_4Pd$. All three of these intermediates gave intramolecular "Heck-vinylation" on reaction with palladium acetate.<89JOC4141>

The 4-thallation product of indole-3-carboxaldehyde has now been shown to be a precursor of 4-nitroindole derivatives. Heating with $NaNO_2$-$CuSO_4$ in DMF at 100° effects the transformation. Similar reaction conditions transform the 7-thallation product of N-acetylindoline to the corresponding nitro compound.<89H643>

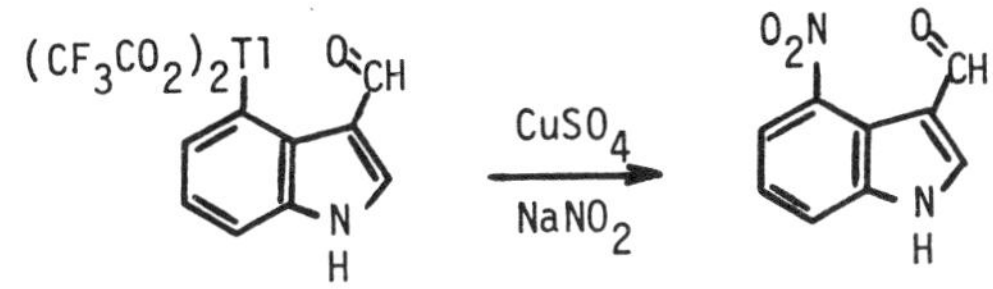

Conditions (NaN_3, $CuSO_4$), were also found for converting the thallium intermediates directed to azidoindoles.

Carbene complexes of both (2-indolyl)methoxymethylene and (3-indolyl)-methoxymethylene have been prepared by addition of an appropriate lithioindole to chromium hexacarbonyl followed by methylation.

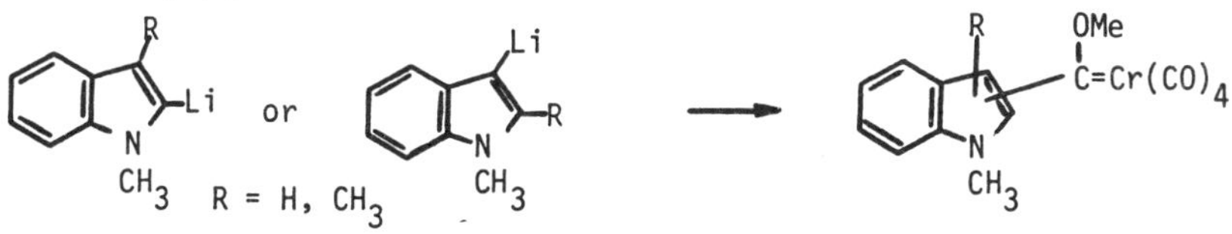

These complexes undergo a characteristic cyclization with acetylenes to provide oxygenated carbazoles. When R = H these can aromatize and are isolated as quinones after oxidation, but the R = CH_3 derivatives are blocked and are iolated as cyclohexadienones.<89JOC3249>

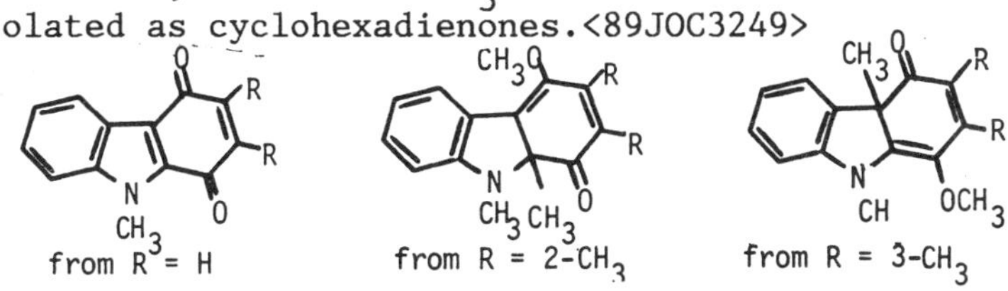

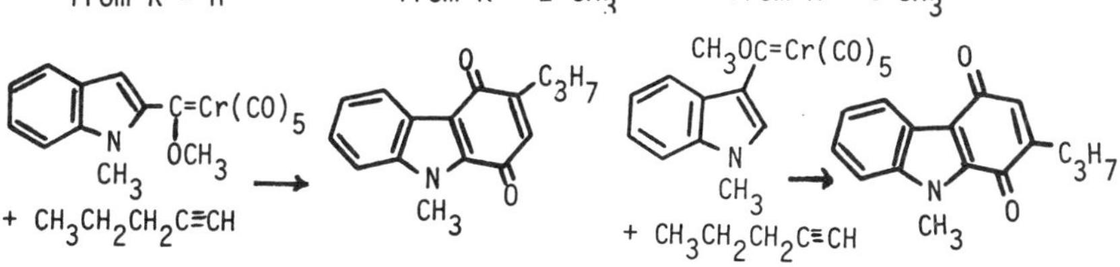

The regioselectivity of the reaction with unsymmetrical acetylenes is such as to place the larger substituent adjacent to the carbonyl group in the initial adduct. By selection of the 2- or 3-indolyl complex, it is therefore possible to access either of the regioisomeric carbazole derivatives.

The metal-carbene complexes derived from vinyldiazo compounds and rhodium acetate react with 1-methoxycarbonylpyrrole to give 2-azabicyclo[3.2.1]octa-2,6-dienes (tropane skeleton).<89TL4653>

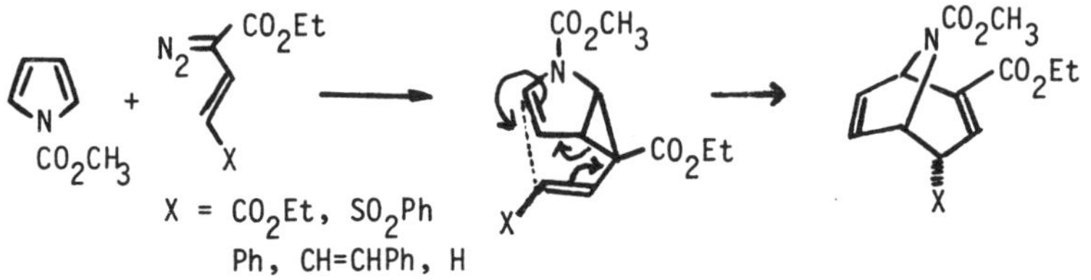

With pyrrole itself or N-methyl pyrrole a mixture of 2- and 3-substitution products are observed. The reaction is presumed on the basis of related chemistry of dienes to proceed by sigmatropic rearrangement a divinylcyclopropane intermediate. A <u>bis</u>-cyclopropanation product was isolated in one instance.

The utility of 2-vinylindoles as reactants in cycloaddition reactions has prompted continued investigation of general synthetic routes to these compounds. One approach involves reaction of 2-aminobenzyltriphenylphosphonium bromide with α,β-unsaturated acyl halides. The resulting amides give 2-vinylindoles via an intramolecular Wittig reaction.<89S364>

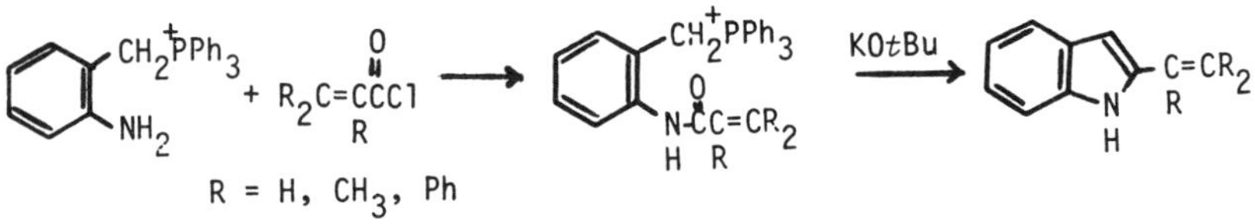

A number of 2-vinylindoles with acceptor substituents in the vinyl group were prepared from 2-formylindoles by a classical Wittig reaction. In a few cases formation of some of the <u>Z</u>-substituted product was observed but the <u>E</u>-isomer was always the major product. Several 3-vinylindoles were prepared by Wittig reactions of N-substituted 3-formyl and 3-acetylindoles.<89M157>

A new approach to 2-vinylindoles is based on lithiation at the 2-methylene group of 1-substituted 2-alkylindole-3-carboxylate esters. On addition of a ketone a pyrano[4,3-b]indole is formed.<89JOC4785>

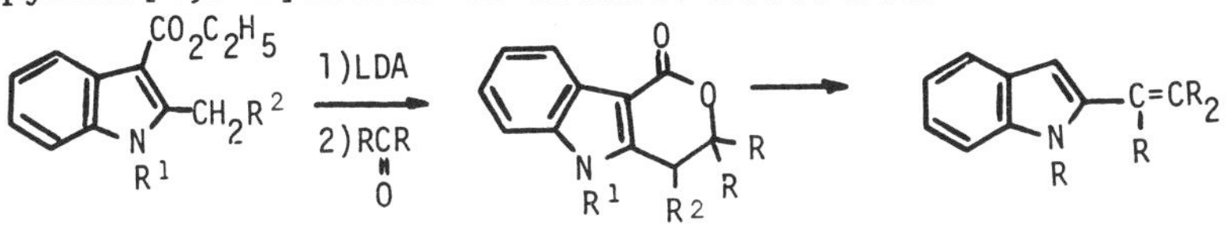

In the presence of additional base, opening of the lactone ring occurs. The corresponding carboxylic acids undergo decarboxylation at 150-160°. With a t-butoxycarbonyl group as the indole-1 substituent, an alternative cyclization occurred, which leads eventually to a 2-vinylindole-3-carboxylate ester.

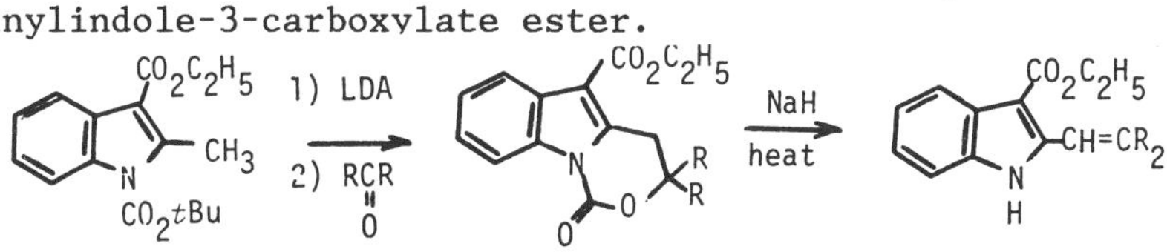

The isolation of natural products having a variety of B-ring alkyl substitution patterns has prompted exploration of methods for synthesis of such compounds. One approach is the annulation of pyrroles. Trimethylsilyl ethers of α,β-unsaturated aldehydes effect substitution of the endoperoxide of 1-methoxycarbonylpyrrole at the 2-position to give a mixture of E- and Z-4(2-pyrrolyl)-but-2-en-1-ones. These intermediates are cyclized to ring-substituted indoles by heating with p-toluenesulfonic acid in benzene.<89H771>

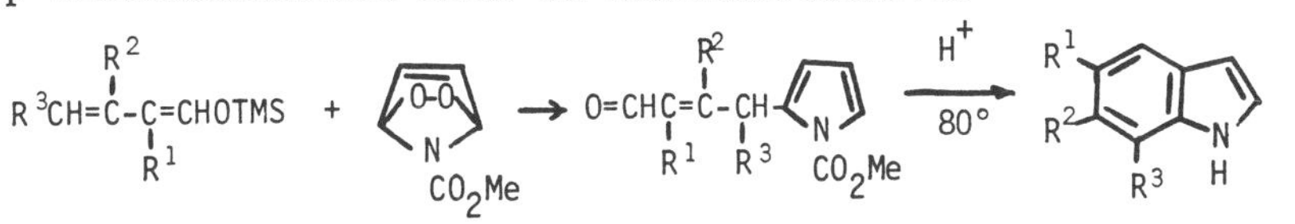

The same group has reported an intramolecular Vilsmeier-Haack reaction which affords a 4-aminoindole derivative which is a model of the teleocidin structure.

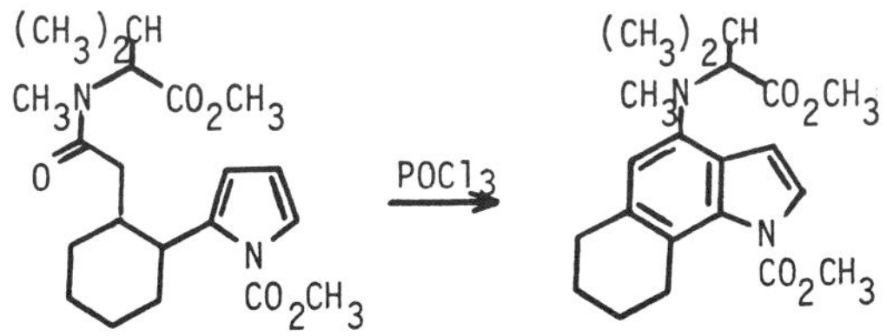

Another approach to construction of a side chain appropriate for annulation of pyrroles to indoles involves addition of 2-(2-dioxolanyl)ethylmagnesium bromide to an N-protected pyrrole-2-carboxaldehyde. Acid-mediated cyclization gives the corresponding N-protected indole.

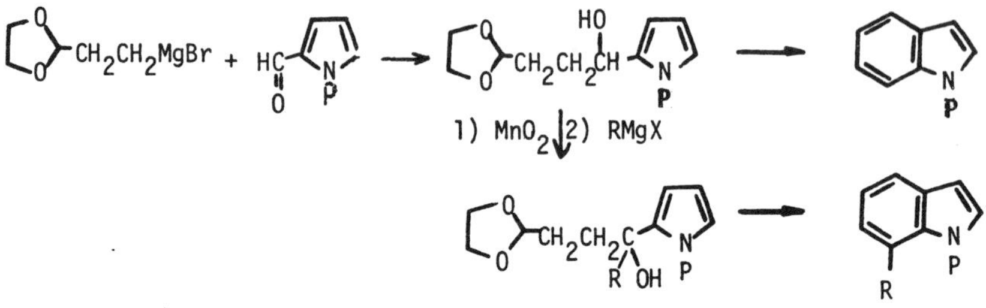

This route can be modified to provide for introduction of a 7-substituent by oxidation of the adduct followed by organometallic addition. This strategy was applied to introduce the tertiary substituent found in the teleocidins. The cyclization failed when the p-tosyl protecting group was in place. However, when the deprotected intermediate was subjected to acid cyclization conditions a 46% yield of two 7-substituted indoles was obtained.<89H783>

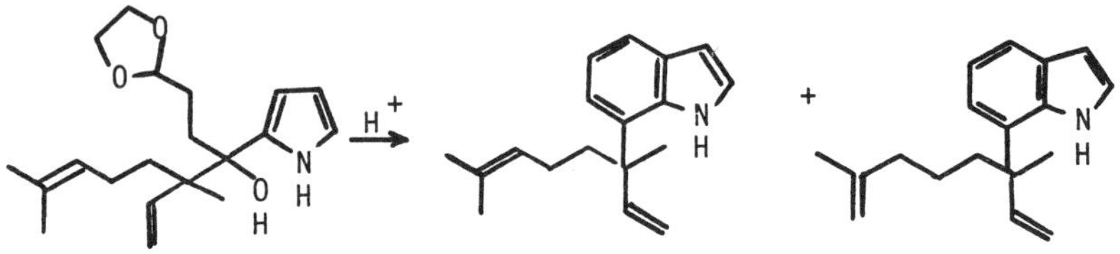

A related strategy was used to construct the cyclopenta[g]indole structure found in the trikentins, which are isolated from a marine sponge. In this case the endoperoxide of 1-phenylsulfonylpyrrole reacts with the trimethylsilyl enol ether of 2,4-dimethylcyclopentanone. After addition of the lithium anion of 2-butanone N,N-dimethylhydrazone, cyclization afforded a 4:3 mixture of <u>cis</u>-and <u>trans</u>-product in 73% yield.<89TL5771>

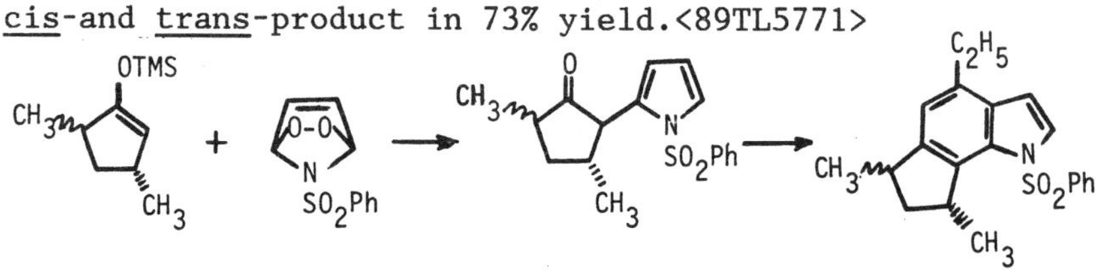

4.2.3 Cycloaddition Reactions

Pyrroles and indoles, and intermediates derived from them, can give a variety of cycloaddition reactions. 2-Vinyl and 3-vinylindoles can function as dienes, as can indole-2,3-quinodimethanes. There have been recent extensions of these reactions, as well as the development of [2+3] dipolar cycloadditions.

3-Methylthio and 3-phenylthiopyrroles with N-carbomethoxy substitution underwent Diels-Alder additions at 12 kbar. N-Phenylmaleimide, methyl acrylate and phenyl vinyl sulfone reacted well. Several more highly substituted dienophiles were ineffective.<89H79>

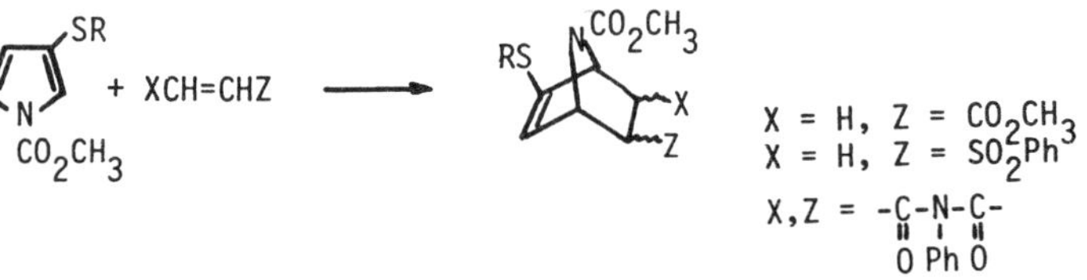

Endo-product was formed preferentially for N-phenylmaleimide and phenyl vinyl sulfone but the addition with methyl acrylate showed little stereoselectivity. Qualitative comparison with the reactivity of N-carbomethoxypyrrole suggests that the thio substituents are activating, with the methylthio group being more effective than phenylthio.

3-(1-Methylthiovinyl)indole, which can be readily prepared from 3-thioacetyl indole, proved to be a good diene toward a range of electrophilic dienophiles. The adducts can be oxidized to carbazoles. With acetylenic dienophiles, carbazoles are formed directly. In a few cases 4-oxotetrahydrocarbazoles formed by hydrolysis of the intermediate adducts were observed.<89CPB1999>

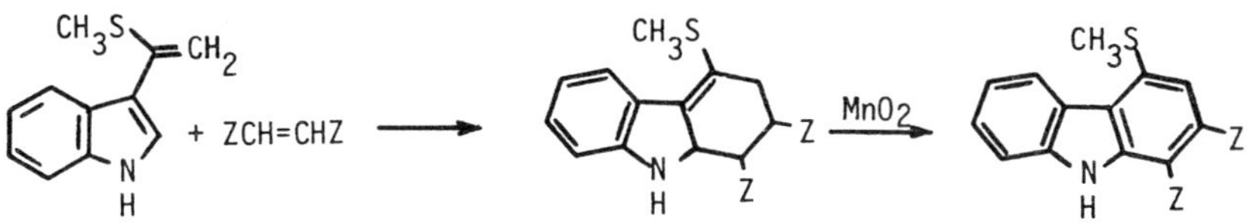

The regioselectivity of Diels-Alder reactions of vinyl indoles has been the subject of continuing study. 2-Vinylindole is regiospecific in its reactions with methyl acrylate and the same pattern of regioselectivity is observed in more substituted systems. Alder rule stereoselectivity is observed.<89H11>

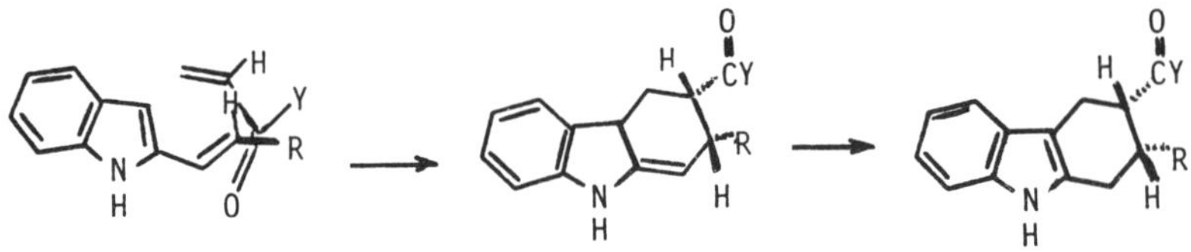

The regiochemistry of the cycloaddition of pyrano[3,4-b]indol-3-ones with a series of dienophiles was examined. The products are carbazoles formed by decarboxylation of the initial adducts. The yields are modest and the selectivity low for R = H, but yields are higher and selectivity is >30:1 with R = CH_3.<89H11> This is attributed to the increased energy of the HOMO of the methyl substituted case.

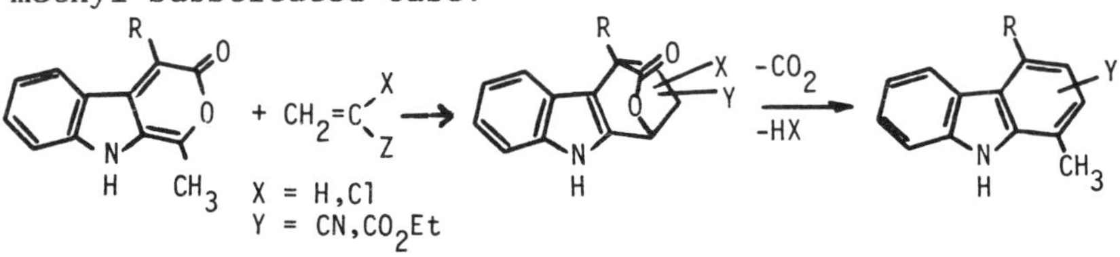

The carbazole alkaloids carbazomycin A and B were synthesized via the intermediate deoxycarbazomycin, which was obtained by Diels-Alder reactions of pyrano[3,4-b]indol-3-one synthon. A similar route yields the marine alkaloid hyellazole.<89JCS(P1)376>

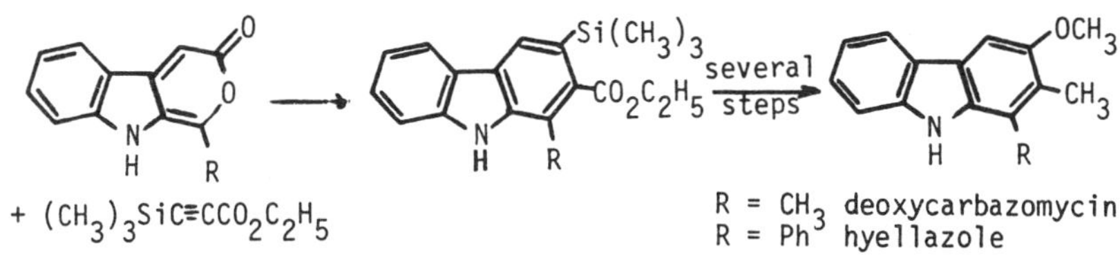

N-Acetylindole-2,3-quinodimethane can be generated as an intermediate from 1-acetyl-2,3-bis-(bromomethyl)indole by reaction with iodide ion. The intermediate is trapped by good dienophiles such as N-phenylmaleimide, dimethyl fumarate and dimethyl acetylenedicarboxylate. 2,3-Quinodimethane intemediates can also be generated by thermolyzing N-substituted pyrano[3,4-b]indol-3-ones. Unsymetrical dienophiles such as methyl acrylate and methyl vinyl ketone gave both regioisomers of the expected tetrahydrocarbazoles.<89TL7289>

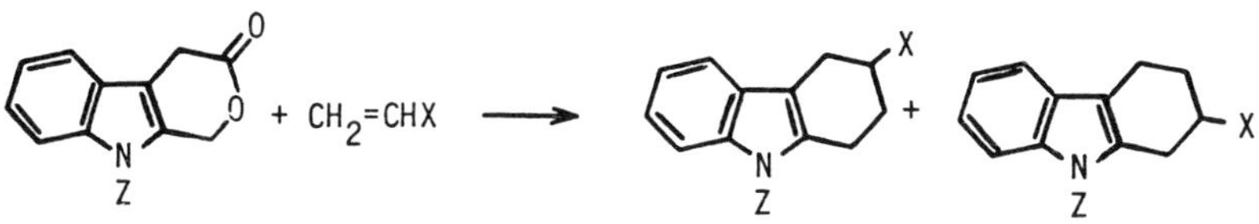

Z = $COCH_3$, $COCF_3$, $O_2COC(CH_3)_3$, SO_2CF_3

In a single case, a reactant with a C-1 methyl group showed enhanced regioselectivity with a 99:1 preference for the 1,2-disubstituted carbazole.

The magnesium derivative of indole undergoes cycloaddition with 2-phenylsulfonyldienes, leading to 2-phenylsulfonyl-1,4,4a,9a-tetrahydrocarbazoles.

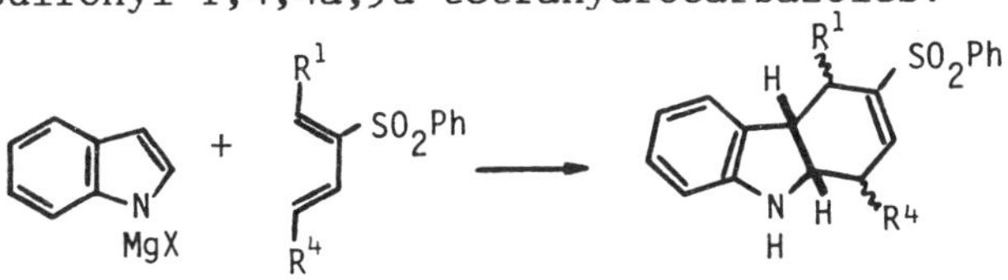

The reaction is not entirely stereospecific with respect to the diene, suggesting a stepwise mechanism.<89TL2589>

Intramolecular Diels-Alder additions in which the indole-2,3-double bond serves as a dienophile have been explored. Successful reactions are observed for a series of N-(hexa-2,4-dienyl)carbamoylindoles. The best yields were obtained when the 3-substitutent is formyl but the 3-phenylsulfinyl and unsubstituted compound also reacted.<89JOC2425>

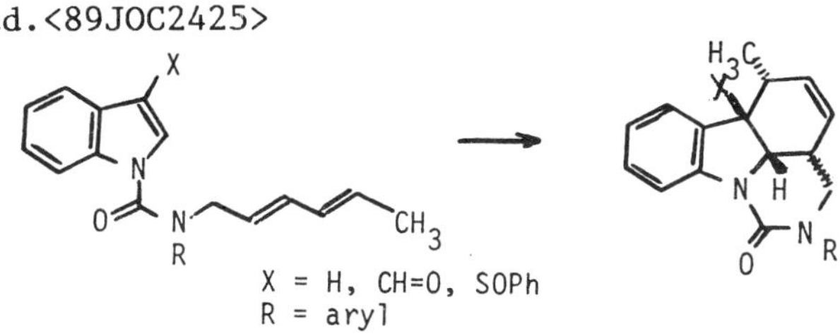

Interestingly the 3-formyl alkoxycarbonyl compound (O for NR) was unreactive.

Methods for construction of pyrrole[1,2-a]indoles have been of longstanding interest because of the occurrence of this structure in the mitomycins. A route to this skeleton through intramolecular cycloaddition of the nitrone **6a** and nitrile oxide **6b** have been accomplished.<89TL1421>

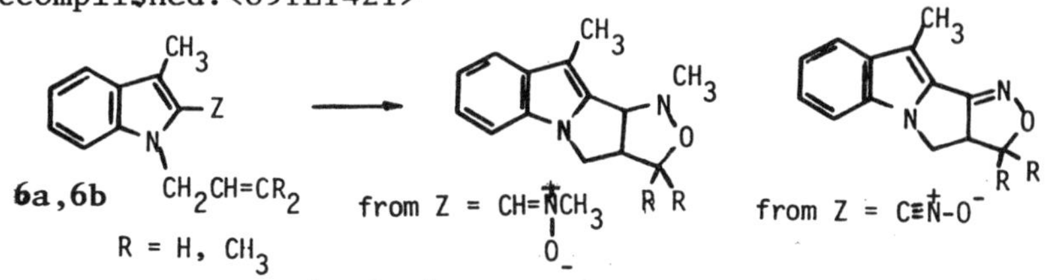

The pyrrolo[1,2-a]indole skeleton can also be constructed by a cycloaddition in which the indole ring serves as the 1,3-dipole. The indole **7** in the presence of silver fluoride generates a methylene azomethine ylide which undergoes cycloaddition with electrophilic dipolarophiles.<89JOC644>

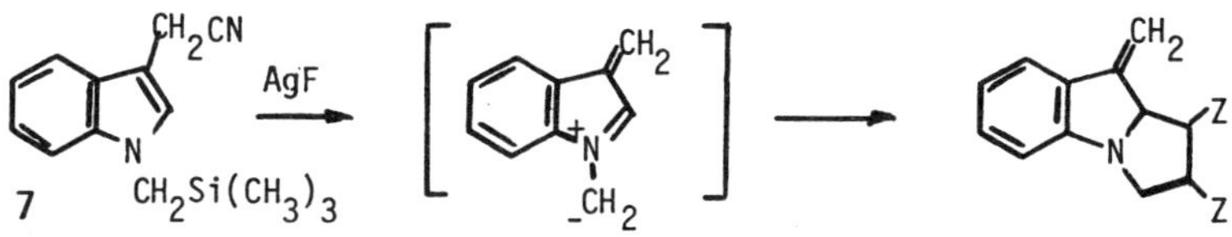

The reaction was successful with such dipolarophiles as maleic anhydride, N-phenylmaleimide, 1,2-bis(phenylsulfonyl)ethene, dimethyl maleate, dimethyl fumarate and acrylonitrile. The stereospecificity and regioselectivity exhibited by the reaction are consistent with a concerted cycloaddition mechanism.

REFERENCES

<88AG(E)1113> L. S. Hegedus, Angew. Chem. Int. Ed. Engl., 1988 **27**, 1113.

<88H2225> T. Sakamoto, Y. Kondo and H. Yamanaka, Heterocycles, 1988, **27**, 2225.

<89CL737> S. Igarashi, Y. Haruta, M. Ozawa, Y. Nishide, H. Kinoshita, and K. Inomata, Chem. Lett., 1989 737.

<89CPB1999> M. Murase, T. Hosaka, T. Koika, and S. Tobinaga, Chem. Pharm. Bull., 1989, **37**, 1999.

<89CR1681> U. Pindur and H. Erfanian-Abdoust, Chem. Rev., 1989, **89**, 1681.

<89H11> U. Pindur, M. Eitel and E. Abdoust-Houshang, Heterocycles, 1989, **29**, 11.

<89H79> J. Keijsers, B. Hams, C. Kruse and H. Scheeren, Heterocycles, 1989, **29**, 79.

<89H415> S. Rajeswari, K. J. Drost and M. P. Cava, Heterocycles, 1989, **29**, 415.

<89H643> M. Somei, F. Yamada, H. Hamada and T. Kawasaki, Heterocycles, 1989, **29**, 643.

89H771> H. Muratake and M. Natsume, Heterocycles, 1989, **29**, 771.

<89H783> H. Muratake and M. Natsume, Heterocycles, 1989, **29**, 783.

<89JA776> S. L. Buchwald, M. W. Wannamaker and B. T. Watson, J. Am. Chem. Soc., 1989, **111**, 776.

<89JCS(P1)376> C. J. Moody and P. Shah, J. Chem. Soc. Perkin Trans. **1**, 1989, 376.

<89JHC1405> A. Kasahara, T. Izumi, S. Murakami, K. Miyamoto and T. Hino, J. Heterocycl. Chem., 1989, **26**, 1405.

<89JOC644> A. Padwa, G. E. Fryxell, J. R. Gasdaska, M. K. Venkatramanan and G. S. K. Wong, J. Org. Chem., 1989, **54**, 644.

<89JOC2425> G. A. Kraus, D. Bougie, R. A. Jacobson and Y. Su, J. Org. Chem., 1989, **54**, 2425.

<89JOC3249> W. E. Bauta, W. D. Wulff, S. F. Pavkovic and E. J. Zaluzec, J. Org. Chem., 1989, **54**, 3249.

<89JOC3394> M. Kawase, T. Kitamura, and Y. Kikugawa, J. Org. Chem., 1989, **54**, 3394.

<89JOC4141> L. S. Hegedus, M. R. Sestrick, E. T. Michaelson and P. J. Harrington, JOC, 1989, **54**, 4141.

<89JOC4736> E. J. Roskamp, P. S. Dragovich, J. B. Hartung, Jr., and S. F. Pederson, J. Org. Chem., 1989, **54**, 4736.

<89JOC4785> J. E. Macor, K. Ryan, and M. E. Newman, J. Org. Chem., 1989, **54**, 4785.

<89JOC84801> J. A. Ganske, R. V. Pandey, M. J. Postich, K. M. Snow and R. M. Smith, J. Org. Chem., 1989, **54**,4801.

<89JOC5856> D. E. Rudisill and J. K. Stille, J. Org. Chem., 1989, **54**, 5856.

<89M157> U. Pindur and L. Pfeuffer, Monatshft. Chem., 1989, **120**, 157.

<89S337> G. H. Walizei and E. Breitmaier, Synthesis, 1989, 337.

<89S364> M. Eitel and U. Pindur, Synthesis, 1989, 364.

<89TL297> R. B. Miller and S. Dugar, Tetrahedron Lett., 1989, **30**, 297.

<89TL1135> B. Burns, R. Grigg, V. Sridharan, P. Stevenson, S. Sukirthalingam and T. Worakun, Tetrahedron Lett., 1989, **30**, 1135.

<89TL1421> P. J. Bhuyan, R. C. Boruah and J. S. Sandhu, Tethredron Lett., 1989, **30**, 1421

<89TL2581> A. Arcadi, S. Cacchi and F. Marinelli, Tetrahedron Lett., **30**, 2581.

<89TL2589> J.-E. Bäckvall, N. A. Plobeck and S. K. Juntunen, Tetrahedron Lett., 1989, **30**, 2589.

<89TL4017> C. J. Moody, A. L. Beck and W. J. Coates, Tetrahedron Lett., 1989, **30**, 4017.

<89TL4653> H. M. L. Davies, W. B. Young and H. D. Smith, Tetrahedron Lett., 1989, **30**, 4653.

<89TL4795> H. G. Chen, C. Hoechstetter and P. Knochel, Tetrahedron Lett., 1989, **30**, 4795.

<89TL5771> H. Muratake and M. Natsume, Tetrahedron Lett., 1989, **30**, 5771.

CHAPTER 4.3
Five-Membered Ring Systems: Furans and Benzo Derivatives

C W BIRD
King's College, London, UK

4.31 INTRODUCTION

The most notable features of furan chemistry reported during this period are the generation of 3,4-dimethylenefuran (1) and the exploration of its chemistry <89 JA3659, 89JA3671, 89JA8732>, and the first syntheses of phenaleno[1,9-bc]furan (2) <88JCS(P1)3169> and the cyclohepta[cd]benzofuran (3) <89JHC365. A timely survey has appearedof the chemistry of isobenzofurans and their application to the synthesis of natural products and polyaromatic hydrocarbons <88T2093>.

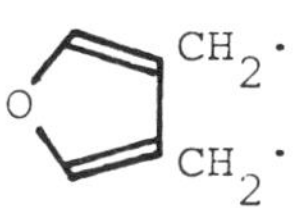

(1)

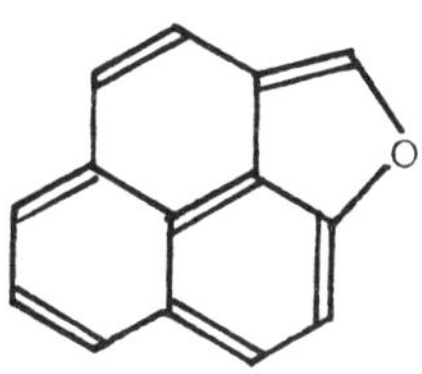

(2)

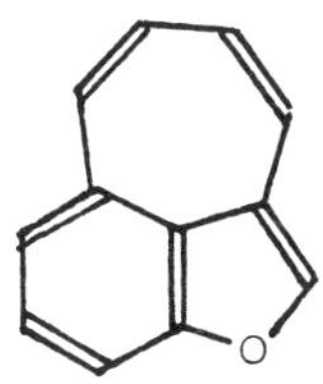

(3)

4.32 REACTIVITY AT RING ATOMS

4.321 Photochemical Reactions

The tri-(*t*-butyl)benzo[c]furan (**4**), which is markedly stable at room temperature, is converted photochemically first to the cyclopropabenzene (**5**) and subsequently isomerised to the benzo[b]furans (**6**) and (**7**) <89TL103>.

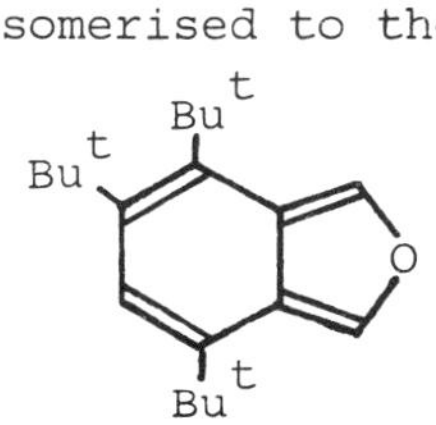

(4)

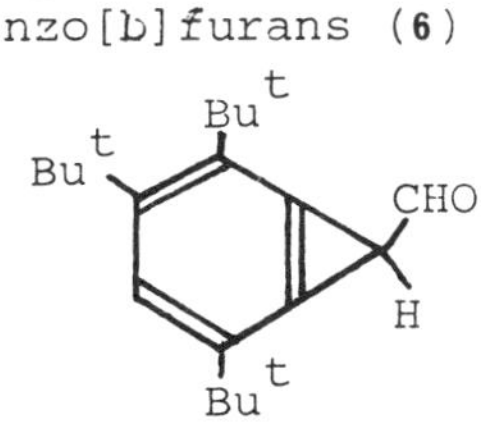

(5)

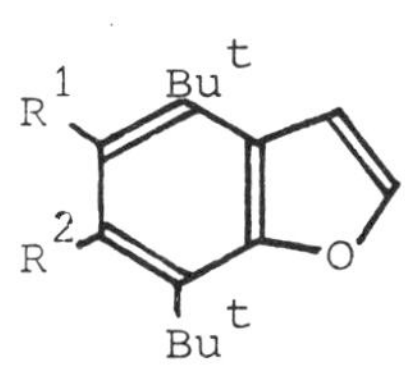

(6) R^1=H;R^2=Bu^t

(7) R^1=Bu^t;R^2=H

4.322 Reactions with Electrophiles

Furans are converted to 2-dialkylaminomethyl derivatives by reaction with *N,N*-dialkylmethyleneimmonium chlorides <88TL2377>. 3-Dialkylaminobutenolides (8) react with Eschenmoser's salt to give the dimethylaminomethylene derivatives (9), which can be further converted to the nitriles (10) <88H(27)1907>. Trimethylsilyloxyfuran with primary alkyl iodides and silver trifluo-

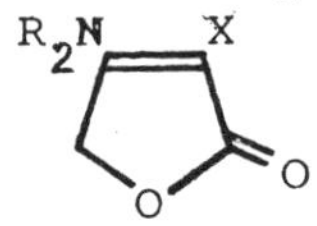

(8) X=H

(9) $X=CH_2NMe_2$

(10) $X=CH_2CN$

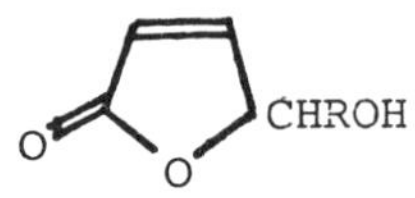

(11)

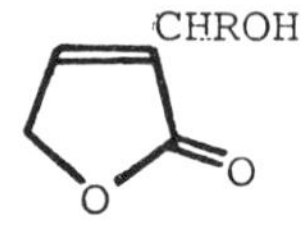

(12)

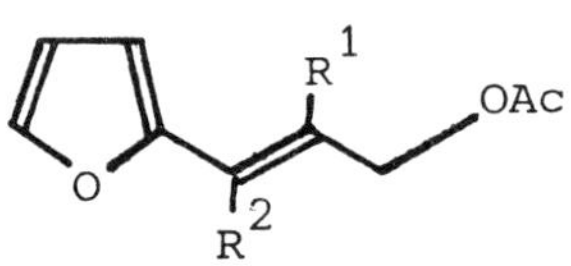

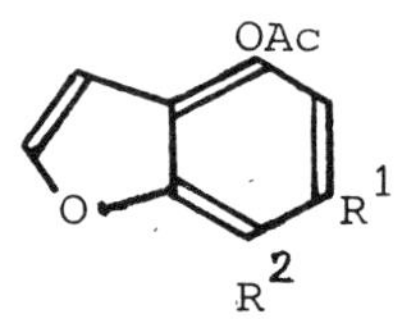

(13) $R^1=H, R^2=Me; R^1=Me, R^2=H; R^1=R^2=H$ (14)

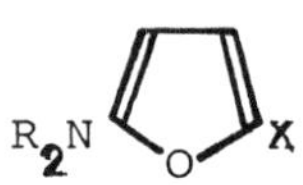

(15) $X=CH=\overset{+}{N}R_2$

(16) X=N=NAr

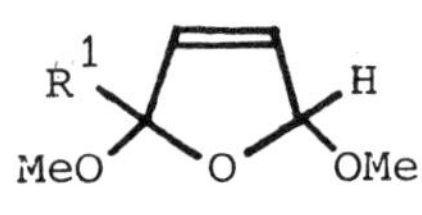

(17) $R^1=H,Me$

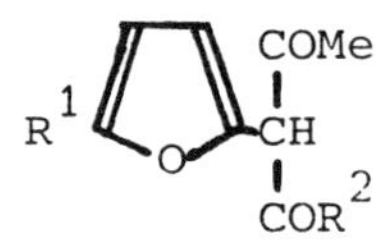

(18) $R^2=Me,OEt$

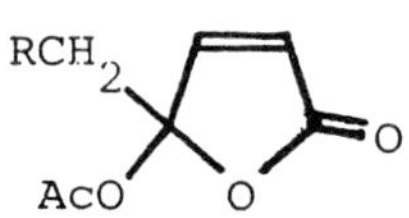

(19) $R=H,Me,PhCH_2CH_2$

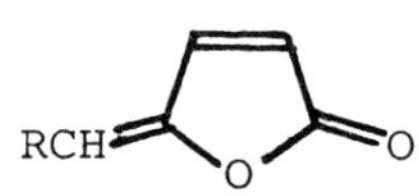

(20)

roacetate gives 4-alkylbut-2-en-4-olides <88CC364>. In cotrast to the reaction of 2-trimethylsilyloxyfuran with aldehydes at C5 forming (11), the corresponding boron and tin furanolates react preferentially at C3 yielding (12) <88CC1595; 89TL1237>. Intramolecular acylation is a key step in the conversion of furylallyl acetates (eg. 13) to benzofurans (14) by reaction with carbon monoxide in the presence of bis(triphenylphosphine)palladium dichloride <89TL95>. *Ipso*-substitution has been observed in the coupling of the *N,N*-dialkyl-iminium salts (15) with arenediazonium salts leading to the azo compounds (16) <88CC1053>. 2,5-Dihydro-2,5-dimethoxyfurans (17) undergo zinc chloride mediated reaction with 1,3-dicarbonyl compounds forming the furans (18) <88JOC5540>.

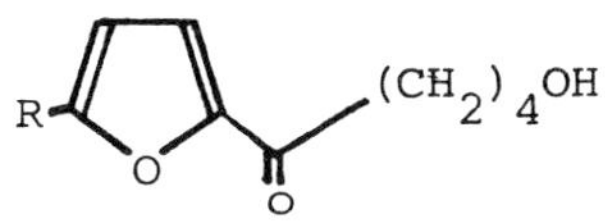

(21) R=H,Me

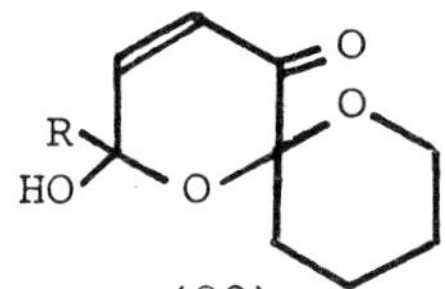

(22)

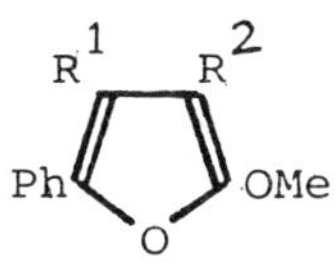

(23)

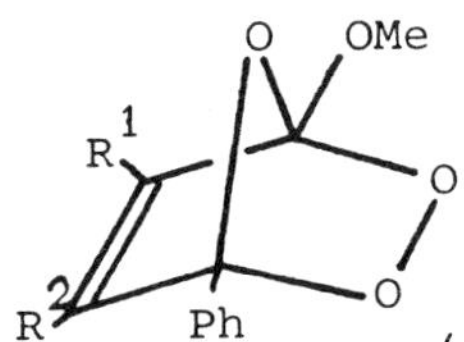

(24)

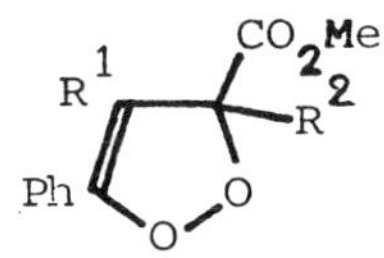

(25)

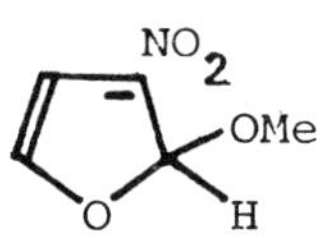

(26)

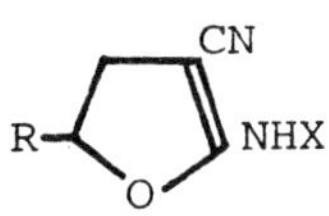

(27) X=H

(28) X=Me

(29) X=COPh

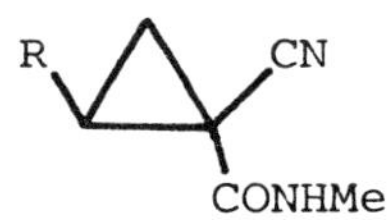

(30) R=H,Ph

4.323 Reactions with Oxidising Agents

Lead tetraacetate oxidation of 2-tributylstannylfurans provides the butenolides (19), which undergo acid catalysed elimination to the corresponding (20)<88CC560>. Oxidation of the furans (21) with aqueous *N*-bromosuccinimide yields the spiroketals (22) <89JOC2044>. The furans (23) bearing electron withdrawing groups at C4 undergo addition of singlet oxygen forming the adduct (24) which in part is converted to the labile dioxolene (25), whereas furans not so substituted at C4 give rise to carbonyl oxides <89JCS(P1)241>.

4.324 Reactions with Nucleophiles

The Meisenheimer adduct (26) is formed by treatment of 3-nitrofuran with methoxide anion <88CS349>. Trialkylaluminiums convert 2-nitrofuran mainly to the 4-alkylbut-2-en-4-olides while 5-methyl-2-nitrofuran yields predominantly the 2-alkyl-4-methylbutenolide <88JCR(S)228>. Bromofuran-2-carboxylic acids can be reductively debrominated with Raney copper in aqueous alkali <88JCR(S)136>. The 2-amino-4,5-dihydrofuran-3-carbonitriles (27) are converted to (28) by methylamine hydrochloride in pyridine, whereas in *N*,*N*-dimethylformamide (28) are rearrranged to the cyclopropanes (30) <89H(29)5>. Treatment of (29) with sodium iodide in DMF generates the pyrrolidones (31).

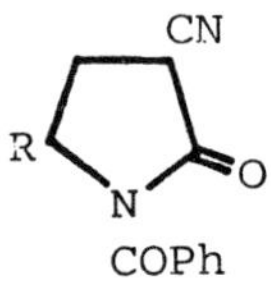

(31)

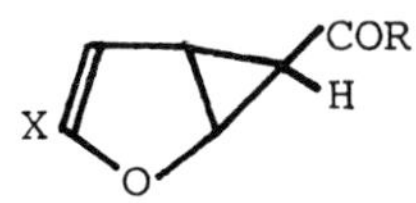

(32) X=CO_2Et;R=Me,MeO

(33) X=CN; R=Me,MeO

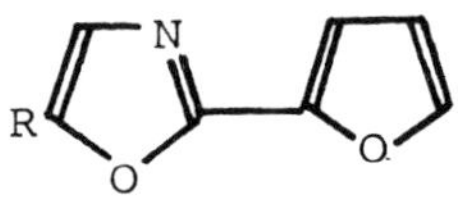

(34)

R$(CH_2)_2COCHN_2$

(35) R=2-furyl

(36) R=3-furyl

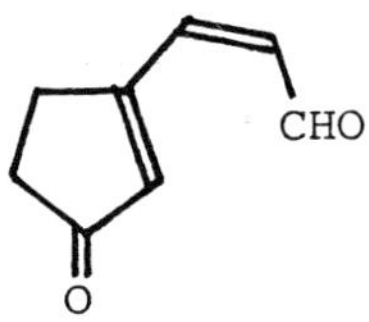

(37)

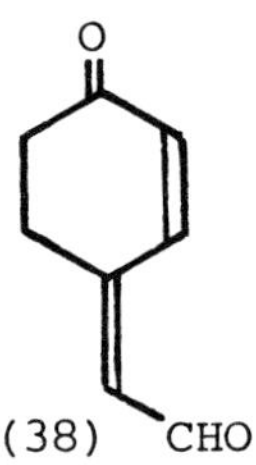

(38)

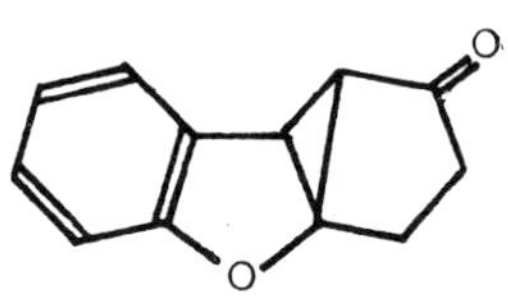

(39)

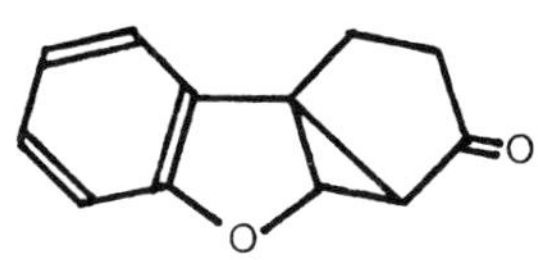

(40)

$RCH_2SO_2CN_2CO_2Et$ (41) R = 3-furyl ; (42) R = 2-furyl

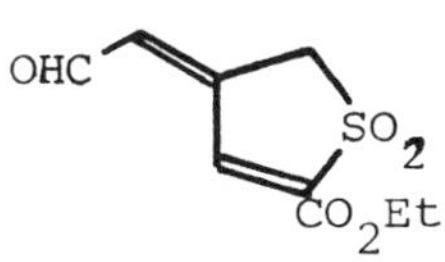

(43)

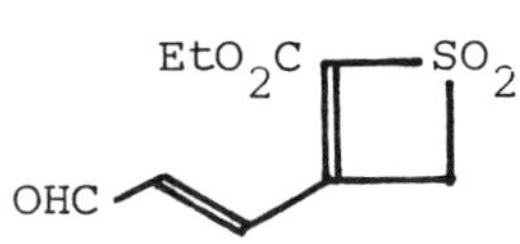

(44)

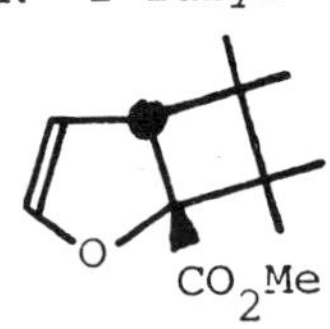

(45)

4.325 Reactions with Carbenes

The rhodium (II) diacetate mediated reaction of α-diazocarbonyl compounds with ethyl furoate provides the adducts (32). The corresponding reaction with furan-2-carbonitrile gives (33) and a larger quantity of (34) <88IZV2833>. Similarly catalysed decomposition of (35) gives (37) and the 3-furyl isomer (36) provides (38). The benzo[b] analogues of (35) and (36) yield (39) and (40) respectively <89JOC299>. The sulphones (41) and

(42) are likewise converted to the thiophene (43) and thietane (44) derivatives respectively <89CJC1071>.

4.326 Cycloaddition Reactions

The photochemically promoted addition of 2,3-dimethylbut-2-ene to methyl furoate gives (45) and (46) as the main products <89JOC135>. Analogous addition of methyl benzoate to furan generates the adduct (47) which can be further converted to (48) <89JOC140>.

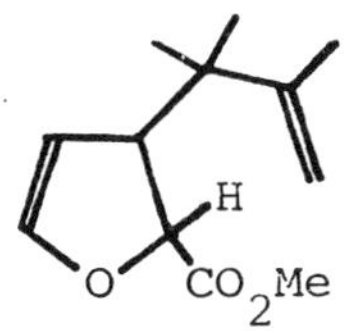

(46)

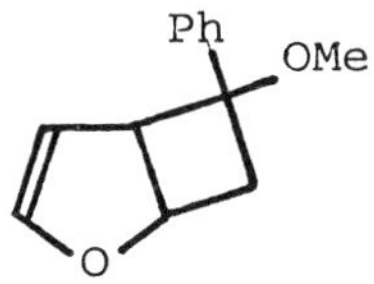

(47)

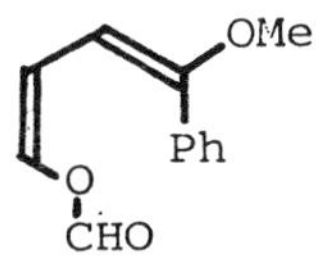

(48)

The cycloaddition of α,β-unsaturated carbonyl compounds to furan may be promoted by adsorption on chromatographic media <88TL175>. Intramolecular cycloadditions to furan as the 4π-component continue to receive much attention both with respect to the effect of substituents and conditions on the rate of reaction<88TL2429; 88TL2493; 88TL5017; 88TL5825; 89JA5469>, and their applications to the synthesis of natural products <88TL6059; 88TL 6501>. The construction of furanodecalins by the intramolecular Diels-Alder reactions of 2- and 3-vinylfurans is exemplified by the conversion of (49) to (50) <88TL2107>.

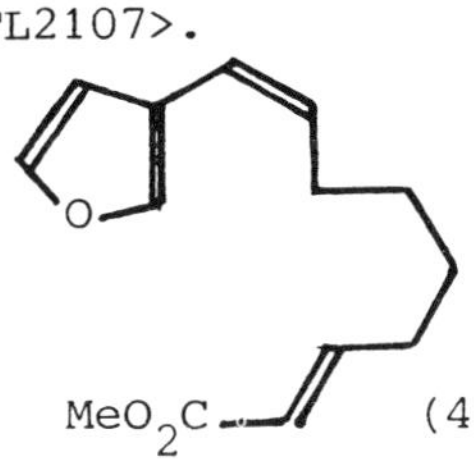

(49)

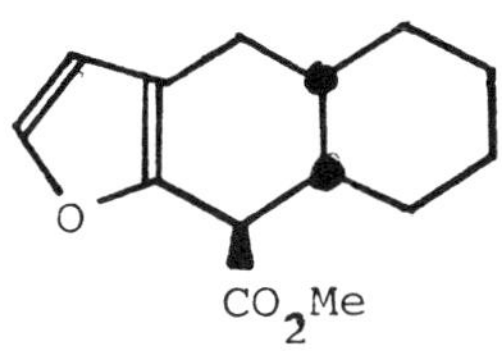

(50)

4.33 REACTIVITY OF SUBSTITUENTS

4.331 Metallo Substituents

2-Lithiobenzo[b]furan-TMEDA is dimeric both in the solid state and in solution <89MI6300>. The reactions of the cuprate $(2\text{-furyl})_2Cu(CN)Li_2$ with enones, epoxides, aldehydes and halides have been surveyed <88TL3045>. 2,5-Dilithiofuran can be generated from furan using the butyllithium/potassium *t*-butoxide/ TMEDA combination <88S316>. A synthesis of the anti-inflammatory manoalide (51) exemplifies the application of a range of furan organometallic derivatives <88TL 1173>. The synthetic applications of [(furyl)methoxymethylene]pentacarbonylchromium continue to be explored<89JOC3625; 89JOC4481>.

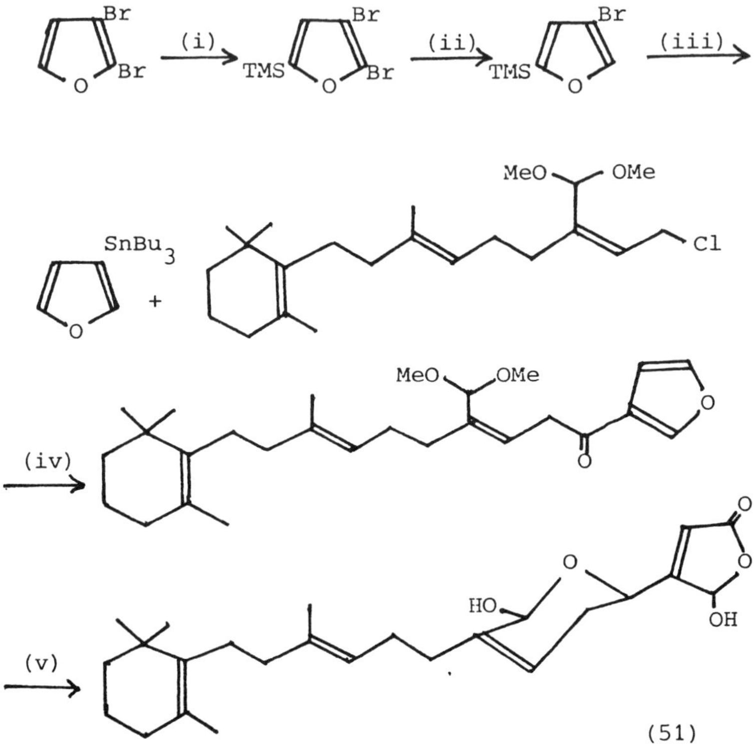

(51)

(i) LDA/TMSCl; (ii) *n*-BuLi/H_2O; (iii) *s*-BuLi/Bu_3SnCl; (iv) CO/Pd(0); (v) $LiAlH_4$; $^1O_2$; *hν*.

4.34 RING SYNTHESIS

4.341 Formation of One Bond

4.3411 Adjacent to Oxygen

Propargyl ketones (52) undergo palladium(0) catalysed cyclisation to the furans (53) <87S1022>. 2-Tosylbut-3-enols (54), readily obtained from the tosylpropene anion and aldehydes, are converted by palladium (II) chloride and copper (II) chloride to the tetrahydrofurans (55), and subsequently by acid treatment to the corresponding 2,5-disubstituted furans <89CL737>.

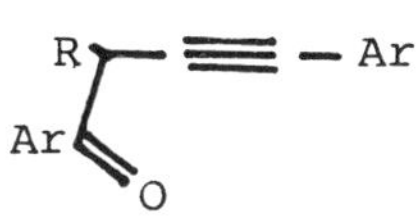
(52) R=H,alkyl

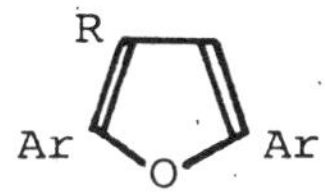
(53)

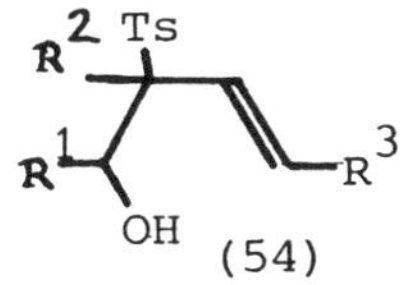
(54)

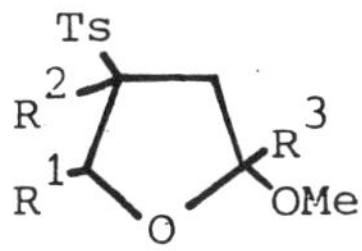
(55)

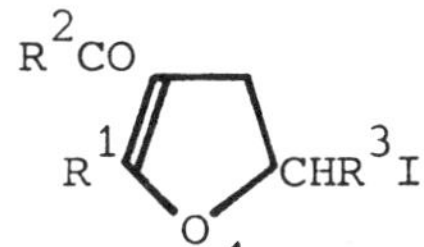
(56) R^1=alkyl,Ph
R^2=OEt,Me,Ph
R^3=H,Me

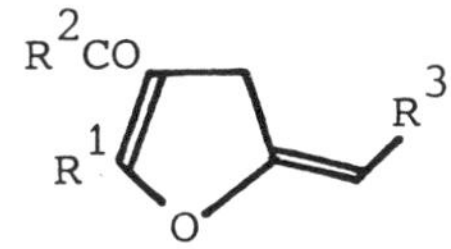
(57)

Iodine treatment of 2-allyl-1,3-dicarbonyl compounds provides the dihydrofurans (56), which can be dehydroiodinated to (57) and subsequently isomerised to (58) <88TL4987>. The cyclic carbonates (59), obtained from RCOC≡CCMe$_2$OH and carbon dioxide in the presence of bis(triphenylphosphine)palladium dichloride and cuprous iodide, undergo thermal conversion into the 2*H*-furan-3-ones (60) <88TL5941>.

From the continuing plethora of papers dealing with the construction of reduced furans attention may be drawn to the directing effect of the benzoate group in

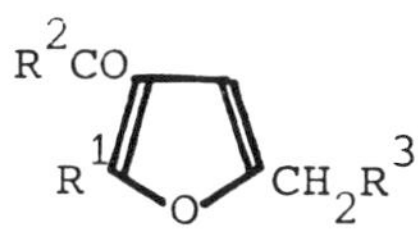
(58)

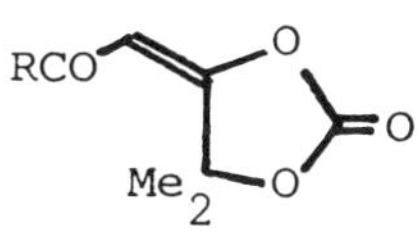
(59) R=alkyl,aryl

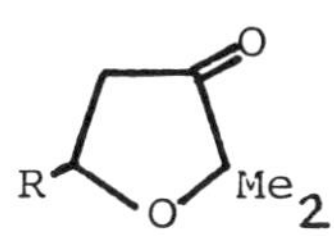
(60)

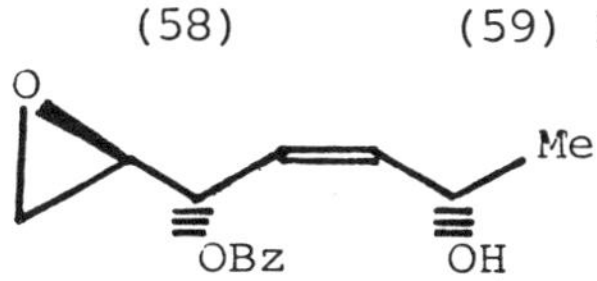
(61)

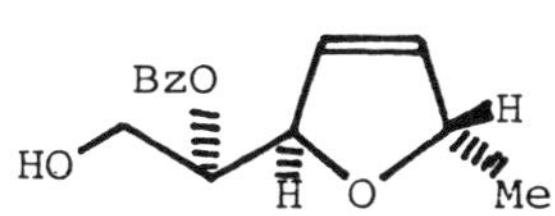
(62)

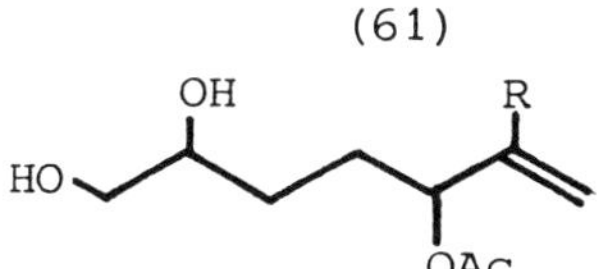
(63)

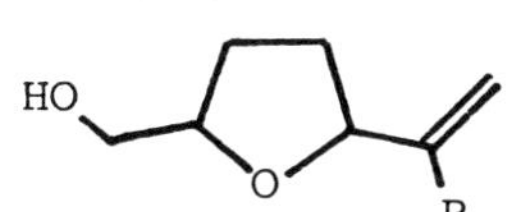
(64)

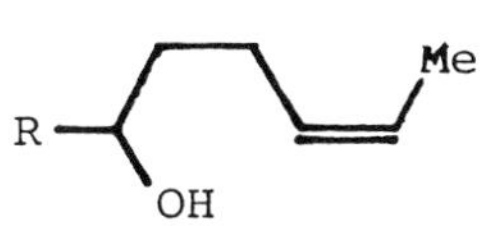
(65)

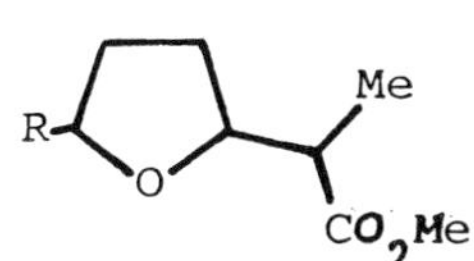
(66)

the boron trifluoride catalysed cyclisation of (61) to (62) <88TL1979>, the palladium(0) catalysed cyclisation of such substrates as (63) to (64) <88TL2927>, and the palladium catalysed intramolecular alkoxycarbonylation of 5-hydroxypentenes (65) to the tetrahydrofurans (66) <89JOC4483>.

4.3412 β to Oxygen

Dipropargyl ethers (67) are cyclised to the furans (68) under the conditions of aminomercuriation <88TL 5029>. A variety of tetrahydrofuran derivatives have been synthesised by treatment of α-phenylthioalkyl but-3-enyl ethers with lithium naphthalenide as in the ring closure of (69) to (70) <89JA2981>.

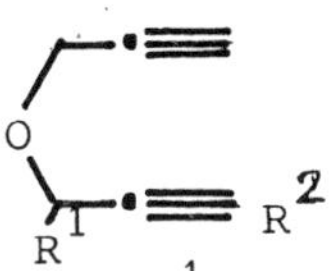

(67) R^1=H,Me,Et
R^2=H,Me,Ph

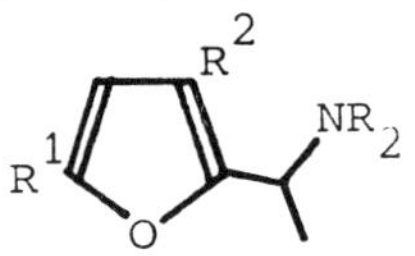

(68)

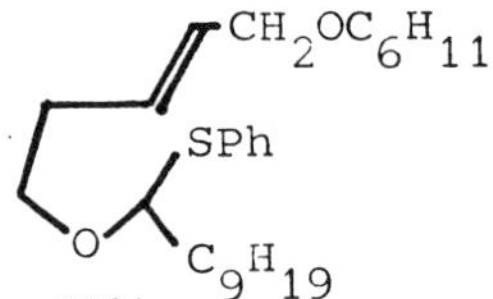

(69)

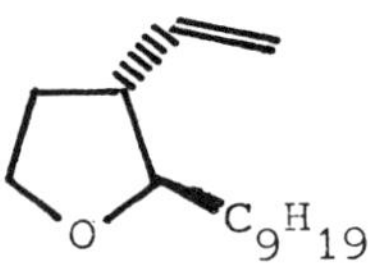

(70)

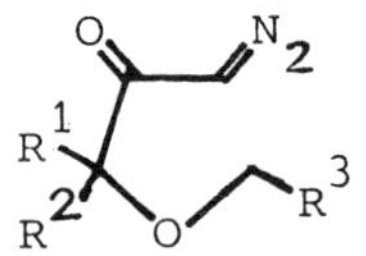

(71) R^1,R^2=H,Me
R^3=alkyl,Ph

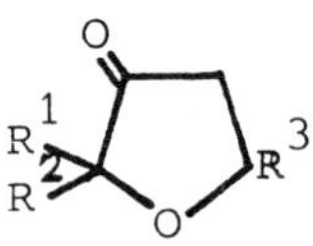

(72)

Tetrahydrofuran-3-ones (72) can be obtained by rhodium (II) diacetate catalysed cyclisation of (71). The initial products may be dehydrogenated with selenium dioxide to the 2*H*-furan-3-ones <89TL1749>. This type of ring closure has been applied to the synthesis of muscarine <89TL1753>.

4.3413 γ to Oxygen

Free radical initiated cyclisations continue to attract attention. Notable developments are the incorporation of activated olefins as in the conversion of (73) to (74) <89H(28)373>, and ring closures of the type illustrated by the benzoyl peroxide promoted cyclisation of (75) to (76) <89JCS(P1)313>. Allyl *o*-iodoaryl ethers are cyclised to the corresponding 3-substituted benzo[b]furans by palladium (0) catalysts <88TL4687; 89H(28)55>. The ring closure of allyl propargyl ethers to methylenetetrahydrofurans is mediated by bis(triphenylarsine)palladium (II) acetate as in the cyclisation of (77) to (78), a key step in the synthesis of phyllanthocin <89JOC4489>.

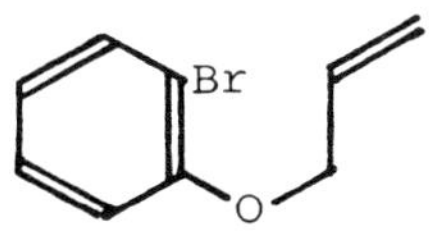

(73)

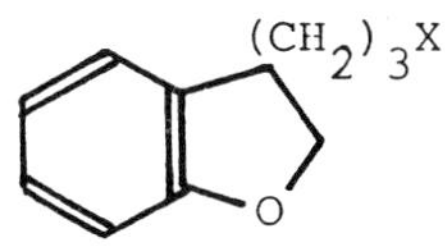

(74) X=CO_2Et,CN,Ph

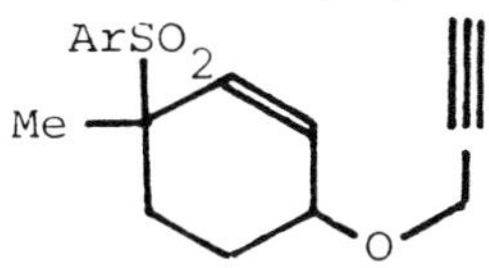

(75)

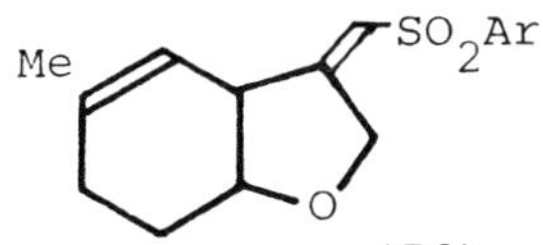

(76)

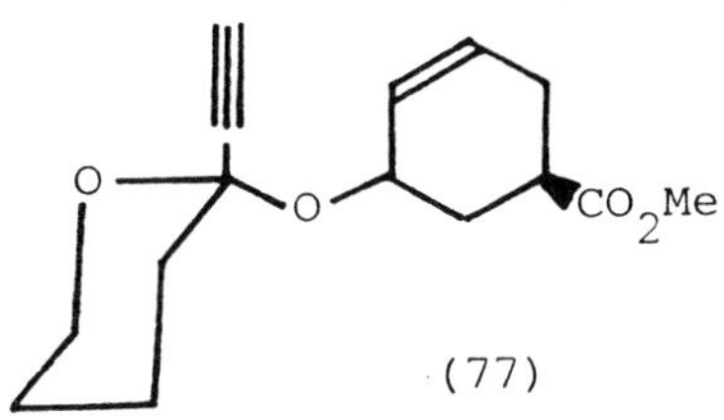

(77)

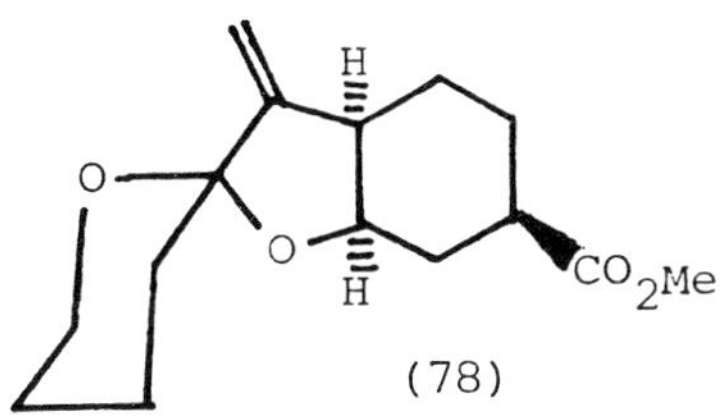

(78)

4.342 From [3+2]Atom Fragments

4.3421 $C_{\alpha}O$ plus C_3 unit

Addition of zirconocenechlorohydride to a protected propargyl alcohol gives (79) which undergoes carbonylation to (80); subsequent treatment with iodine generates the butenolide (81) <88TL3445>. Reaction of the sulphonylepoxide (82) with methyllithium generates the anion (83) which adds to aldehydes forming (84). This undergoes acid catalysed cyclisation to the 2,5-dihydrofuran (85), or oxidative cyclisation with pyridinium chlorochromate to the furan (86) <89JOC1491>. Reaction of the organozinc bromide (87) with a ketone provides the intermediate (88) whose cyclisation to (89) is catalysed by tetrakis(triphenylphosphine)palladium (0) <88TL3579>. 3-Methylenetetrahydrofurans (91) have also been synthesised from (90) and aldehydes under catalysis by tributyltin acetate and tetrakis(triphenylphosphine)palladium (0) <89JA5902>. Acylation of the allene (92) with acyl chlorides in the presence of aluminium chloride generates the furans (93) which can be desilylated with pyridinium hydrofluoride <89JA4407>.

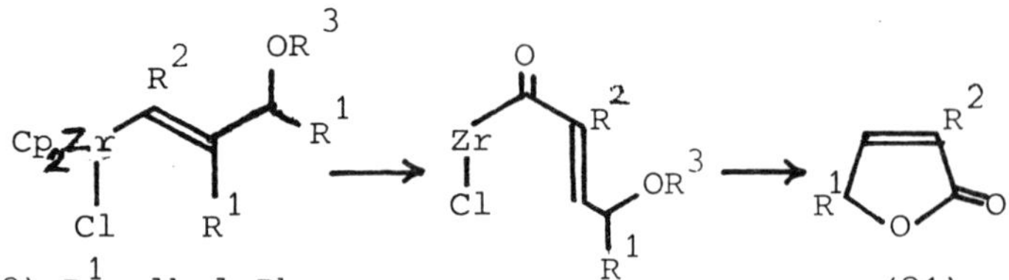

(79) R^1=alkyl,Ph
R^2=H,alkyl

(80)

(81)

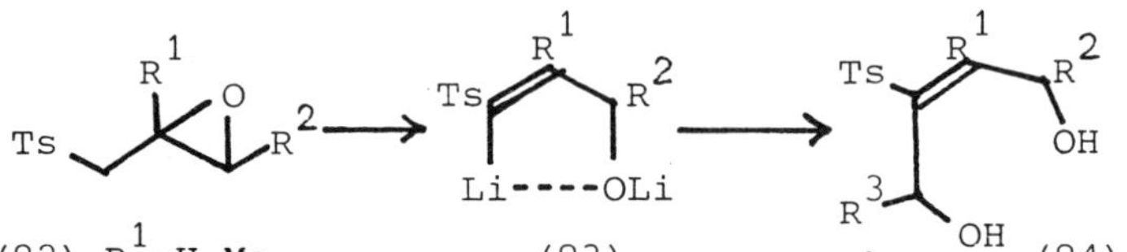

(82) R^1=H,Me
R^2=Me,H
R^3=i-Pr,Ph

(83)

(84)

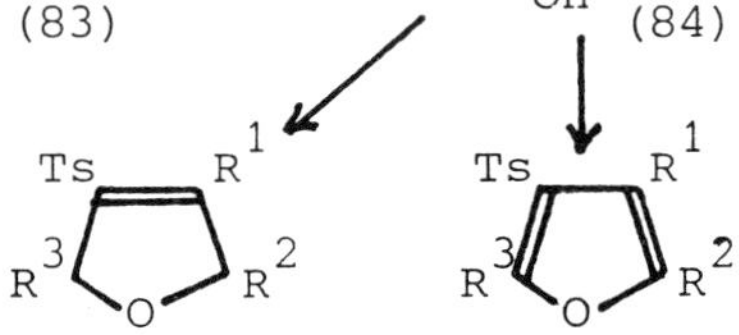

(85)

(86)

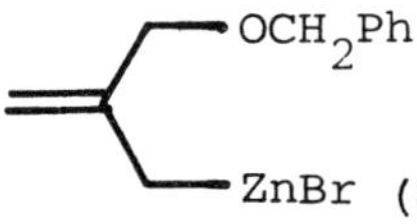

(87)

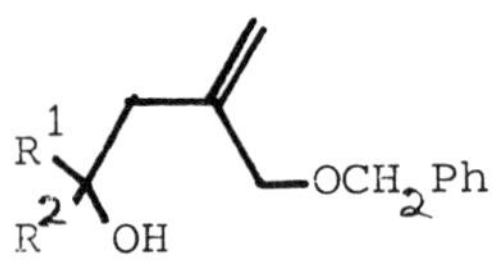

(88)

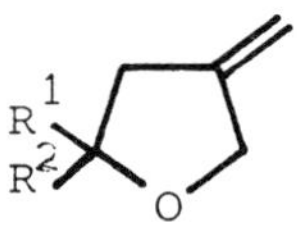

(89)
(91) R^2=H

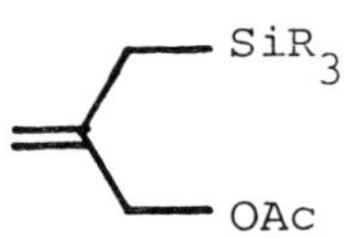

(90)

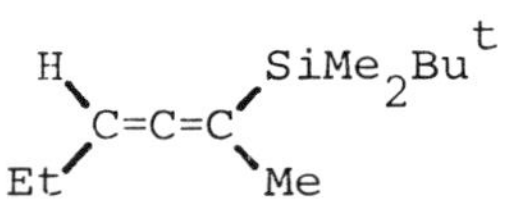

(92)

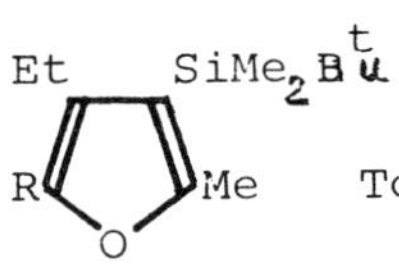

(93)

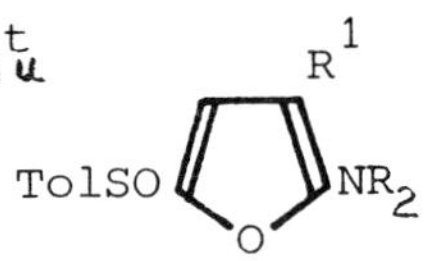

(94) R^1=Me,Ph,acyl

4.3422 $C_\alpha C_\beta$ plus C_2O unit

The reaction of alk-1-ynyldialkylamines, $R_2NC{\equiv}CR'$, with p-toluenesulphonylacetylene yields the sulphinylfurans (94) <88CB2163>. The butenediol derivative (95) has been introduced as a useful synthon for tetrahydrofuran synthesis as in the generation of (96) <89JOC4246>. Oxidative addition of 1,3-dicarbonyl compounds to vinyl acetates mediated by ceric ammonium nitrate provides the dihydrofurans (97), which undergo acid catalysed elimination of acetic acid to yield the corresponding furans <88SC1841>. Condensation of 2,2-diethoxyvinyltriphenylphosphonium tetrafluoroborates (98) with α-hydroxyketones in the presence of sodium hydride yields the furans (99) <88T5095>.

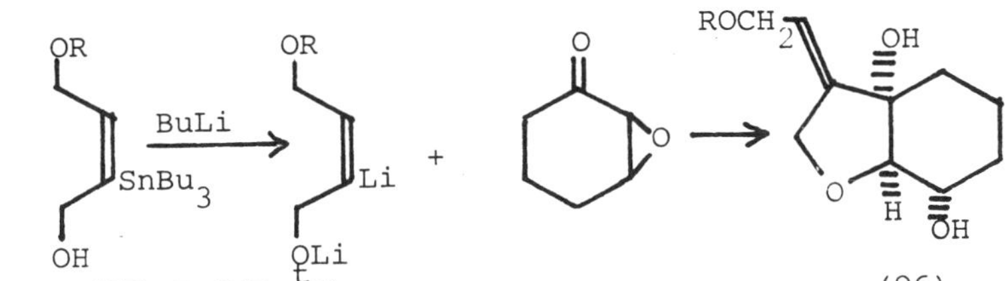

(95) R=SiBu^{t}Ph$_2$ (96)

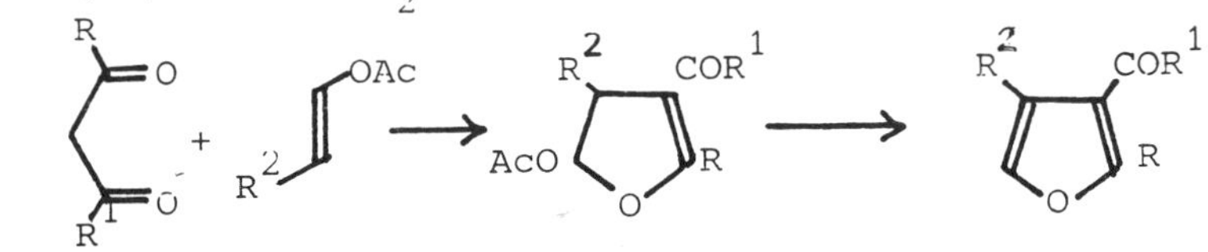

(97)

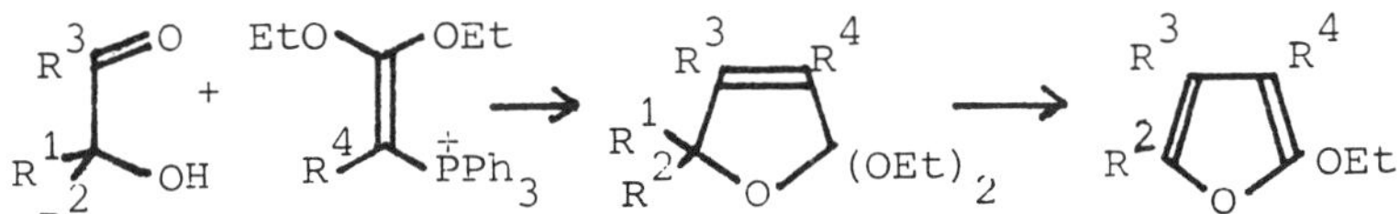

(98) (99)

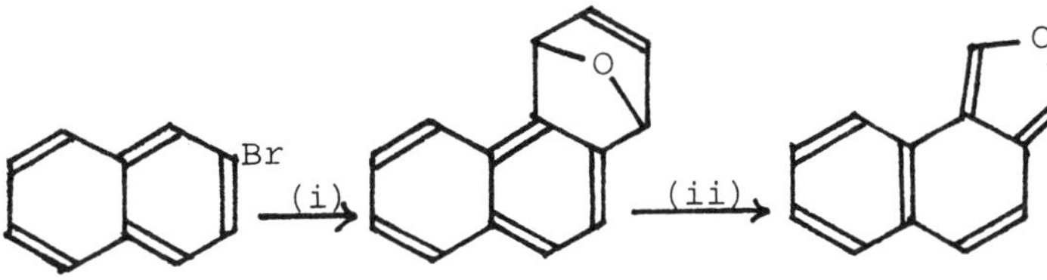

(100)

(i) $NaNH_2$/furan; (ii) 3,6-dipyridyltetrazine.

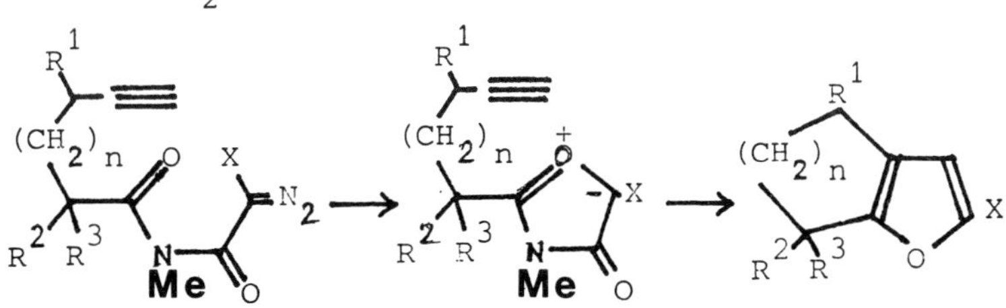

(101) (102)

4.3423 $C_\beta C_\beta$ plus $C_\alpha OC_\alpha$ unit

The route indicated for the generation of naphtho-[1,2-c]furan (100) has been applied to the synthesis of a variety of arene[c]furans <88AJC235>. The use of intramolecular cycloadditions to intermediary munchnones is typified by the conversion of (101) to(102) <89CB1081>.

REFERENCES

87S1022 H. Sheng, S, Lin and Y. Huang, Synthesis, 1987, 1022.

88AJC235 J. Moursounidis and D. Wege, Aust. J. Chem., 1988, 41, 235.

88CB2163 G. Himbert and S. Kosack, Chem. Ber., 1988, 121, 2163.

88CC364 C.W. Jefford, A.W. Sledeski, J. Chem. Soc., Chem. Commun., 1988, 364.

88CC560 M. Yamamoto, H. Izukawa, M. Saiki and K. Yamada, J. Chem. Soc., Chem. Commun., 1988, 560.

88CC1595 C.W. Jefford, D. Jaggi and J. Boukouvalas, J. Chem. Soc., Chem. Commun., 1988, 1595.

88CCC1053 T. Gracza, Z. Arnold and J. Kovac, Coll. Czech Chem. Comm., 1988, 53, 1053.

88CS349 F. Sancassan, C. Dell'Erba, S. Gronowitz, G. Consiglio and D. Spinelli, Chem. Scr., 1988, 28, 349.

88H(27)1907 T. Momose, N. Toyooka, T. Nishi and Y. Takeuchi, Heterocycles, 1988, 27, 1907.

88IZV2833 L.E. Saltykova, A.E. Vasilviskii, V.M. Shostakovskii and O.M. Nefedov, Izv. Akad. Nauk SSSR, Ser. Khim., 1988, 2833.

88JCS(P1)3169 G. Weeratunga, M. Austrup and R. Rodrigo, J. Chem. Soc., Perkin 1, 1988, 3169.

88JCR(S)136 M. Tashiro, S. Mataka, K. Nakayama, H. Tsuzuki and T. Yonemitsu, J. Chem. Res., Synop., 1988, 136.

88JCR(S)228 A. Pecunioso and R. Menicagli, J. Chem. Res., Synop., 1988, 228.

88JOC5540 R. Antonioletti, F. Bonadies and A. Scettri, J. Org. Chem., 1988, 53, 5540.

88S316 B.L. Feringa, R. Hulst, R. Rikers and L. Brandsma, Synthesis, 1988, 316.

88SC1841 E. Baciocchi and R. Ruzziconi, Synth. Commun., 1988, 18, 1841.

88T2093 R. Rodrigo, Tetrahedron, 1988, 44, 2093.

88T5095 R.W. Saalfrank, W. Hafner, J. Markmann and H.J. Bestmann, Tetrahedron, 1988, 44, 5095.

88TL175 V.V. Veselovsky, A.S. Gybin, A.V. Lozanova, A.M. Moiseenko, W.A. Smit and R. Caple, Tetrahedron Lett., 1988, 29, 175.
88TL1173 S. Katsumura, S. Fujiwara and S. Isoe, Tetrahedron Lett., 1988, 29, 1173.
88TL1979 M.T. Nunez, M.L. Rodriguez and V.S. Martin, Tetrahedron Lett., 1988, 29, 1979.
88TL2107 J.A. Cooper, P. Cornwall, C.P. Dell and D.W. Knight, Tetrahedron Lett., 1988, 29, 2107.
88TL2377 H. Heaney, G. Papageorgiou and R.F.Wilkins, Tetrahedron Lett., 1988, 29, 2377.
88TL2429 M.E. Jung and J. Gervay, Tetrahedron Lett., 1988, 29, 2429.
88TL2493 S. Cauwberghs, P.J. De Clercq, B. Tinant and J.P. Declercq, Tetrahedron Lett., 1988, 29, 2493.
88TL2927 B.M. Trost and A. Tenaglia, Tetrahedron Lett., 1988, 29, 2927.
88TL3045 J.S. Ng, J.R. Behling, A.L. Campbell, D. Nguyen and B. Lipshutz, Tetrahedron Lett., 1988, 29, 3045.
88TL3445 S.L. Buchwald, Q. Fang and S.M. King, Tetrahedron Lett., 1988, 29, 3445.
88TL3579 J. van der Louw, J.L. van der Baan, H. Stichter, G.J.J. Out, F. Bickelhaupt and G.W. Klumpp, Tetrahedron Lett., 1988, 29, 3579.
88TL4687 R.C. Larock and D.E. Stinn, Tetrahedron Lett., 1988, 29, 4687.
88TL4987 R. Antonioletti, F. Bonadies and A. Scettri, Tetrahedron Lett., 1988, 29 ,4987.
88TL5017 L.M. Harwood, S.A. Leeming, N.S. Isaacs, G. Jones, J. Pickard, R.M. Thomas and D. Watkin, Tetrahedron Lett., 1988, 29, 5017.
88TL5029 J. Barluenga, F. Aznar and M. Bayod, Tetrahedron Lett., 1988, 29, 5029.
88TL5825 L.M. Harwood, G. Jones, J. Pickard, R.M. Thomas and D. Watkin, Tetrahedron Lett., 1988, 29, 5825.
88TL5941 Y. Inoue, K. Ohuchi and S. Imaizumi, Tetrahedron Lett., 1988, 29, 5941.
88TL6059 M.E. Jung and Vu Chi Truc, Tetrahedron Lett., 1988, 29, 6059.

88TL6501 S.G. Cauwberghs and P.J. De Clercq, Tetrahedron Lett., 1988, 29, 6501.

89CB1081 M.E. Maier and B. Schoeffling, Chem. Ber., 1989, 122, 1081.

89CJC1071 S.D. Babu, M.D. Hrytsak and T. Durst, Can. J. Chem., 1989, 67, 1071.

89CL737 S. Igarashi, Y. Haruta, M. Ozawa, Y. Nishide, H. Kinoshita and K. Inomata, Chem. Lett., 1989, 737.

89H(28)55 E. Negishi, T. Nguyen, B. O'Connor, J.M. Evans and A. Silveira, Heterocycles, 1989, 28, 55.

89H(28)373 H. Togo and O. Kikuchi, Heterocycles, 1989, 28, 373.

89H(29)5 K. Yamagata, H. Maruoko, Y. Hashimoto and M. Yamazaki, Heterocycles, 1989, 29, 5.

89JA2981 C.A. Broka and T. Shen, J. Amer. Chem. Soc., 1989, 111, 2981.

89JA3659 K.J. Stone, M.M. Greenberg, S.C. Blackstock and J.A. Berson, J. Amer. Chem. Soc., 1989, 111, 3659.

89JA3671 M.M. Greenberg, S.C. Blackstock, K.J. Stone and J.A. Berson, J. Amer. Chem. Soc., 1989, 111, 3671.

89JA4407 R.L. Danheiser, E.J. Stoner, H. Koyama, D.S. Yamashita and C.A. Klade, J. Amer. Chem. Soc., 1989, 111, 4407.

89JA5469 M.E. Jung and J. Gervay, J. Amer. Chem. Soc., 1989, 111, 5469.

89JA5902 B.M. Trost, S.A. King and T. Schmidt, J. Amer. Chem. Soc., 1989, 111, 5902.

89JA8732 J.C. Scaiano, V. Wintgens, K. Haider and J.A. Berson, J. Amer. Chem. Soc., 1989, 111, 8732.

89JCS(P1)241 M.L. Graziano, M.R. Iesce, G. Cimminiello and R. Scarpati, J. Chem. Soc., Perkin 1, 1989, 241.

89JCS(P1)313 T.A.K. Smith and G.H. Whitham, J. Chem. Soc., Perkin 1, 1989, 313.

89JHC365 T. Horaguchi, E. Hasegawa, T. Shimizu, K. Tanemura and T. Suzuki, J. Het. Chem., 1989, 26, 365.

89JHC439 M. Tinggaard, P. Hansen, P.K. Mogensen, O. Simonsen and J. Becher, J. Het. Chem., 1989, 26, 439.

89JOC135	T.S. Cantrell and A.C. Allen, J. Org. Chem., 1989, 54, 135.
89JOC140	T.S. Cantrell, A.C. Allen and H. Ziffer, J. Org. Chem., 1989, 54, 140.
89JOC299	A. Padwa, T.J. Wisnieff and E.J. Walsh, J. Org. Chem., 1989, 54, 299.
89JOC1491	C. Najera and M. Yus, J. Org. Chem., 1989, 54, 1491.
89JOC2044	F. Perron and K.F. Albizati, J. Org. Chem., 1989, 54, 2044.
89JOC3625	A. Yamashita, A. Toy and T.A. Scahill, J. Org. Chem., 1989, 54, 3625.
89JOC4246	A.G.M. Barrett, T.E. Barta and J.A. Flygare, J. Org. Chem., 1989, 54, 4246.
89JOC4481	A. Yamashita, A. Toy, N.B. Ghazal and C.R. Muchmore, J. Org. Chem., 1989, 54, 4481.
89JOC4483	M.F. Semmelhack and N. Zhang, J. Org. Chem., 1989, 54, 4483.
89JOC4489	B.M. Trost, E.D. Edstrom and M.B. Carter-Petillo, J. Org. Chem., 1989, 54, 4489.
89MI6300	S. Harder, J. Boersma, L. Brandsma, J.A. Kanters, W. Bauer, R. Pi, P. von R. Schleyer, H. Scholhorn and U. Thewalt, Organometallics, 1989, 8, 1688.
89TL95	M. Iwasaki, J. Li, Y. Kobayashi, H. Matsuzuka, Y. Ishii and M. Hidai, Tetrahedron Lett., 1989, 30, 95.
89TL103	S. Miki, M. Yoshida and Z. Yoshida, Tetrahedron Lett., 1989, 30, 103.
89TL1237	C.W. Jefford, D. Jaggi and J. Boukouvalas, Tetrahedron Lett., 1989, 30, 1237.
89TL1749	J. Adams, M.A. Poupart, L. Grenier, C. Schaller, N. Ouimet and R. Frenette, Tetrahedron Lett., 1989, 30, 1749.
89TL1753	J. Adams, M.A. Poupart and L. Grenier, Tetrahedron Lett., 1989, 30, 1753.

CHAPTER 4.4
Five-Membered Ring Systems: With More Than One N Atom

S A LANG, JR. AND C B ZIEGLER, JR
American Cyanamid Company, Pearl River, NY, USA

4.4.1 INTRODUCTION

Di-, tri- and tetra azoles continue to be highly useful synthetic intermediates for the preparation of a variety of polycyclic and bridged systems. In addition, these rings are often integral parts of molecules which possess biological applications in agricultural or medical areas. A large number of publications have appeared on metal complexes of the title compounds and many focus on their catalytic properties or use in chiral reagents. A major review on 1,2,4-triazolines appeared in 1989 [89AHC(46)169].

4.4.2 PYRAZOLES AND RING-FUSED PYRAZOLES

The development of new methodology or improvements upon existing methods allows the synthesis of pyrazoles with new substituents. 3-Trifluoromethyl substituted pyrazoles (**1**) have been prepared regiospecifically from trifluoroacetyl acetylenes and hydrazine via a 1,2- or 1,4-cycloaddition mechanism [89TL(30)2049]. Other pyrazole syntheses bearing the

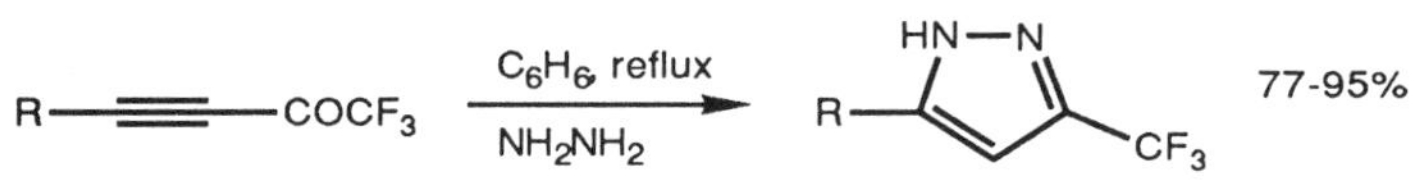

(1)

trifluoromethyl substituent are also noteworthy. As part of a program searching for glycine agonists, a highly functionalized 5-trifluoromethyl analogue (**2**) was prepared from trifluoroacetyl chloride via an oxazole. This method should prove to be general for this substitution class of pyrazole [89JMC(32)2116].

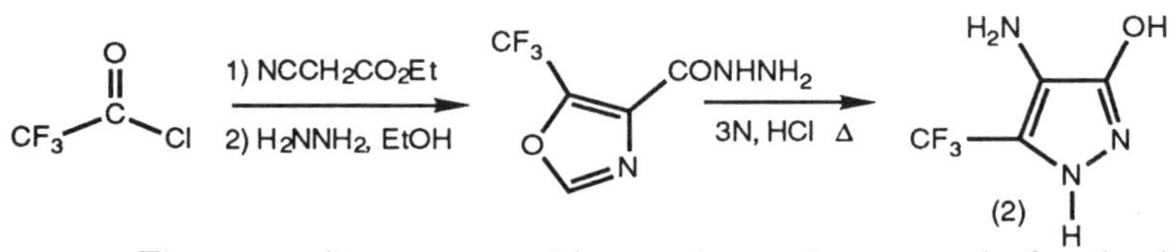

The patent literature provides an interesting example for the first synthesis of 4-trifluoromethylthio-5-aminopyrazoles (**3**) in a one-pot procedure [89EUP320750]. The utilization of a sulfur dioxide extrusion from readily available 1,2,6-thiadiazine 1,1-dioxides provides a convenient synthesis of 2,3,4-trisubstituted systems (**4**) [89H(29)245].

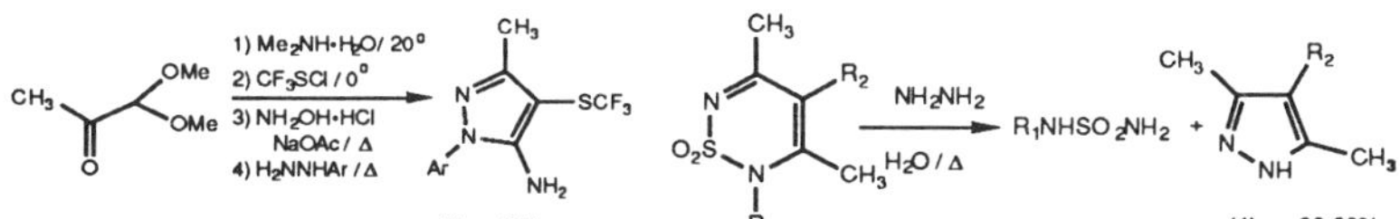

(3) 91% (4) 60-90%

A novel synthesis of 5-chloro-3-methoxycarbonyl-1-arylpyrazoles (**5**) from phosphoranes of the type (**6**) compliment the existing methodology [89JHC(26)241].

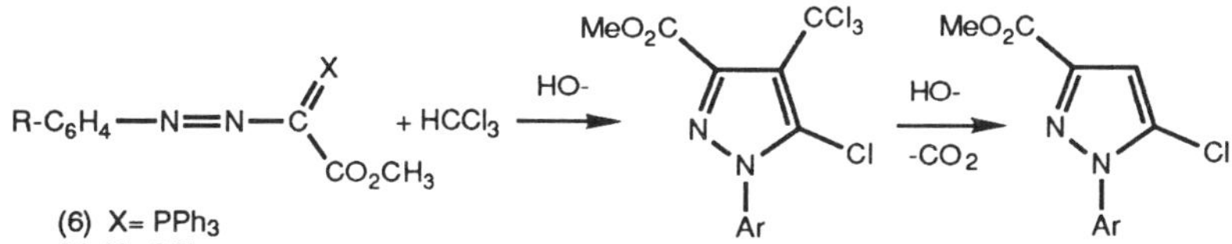

(6) X= PPh_3
(7) X= CCl_2
(8) (5)

Mechanistic details for this synthesis are involved but initially entail dichlorocarbene addition to (**6**) that gives (**7**) followed by the incorporation of two more carbon units from the chloroform solvent. The intermediate (**8**) is stable but is easily converted to the product (**5**). The patent literature provides two useful routes to 4-nitro-5-aminopyrazoles (**9**) as shown [89EUP320765, 320764]. Both routes are high yielding and should be general at least for preparing N-aryl examples.

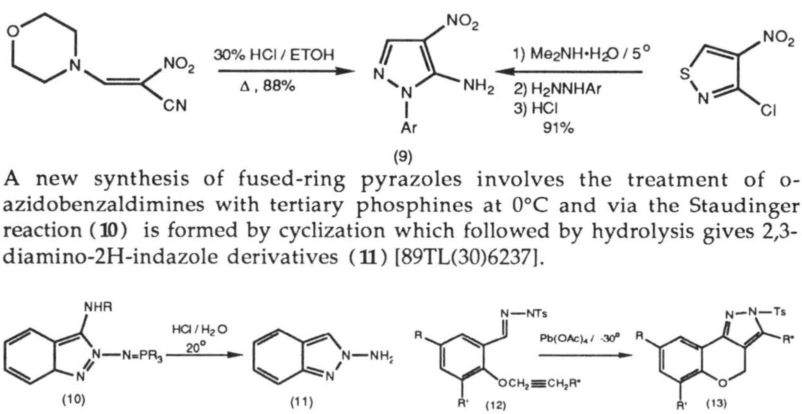

(9)

A new synthesis of fused-ring pyrazoles involves the treatment of o-azidobenzaldimines with tertiary phosphines at 0°C and via the Staudinger reaction (**10**) is formed by cyclization which followed by hydrolysis gives 2,3-diamino-2H-indazole derivatives (**11**) [89TL(30)6237].

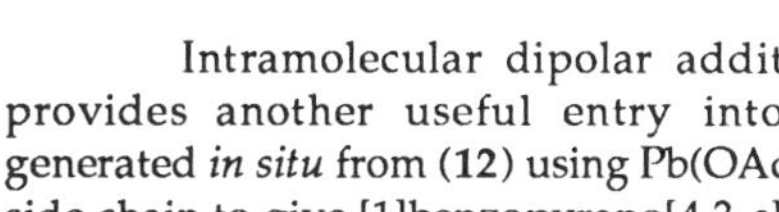

(10) (11)

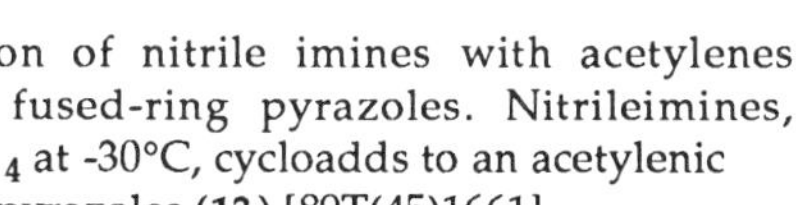

(12) (13)

Intramolecular dipolar addition of nitrile imines with acetylenes provides another useful entry into fused-ring pyrazoles. Nitrileimines, generated *in situ* from (**12**) using $Pb(OAc)_4$ at -30°C, cycloadds to an acetylenic side chain to give [1]benzopyrano[4,3-c]-pyrazoles (**13**) [89T(45)1661].

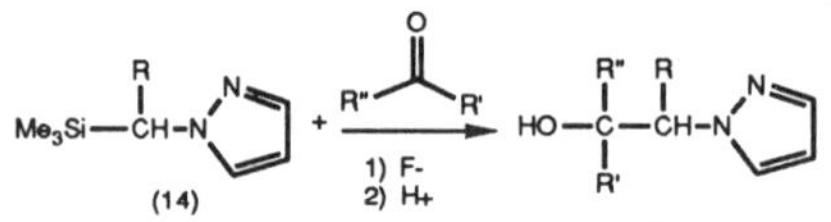

(14) (15)

R = SPh, SMe, OMe

Some interesting N-alkylation studies have focused on the pyrazole nucleus. Phenylthio, methylthio and methoxysubstitute silylmethylpyrazoles act as N-pyrazolymethyl anion equivalents [89T(45)637]. Compounds of the general structure (**14**) were reacted with a variety of carbonyl compounds in the presence of fluoride ions to yield the corresponding pyrazolylethanols (**15**) in yields ranging from 43 to 93%. The reaction conditions were dependent on the R substituent. For instance, when R = SPh the fluoride induced condensation proceeded at -30°C whereas higher reaction temperatures are required when R = SMe, OMe. In addition to N-alkylpyrazoles, the corresponding imidazole and 1-substituted-1,2,4-triazoles work equally well in this alkylation reaction. Similar overall transformations have been demonstrated using the lithio anion of (**14**) R = SPh [89CJC(67)1144].

Regioselective N-alkylation was noted when ethyl 5-cyano-1H-pyrazole-4-carboxylate (**16**) was treated with tertiary carbocations under mild conditions using catalytic H_2SO_4. Under these conditions (**17**) was formed. On the other hand, (**17b**) was formed when boron trifluoride was used as catalyst. The regioselective alkylation reflects either the strong coordination of the Lewis acid at N-1 or, more likely, its bond formation at this center. From this, the authors conclude that the major tautomeric species of (**16**) has its lone electron pair at N-1 [89JHC(26)3]. The synthesis of masked aminopyrazolecarboxylic acid synthons (**18**) and (**19**) has been detailed [89JOC(54)428]. Each was prepared via Curtius methodology from

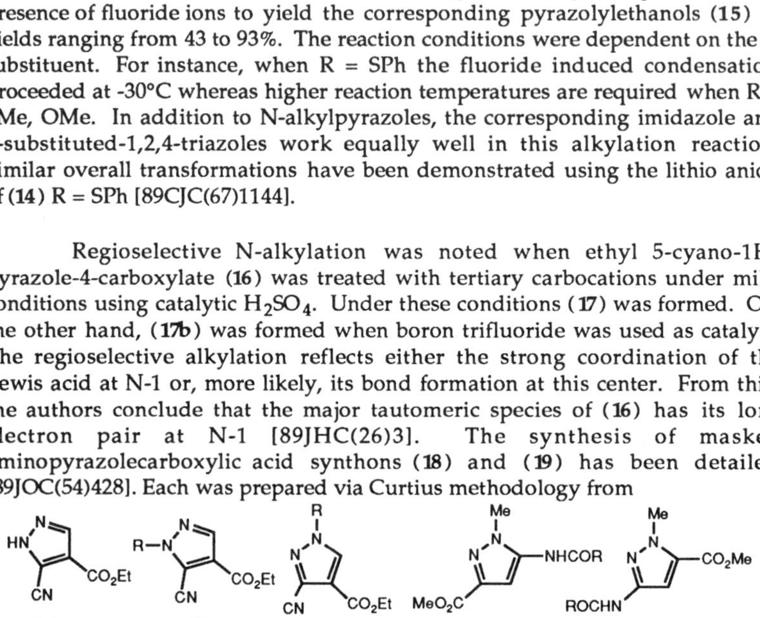

(16) (17) (17b) (18) (19)

the requisite half ester-acid pyrazole. These pyrazoles are useful simpler analogues to netropsin and distamycin A which are DNA minor groove binding antibiotics. As part of a C-nucleoside synthetic program a direct method for double functionalization of the pyrazole ring was developed [89JCS(PI)925]. Treatment of 3(5)-alkylpyrazoles (**20**) with ammonium nitrate in trifluoroacetic anhydride gave (**21**) directly. This procedure, in conjunction with cine-substitution of cyanide to (**22**) offers a direct route to vicinal disubstitution of this sort.

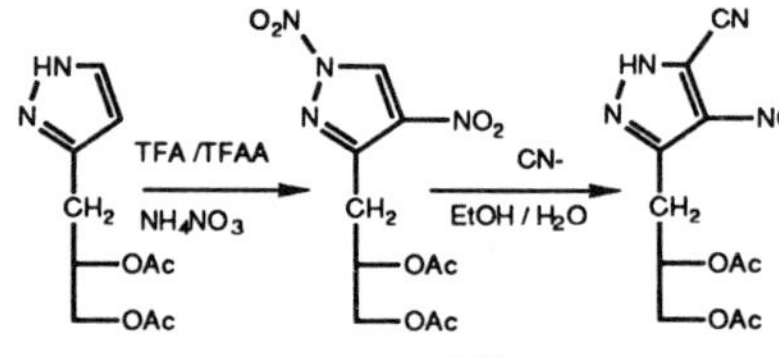

(20) (21) (22)

4.4.3 IMIDAZOLES AND FUSED-RING IMIDAZOLES

New routes to prepare imidazoles and fused-ring imidazoles have appeared. Clever utilization of the imidazole nucleus in synthesis and as mechanistic probes has been advanced. Protonated 1,3-diazabutadienes (23) are very reactive with nucleophilic isocyanates to give 5-iminoimidazolines (24). These [1+4] cycloadditions are rare and are much faster with salts (23) than with the corresponding diazadienes [89JOC(54)1185]. A series of bridged 2,2'-bi-imidazole derivatives have been prepared [89JOC(54)3057]. The conformations of compounds (25) and (26) were studied with respect to their absorption spectra. Two trends were apparent. As the N,N'-bridge became longer for (25) the absorption energy increased which indicated less pi-delocalization, or less coplanarity between the two rings. On the other hand, the corresponding bi-bridged (26) appeared more planar based on its longer absorption wavelength.

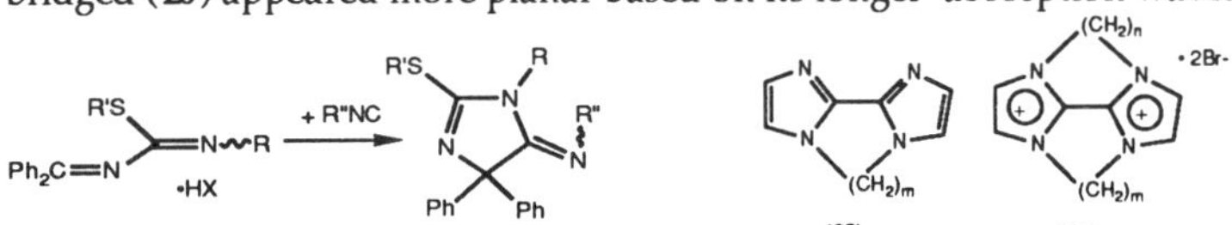

(23) (24) (25) m = 2-4 (26) n = 2-4, m = 2-4

Some useful and interesting fused-ring imidazole syntheses have appeared. The Staudinger reaction serves as the introductory step in a versatile synthesis starting with the azidoimidazole (27). Reaction with triphenylphosphine yields the unisolated iminophosphorane (28) which then undergoes intramolecular Wittig ring closure to (29). On the other hand, the iminophosphorane (30) derived from 2-azidomethylbenzimidazole and triphenylphosphine reacts with aromatic isocyanates to give the corresponding tricyclic species (31) [89T(45)1823]. This is the first example of heterocyclization based on the reaction of a carbodiimide with the NH of an imidazole.

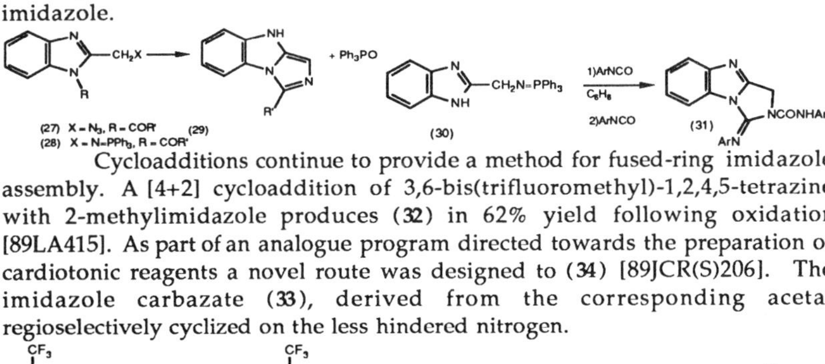

(27) X = N_3, R = COR' (28) X = N=PPh_3, R = COR' (29) (30) (31)

Cycloadditions continue to provide a method for fused-ring imidazole assembly. A [4+2] cycloaddition of 3,6-bis(trifluoromethyl)-1,2,4,5-tetrazine with 2-methylimidazole produces (32) in 62% yield following oxidation [89LA415]. As part of an analogue program directed towards the preparation of cardiotonic reagents a novel route was designed to (34) [89JCR(S)206]. The imidazole carbazate (33), derived from the corresponding acetal regioselectively cyclized on the less hindered nitrogen.

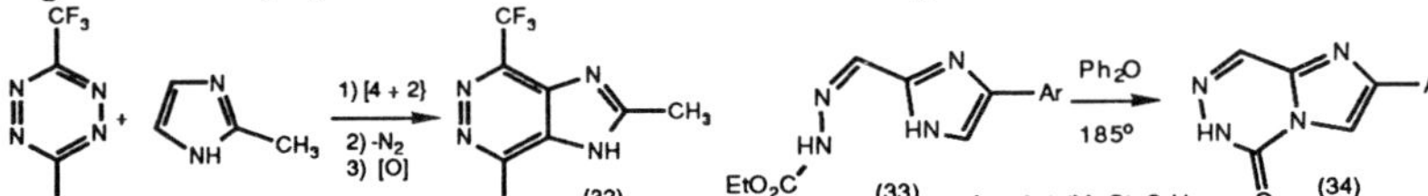

(32) (33) (34) Ar = 2,4-$(MeO)_2C_6H_3$

Imidazole and fused-ring imidazoles are suitable mimics of histidine which plays an important role in the in vivo activation of the antibiotic, bicyclomycin. Most proposals on its mode of action suggest that nucleophilic species in bacterial cell walls significantly contribute to the overall process. In one study various imidazoles were added in Michael fashion to the activated form (**35**) of bicyclomycin. The Michael adduct (**36**) went on to form a diastereomeric mixture of ring opened products [89JA(111)4897].

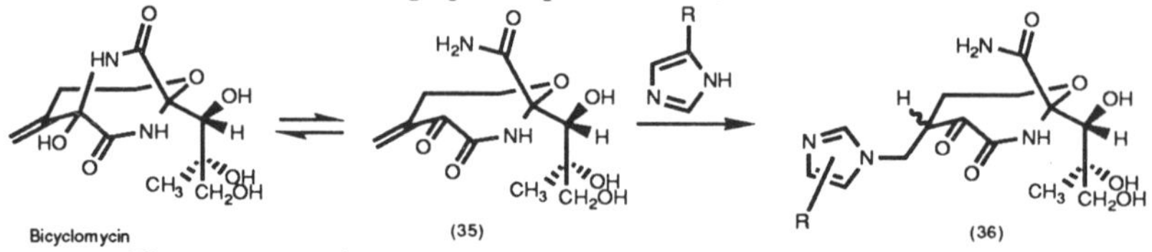

There are a few examples of 2-acylimidazolium salts acting as acylating agents. However, the first instance of an intramolecular version of this reaction to give γ–lactones has been reported. Reaction of the anion (**37**) with ketones and aldehydes proceeds regioselectively to give the enol silyl ether (**38**) which suffers cleavage to the γ-lactone (**39**) after desilyation,

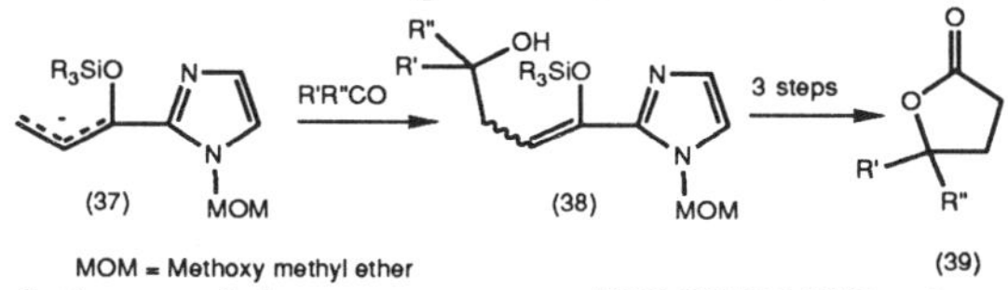

N-methylation and base treatment [89JCS(PI)838]. A general oxidative preparation of converting imidazole-2(3H)-thiones (**40**) to imidazolyl disulfides (**41**) has been reported [89S714]. The electrochemical oxidation produces the disulfides in fair to excellent yields. Substituted imidazoles undergo Bamberger ring cleavage to give some interesting synthetic intermediates. One such case is demonstrated by the synthesis of 5,6-diaminocaproic acid from (**42**) [89LA493] via the bis-N-benzoyl derivative (**43**).

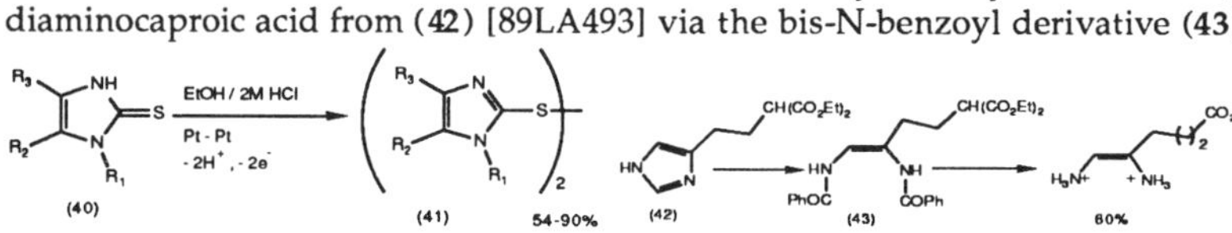

4.4.4 1,2,3-TRIAZOLES AND RING FUSED 1,2,3-TRIAZOLES

Azide cycloadditions or reactions of propargyl halides with azide is the most reported method to generate 1,2,3-triazoles and a number of mechanistic studies of this transformation have been reported. The interconversion of the N-substituent has also been examined. The reaction of methyl 3,3-diazido-2-cyanoacrylates with amines leads to vinyl azides (**44**) a number of which are quite stable [89JOC(54)4356]. The more stable vinyl azides undergo a [1,5] ring closure to 1,2,3-triazoles (**45**) via 4H-1,2,3-triazoles(**46**). In contrast, vinyl azides with a monoalkylamino substitutent in the 4-position produce tetrazolyltriethylammonium salts (**47**) via a [1,5'] closure when treated with triethylamine followed by treatment with HCl to produce the vinyl triazole (**48**). Treatment of (**44**) with 1,3-propanediamine or 1,4-butanediamine produced (**49**).

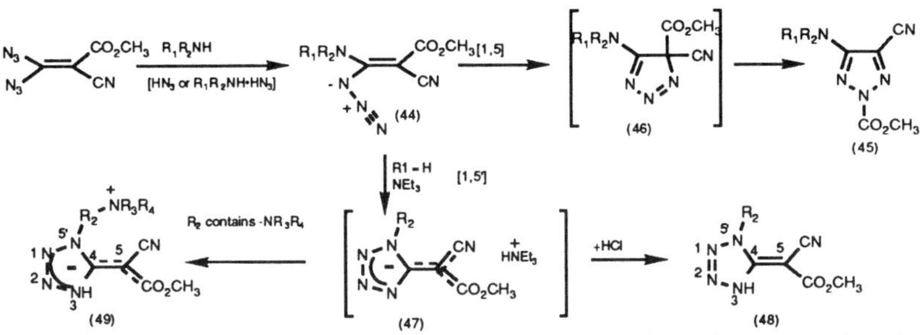

Treatment of propargyl halides with azide afforded propargylazides which rearrange at room temperature to produce 1,2,3-triazoles [89CB(122)911; 89CB(122)1175]. Short-lived allenyl azides (50) and triazafulvenes (51) were proposed intermediates. The reaction of 1,4-dichloro-2-butyne and azide yielded a monoazide which rearranged to (52) via an azidobutatriene.

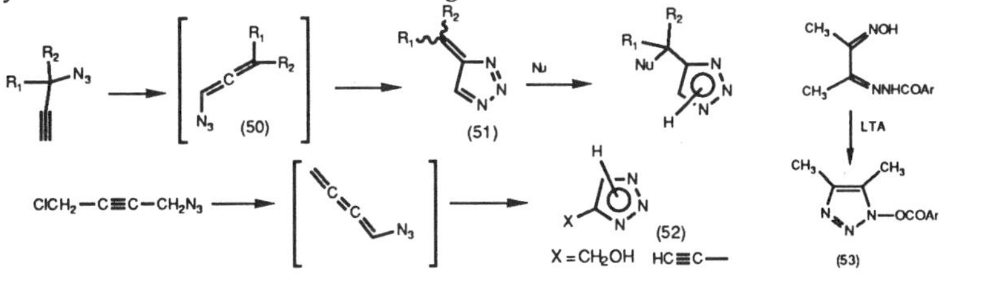

1,2,3-Triazoles were also prepared by the catalytic cyclization of propargylamines or 1,2-azidoaminoethanes [89JOC(54)5302] or by the reaction of ketenimines with trimethylsilyldiazomethane [89H(29)133]. The latter reaction yielded silyl-1,2,3-triazoles and mechanistic discussions were included. Lead tetraacetate oxidation of α-hydroxyimino-aroylhydrazones of biacetyl produced a mixture containing triazole oxides (53). The mechanism of formation is discussed and the structure was confirmed by X-ray studies [89JCS(P1)619]. 5-Amino-1,2,3-triazoles with an electron-withdrawing N-1 moiety cannot be diazotized directly to the corresponding 5-azido compound. Diazo transfer with tosylazide produced the 5-azido compounds (54) which were thermalized to produce diazotetrazoles (55) or triazenes depending upon the nature of the substituent [89T(45)749].

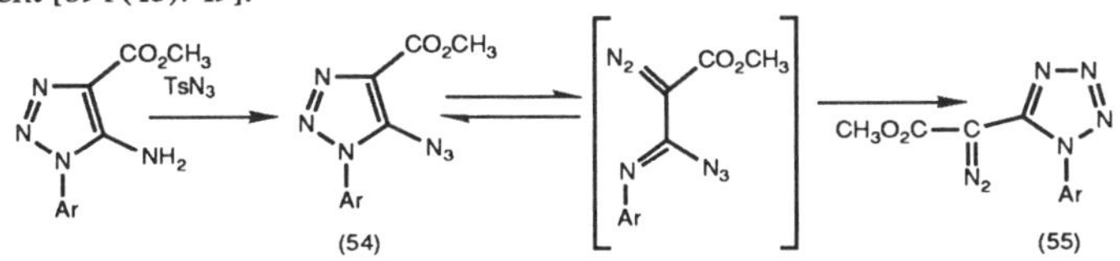

Arylazides react with ethyl cyanoacetate to produce 5-amino-1,2,3,-triazoles (56) which undergo a Dimroth rearrangement on treatment with 2-propyl or 2-butylisocyanate to give positional isomers (57) [89CZ(113)11]. A ^{13}C NMR study of the rearrangement was a part of the research effort. Modelization techniques were applied for the diazomethane addition to 2-cyanochromone reported in the last volume of PHC [89MI73].

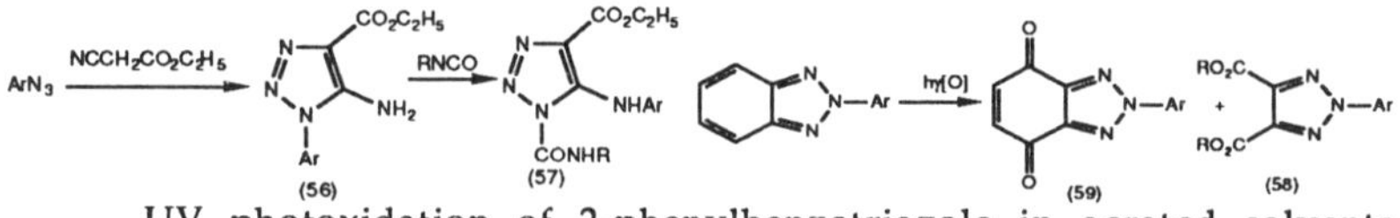

UV photoxidation of 2-phenylbenzotriazole in aerated solvents generated 2-phenyl-1,2,3-triazole-4,5-dicarboxylic acid (**58**) via a proposed 2-phenylbenzotriazol-4,7-dione (**59**) intermediate. The proposed intermediate generates (**58**), the same product under the reaction conditions when photostabilizer Tinuvin P is ulitized [89AJC(42)441].

Ab initio studies on the tautomerism and aromaticity of 1,2,3-triazoles and their benzoanalogues provide a theoretical basis for lone pair and aromaticity studies that were correlated to the thermodynamically observed differences [89JA(111)7348]. Ribosylation of several benzotriazoles in the presence of NAD and NAD-nucleosidase produced exclusively the N-1 dinucleotides as based on UV and ^{15}N NMR analysis. In contrast, the respective indazole analogs yielded the N-2 dinucleotides [89JHC(26)339]. A number of 2-phenyl-5-aminobenzotriazines were synthesized and used as fluorescent probes to examine affinity for basic drug glycoprotein binding studies [89CPB(37)1009]. Pharmaceutically active, highly branched tertiary amines were prepared by the reaction of Grignard reagents with adducts of cyclic ketones, benzotriazole and secondary amines as an attractive alternative to available routes [89S66]. 7-Aminotriazolo-(1,2,3)(1,5-a)-pyridines (**60**) were prepared from lithio triazolopyridines (**61**) or 6-aminopyridines [89T(45)7041]. A variety of 1-carbonylmethyl-1H-benzotriazoles were synthesized and converted into thiadiazolo and thiadiazine derivatives of benzotriazole. The usual methodologies were employed to confirm structural features [89JHC(26)177].

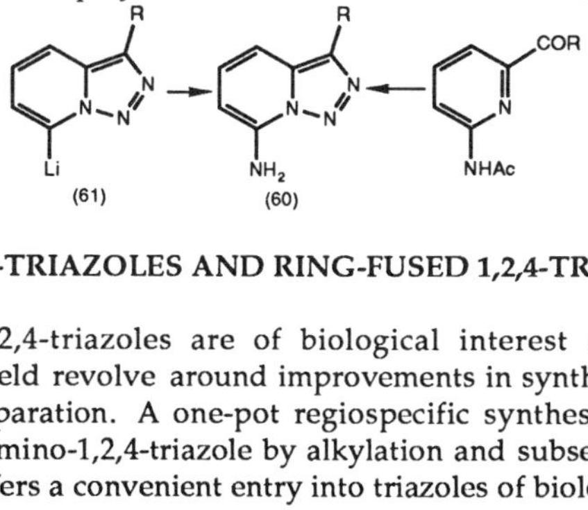

4.4.5 1,2,4-TRIAZOLES AND RING-FUSED 1,2,4-TRIAZOLES

Many 1,2,4-triazoles are of biological interest and much of the activities in the field revolve around improvements in synthetic methodologies and analogue preparation. A one-pot regiospecific synthesis of 1-alkyl-1,2,4-triazoles from 4-amino-1,2,4-triazole by alkylation and subsequent deamination [89JOC(54)731] offers a convenient entry into triazoles of biological interest. The intramolecular cyclodehydration of hydrazinothiazol-4-ylacetic acid esters (**62**) yield the desired bicyclic thiazolo[2,3-c]-s-triazoles (**63**) together with the trisubstituted 1,2,4-triazoles (**64**) [89LA75].

A general synthesis of 1-amino-1H-1,2,4-triazoles from their 4-amino-4H-counterpart via a 1,4-diamino-1,2,4-triazolium salt was described. The parent compound itself cannot be prepared by the described one-pot methodology but could be synthesized by monoamination of the 1,2,4-triazolide anion [89S269]. A convenient synthesis of 5-(1-aminoalkyl)-1H-1,2,4-triazoles (**65**) from t-BOC-α-aminoacids and triphenylhydrazonoiminophosphorane derivatives was also described [89S399]. In this case chromatographic separation of the desired triazole from triphenylphosphine was necessary.

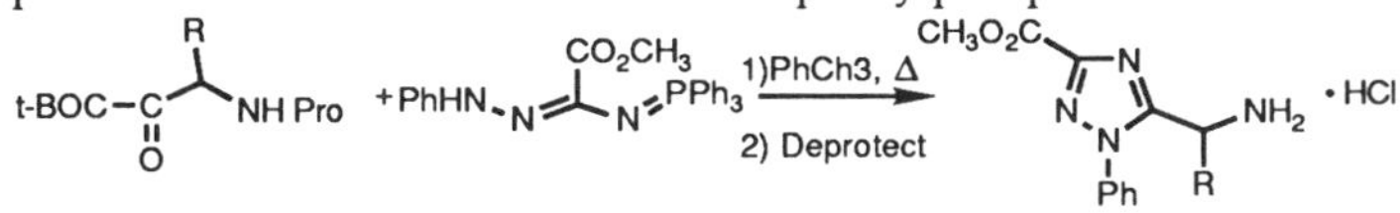

(65)

Analogously, triphenylphosphiniminohydrazones react with acetic, maleic, fumaric and phthalic anhydrides to produce, in refluxing benzene, 5-alkyl-1H-1,2,4-triazoles (**66**) and this procedure represents an improvement over similar methodologies. Mechanistic possibilities were also discussed [89JCR(S)16].

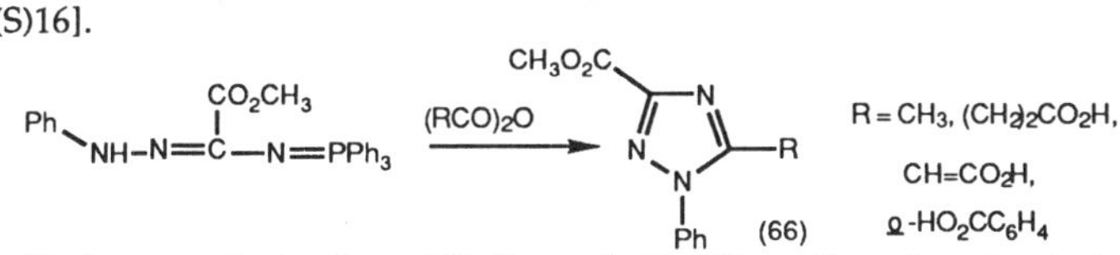

Hydrazone derivatives (**67**, R_1=aryl, R_2=H) cyclize when treated with acid to form bis-triazoles (**68**) and triazoles (**69**) via a protonated intermediate. In (**67**, R_1=aryl, R_2=alkyl), only the N-alkenyltriazoles (**70**) were isolated [89JHC(26)327, 763].

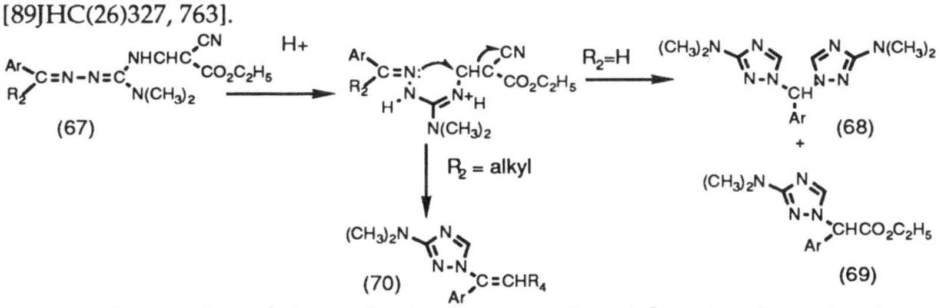

Conversion of the carboximido group in a 1,2,4-triazole containing a furanose generated a mixture of compounds containing nitrile, imidate and amidine analogs [89JMC(32)1447]. Reaction of 2,5-bis(perfluoroalkyl)-1,3,4-oxadiazoles (**71**) with primary amines generated 4-substituted-3,5-bis(perfluoroalkyl)-4H-1,2,4-triazoles (**72**) [89JHC(26)225].

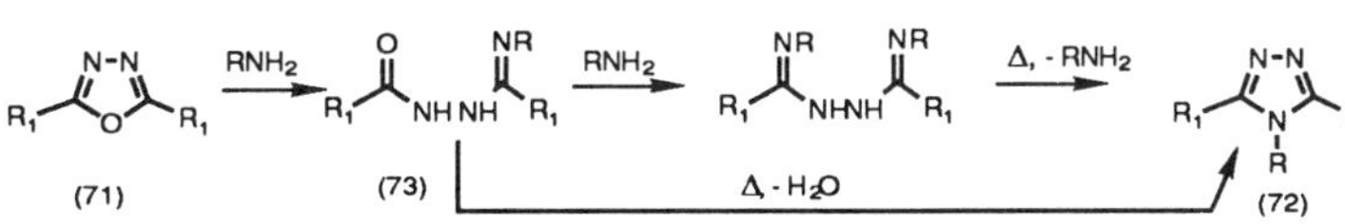

The reaction of 2,5-perfluoroalkyl-1,3,4-oxadiazoles (**71**) with hydrazine in methanol at 0°C was reported to yield hydrazine derivatives (**73**, R = NH_2) which when treated with acetic acid afforded 4-aminotriazoles (**72**, R

= NH_2). Deamination to known triazoles was cited as structural proof [66JOC(31)781]. The reaction of 3,5-bis(trifluoromethyl) analog (**71**, $R_1 = CF_3$) at 0°C in ethanol produced a dihydrotetrazine, which on subsequent oxidation yielded a known tetrazine [78JCP(P1)378]. A re-examination of the reaction of hydrazine and (**71**, $R_1 = CF_3$) in methanol at -42°C yielded (**73**, $R_1 = CF_3$, R = NH_2) rather than the expected dihydrotetrazine. The ^{19}F NMR study of the reaction product clearly showed only (**73**) and no dihydrotetrazine and further experiments ruled out the production of (**72**, $R_1 = CF_3$) at higher temperatures from a dihydrotetrazine [89JOC(54)1760].

5-Aryl-3-mercapto-4-substituted-triazoles were prepared by the cyclization of benzoylthiosemicarbazides with sodium hydroxide [89IJC(28B)83]. Cycloaddition of 4-phenyl-4H-1,2,4-triazole-3,5-dione to fulvene generated a number of polycyclic benzo-norbornadienes some containing a triazole unit [89CB(122)133]. The flash vacuum pyrolysis of 1-acetyl-1,2,4-triazole (**74**) yielded 5-methyloxazole and nitrogen. Additional acyltriazole analogues likewise generated functionalized oxazoles in yields of 10-90% depending upon R. In contrast, N-acetylimidazole yielded a mixture of 2- and 4-acetylimidazole and N-acetylpyridazine produced pyridazone and ketene [89H(29)103].

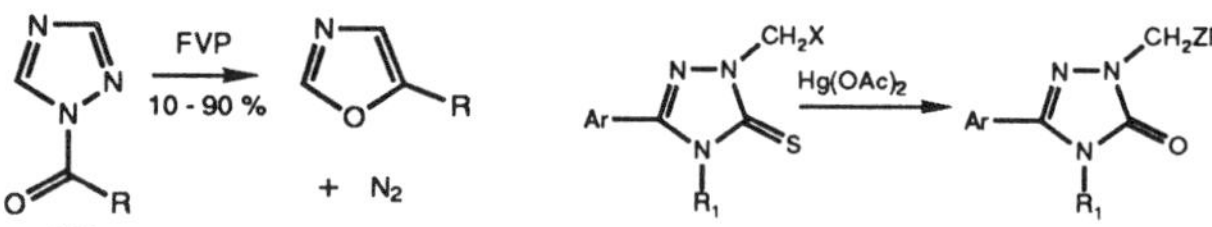

(74) (78) X = Cl, SR, OR (79)

Silylmethyl-1,2,4-triazoles were prepared by the reaction of 1,2,4-triazole with trimethylsilylmethyl chlorides. De-silylation with fluoride and reaction of the generated anion with ketones selectively produced a variety of 2-(1,2,4-triazolyl)ethanols. In contrast to the derivatives containing methylthio, phenylthio or methoxy, the unsubstituted 1-[(trimethylsilyl)methyl]-1,2,4-triazole (**76**) afforded the desired ethanol (**75**) along with an unexpected methanol (**77**) [89T(45)637]. The chloro of 2-chloromethyl-2H-1,2,4-triazole-3-thiones (**78**) is replaced by thiolates or alkoxides and subsequent de-sulfurization by mercuric acetate yielded the oxygen analogs (**79**) [89MI261].

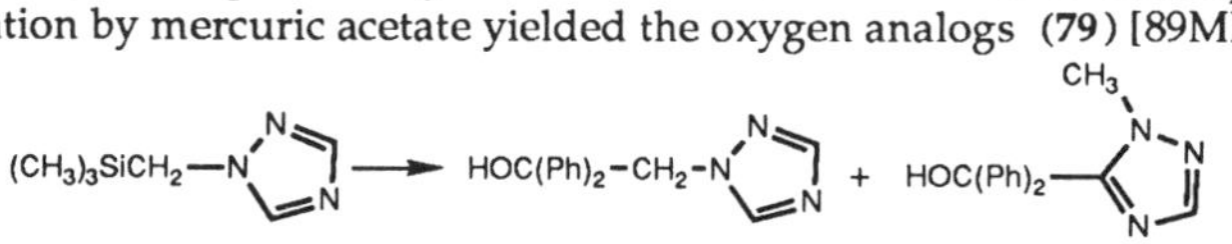

(76) (75) (77)

The reaction of ^{14}C methyl-labeled acetyl chloride with amidohydrazones produced 1,2,4-triazoles labeled in the 3-position [90TL(30)6469]. Phosphoric triazolides have been employed in the chiral synthesis of diacylglycerol analogs [89LA59] The enamtiomeric synthesis of antifungal triazoles was accomplished by the utilization of chiral oxathiane derivatives [89AG(101)507]. The biological implications of guanidinium type resonance stabilization energy, was investigated using *ab initio* calculations on a series of 3-amino-1,2,4-triazoles [89MI35].

The importance of thione versus thiol tautomeric forms of 4-amino-1,2,4-triazoles substituted in the 5-position by sulfur was investigated [89CJC(67)198]. The alkylation of 5-phenyl-1,2,4-triazole-5-thione with N-substituted- quinazolinoneacetamides occurred at the N-2 position rather than the sulfur [89AP(322)49]. 4-Aminopyrazolo-1,2,4-triazoles were obtained by the cyclization of triaminoguanidines with formic and acetic acids and the compounds were tested for antimicrobial activity [89AP(322)63].

Two fused 1,2,4-triazole thiol derivatives utilized as substituents on a cephalosporin or carbapenem, respectively 1,2,4-triazolo[1,5-a]pyrimidine-7-thiols (**80**) [89EUP302633] and (6,7-dihydro-5H-pyrazolo[1,2-a][1,2,4-triazole-6-thiols (**81**) [89JAP25779] were prepared. The latter nucleus (**81**) was prepared *in situ* by formation of the pyrazolothio (**82**) followed by cyclization with forimidate (or formamidine) salts with the proposed intermediacy of the bis-formamide (**83**). 3-Aryl-4-alkyl-1-methyl-1,2,4-triazol-5H-one derivatives as exemplified by (**84**) are seizure antagonists and thus have potential for therapeutic use in neurodegenerative disorders such as Huntington's disease, Parkinsonism and Alzheimers syndrome [88EUP273309].

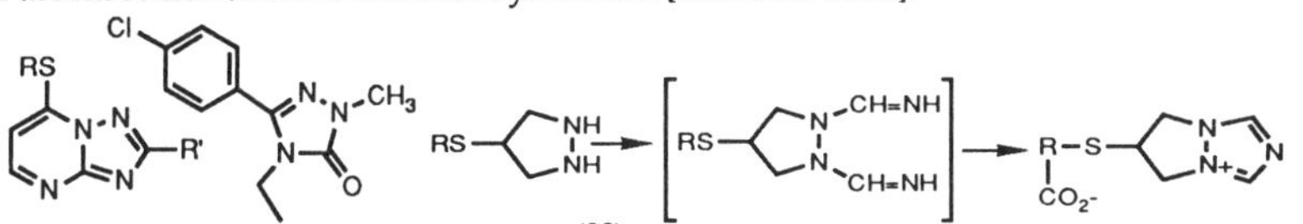

(80) (84) (82) R = 2'-Carbapemem- (83) (81)

In an improved synthesis, 2,4-dihydro-1-oxo-(1,2,4-triazolo)(3,4-c)(1,4)-benzoxazines and (1,4)-benzothiazines (**85**) were prepared in an overall yield of 80% in two steps from benzoxazines or thiazines by formation of the carbohydrazides with subsequent cyclization by reaction with potassium carbonate and PEG4000 in refluxing ethanol [89CI(L)227]. As an example for a simple procedure for the oxidation of thiols to disulfides in 75-97% yields using ethyl dichlorophosphate, thiadiazole-3-thiols were converted into their disulfides in greater than 90% yield [89CI(L)282].

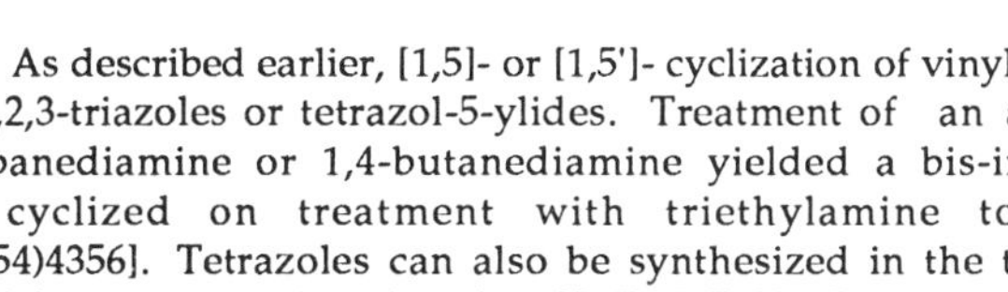

(85)

4.4.6 TETRAZOLES AND RING-FUSED TETRAZOLES

As described earlier, [1,5]- or [1,5']- cyclization of vinyl azides produced either 1,2,3-triazoles or tetrazol-5-ylides. Treatment of an acryloazide with 1,3-propanediamine or 1,4-butanediamine yielded a bis-intermediate (**86**) which cyclized on treatment with triethylamine to produce (**87**) [89JOC(54)4356]. Tetrazoles can also be synthesized in the thermolysis of 5-azido-1,2,3-triazoles as listed earlier [89T(45)749]. The reaction of tributyltin azide and (3-cyanopropyl)triphenyltin at 140°C for 3 hr generated the anticipated 2-(tributylstannyl)-5-(3'-triphenylstannylpropyl)tetrazole (**88**).

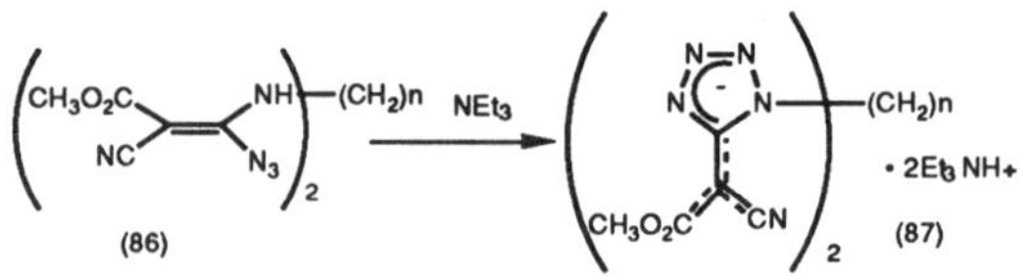

Prolonged heating for 21 hr yielded 2,3,4,5-tetraaza-6-diphenylstannyl[3,4]-bicyclonona-1,3-diene (**89**) in a thermal cyclization process [89JOM(C5)361].

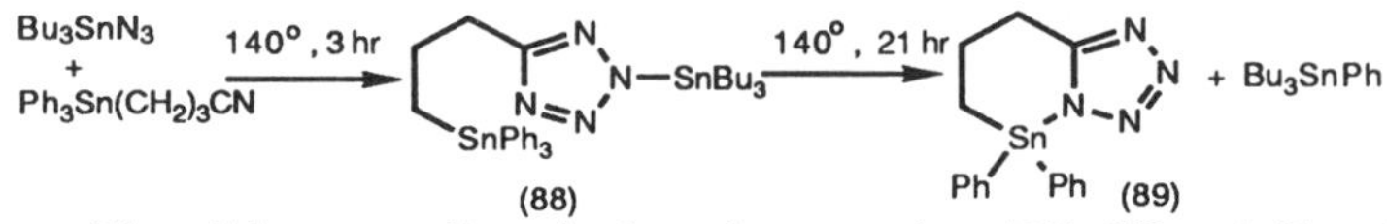

The 5-heterocyclic-substituted tetrazoles (**90**) [Het=4-(5-amino-pyrazolo)- or 5-(4-amino-3-alkylthiothiadiazolo)-] react with nitriles, ortho esters and a variety of other reagents to produce 6 distinct tricyclic heterocyclic ring systems containing fused tetrazoles [89LA83] two of which (**91,92**) contain, in effect, a linear string of 6 consecutive nitrogen atoms.

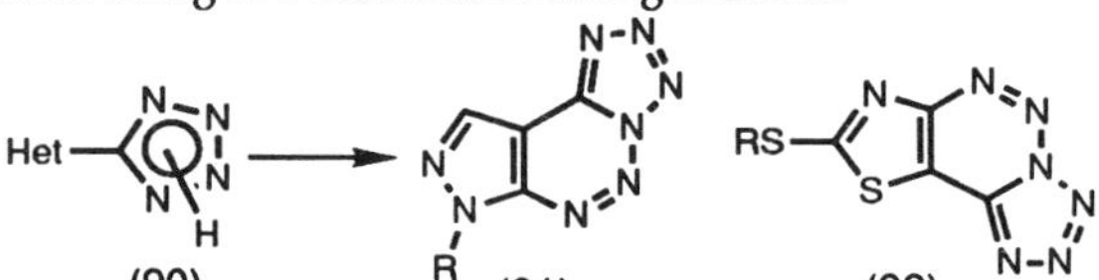

The reaction of 5-(chloromethyl)tetrazoles with hexamethyltetramine generated the dimer (**93**) which contains 10 nitrogens. When N-substituted derivatives are employed, 1,3,5-tri-(5H-tetrazolo-5-yl)methyl-s-triazines were isolated. Methylaminotetrazole (**94**) also produced (**93**) on treatment with formaldehyde in a reversible reaction [89LA689]. Isomeric derivatives of (**93**) were ruled out by spectral and chemical considerations.

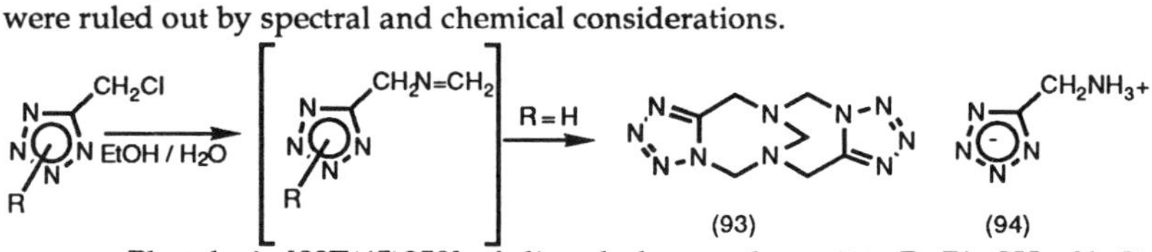

Photolysis [89T(45)259] of dimethyltetrazolone (**95**, R=R'=CH_3, X=O) at 12° K in Ar and an N_2 matrices yielded diaziridinones identical to those obtained in solution photolysis [75AG(87)422; 81CB3253]. Methyl isocyanate was observed and methylazide was speculated as an intermediate. Iminotetrazolones (**95**, X=NH) generated the respective carbodiimides and iminodiaziridines (**96**, X=NH). Tetrazolonothiones (**95**, X=S) also produce carbodiimides. Photolysis of the vinyl analog (**95**, R=t-Bu, R'=vinyl) generated in a 12K matrix the vinyldiaziridinone (**96**, R=t-Bu, R'=vinyl) while in fluid solutions the imidazolone (**97**, R =t-Bu, R'=H) was produced [89T(45)259].

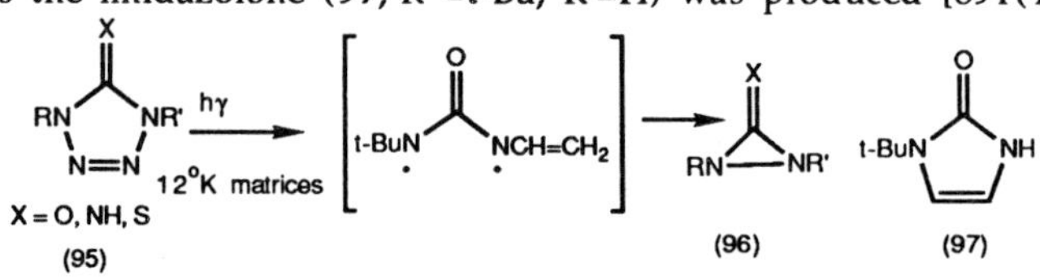

The reaction of nitrile (**98**) with NH_4Cl and NaN_3 in DMF at 115° for 18 hr. followed by treatment with HCl and water generated the expected tetrazole (**99**) but with the unexpected aminolysis of the 5-morpholino group [89JHC(26)713].

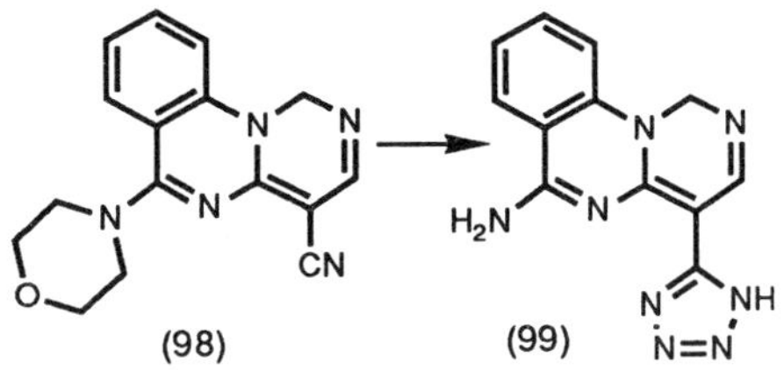

(98) (99)

Pyrrolo analogues of antifungal agent bifonazole have similar therapeutic activity. Substitution of the imidazole group in bifonazole by triazolo or tetrazolo groups produce compounds with reduced activity [89AP(322)369]. Novel fused 2-[t-butlyamino-1-(R)-tetrazolo[1,5-a]pyridin-6yl]ethanols are potent animal growth promotors acting by β-adrenergic blockage [89EUP318092]. 5-Phenoxyalkyl-1H-tetrazoles possess antiallergic activity based on their ability to inhibit leukotriene biosynthesis [89EUP317165]. 5-(1H-Tetrazolo)biphenyls are potent angiotensin II blockers [89EUP324377].

Other tetrazolo compounds have also been prepared for a variety of other medicinal indications such as HMG-CoA reductase inhibition [89JMC(32)2038], anti-ulcer agents [89CPB(37)322], anti-allergy agents [89JMC(32)1360], anti-diabetic indications [89JMC(32)11] and for cardiac stimulating properties [89JMC(32)577]. Studies on the molecular and crystal structure of vinylogous 5-amino-1H-tetrazole derivatives [89ACS(43)298] and on the electronic structures of a series of 5-substituted-1,4-dihydro-5H-tetrazoles appeared in 1989 [89CB(122)985].

4.4.7 METAL COMPLEXES WITH RINGS CONTAINING 2 OR MORE NITROGENS

The reported ligating properties of 5-membered rings with 2 or more nitrogens continues to expand at a rapid pace. These systems are highly important in natural systems, in models of bio-organic synthesis and in electronic transfer media. Aromatic stacking in copper complexes [89IC(28)4066] and spectroscopic properties of copper complexes [89IC1128] of triazoles were of particular interest. Studies related to chromium complexes [89IC(28)2688] and reports on the electrochemical transformations of ruthenium triazoles [89IC(28)1413] point to the continued importance of 5-membered rings containing two or more nitrogens as ligands. Interest in the synthesis of improved metalloenzyme mimics remains high. Some highlights of this research are reported below. The first example of a dinucleating ligand that mimics the characteristics of histidine dominated enzyme cores was reported. The hexaimidazole ligand (**100**) has been shown to complex two Cu(II) ions. Crystallographic data provided conclusive structure proof [89JA(111)4532].

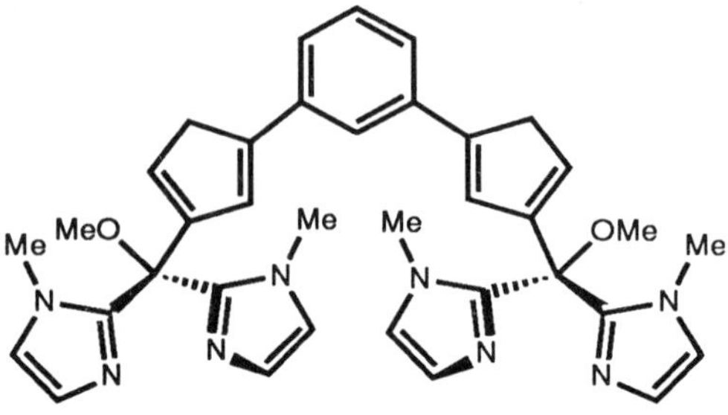

(100)

A new class of tripodent chelating ligand have been synthesized to model carbonic anhydrase. Tris (imidazolyl) ligands with general structure (**101**) were prepared in a one-pot procedure starting from 2-chloromethylimidazole and acetonitrile [89JOC(54)1439].

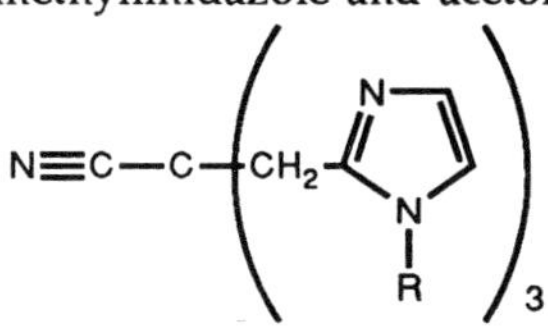

(101)

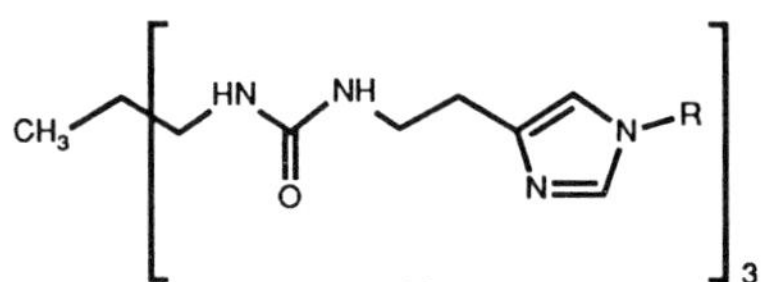

(102)

Another one-step synthesis prepared a tripodent group with general structure (**102**). In (**102**, R=benzyl only mild selectivity was seen for Na, K or Li but unprecedented affinity for hydrazinium ion complexation was noted [89CJC(67)1363]. Reaction of a secondary amine such as 1,4,7-triazacyclononane with 1-methyl-2-chloromethylimidazole in acetonitrile and base presents a general route to nitrogen crown ligands (**103**) with pendant biomimetic imidazole groups [89CC126]. Other interesting crown ethers containing 18-ring atoms (**104**) and 36-ring atoms (**105**) were obtained in approximately 25% yield by 1:1 and 2:2 cyclization of 1-methyl-3,5-bis(chloromethyl)-1H-pyrazole and disodium tetraethylene glycolate under the template effect of CsCl. Transport rates of alkali and ammonium ions are much higher for both compared to earlier prepared ester crown systems [89JOC(54)1391].

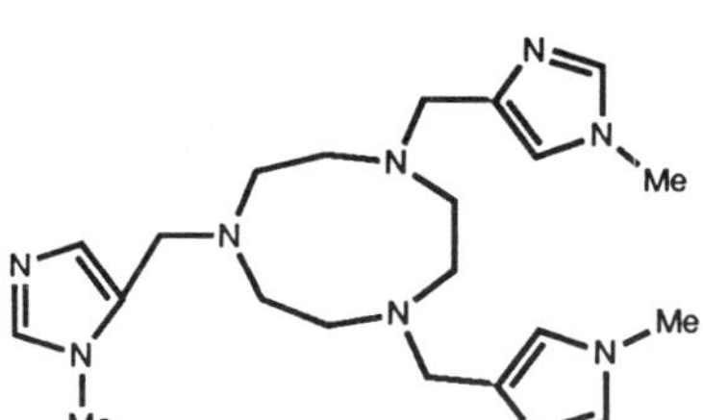

(103)

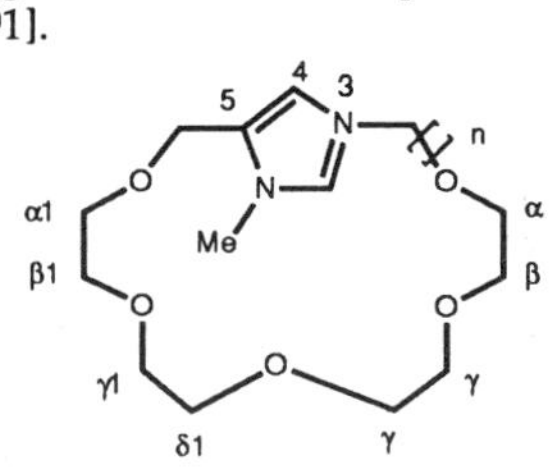

n = 1 (104)
n = 2 (105)

An improved synthesis of a versatile diol (**106**) provides a ready entry into a wide variety of imidazole-containing crown ethers. The key step in this method employs the Swern oxidation of easily prepared imidazoline (**107**) [89JOC(54)1256].

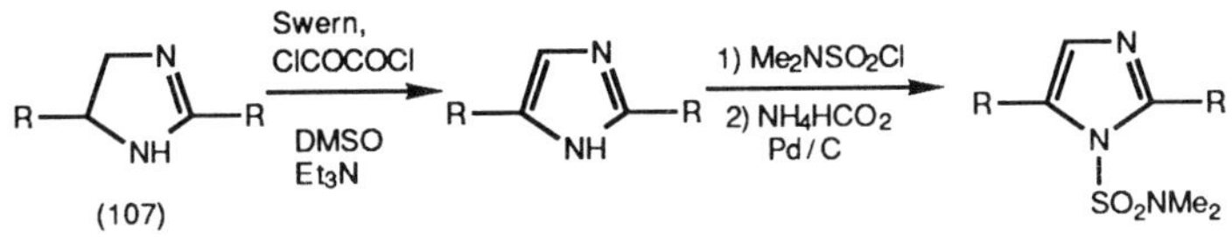

(107) (106)

R = $-CH_2O-(CH_2)_2-OBz$

R = $-CH_2O-(CH_2)_2CH_2OH$

ACKNOWLEDGEMENTS: We wish to thank Ms. Elizabeth Fromwiller-Campisano for her patience and assistance in the preparation and finalization of this document. We also thank Ms. E. Delos Santos for her efforts in obtaining the cited references.

4.4.8 REFERENCES

75AG(87)422 H. Quast and L. Bieber, Angew. Chem., 1975, 87, 422.

89ACS(43)298 J. Sieler, K. M-Moe and G. W. Fischer, Acta Chemica Scand., 1989, 43, 298.

89AG(101)507 Johannes Kaulen, Angew. Chem., 1989, 101, 507.

89AHC(46)169 P. K. Kadaba, Avances in Heterocyclic Chemistry,1989,46,pp. 169-281, A. R. Katritzky, Ed.,Academic Press, 1989.

89AJC(42)441 R. G. Amiet, N. A. Evans and P. M. Wynne, Aust. J. Chem., 1989, 42, 441.

89AP(322)49 S. Büyüktimkin, N. Büyüktimin , O. Ozdemir and S. Rollas, Arch. Pharm. (Weinheim), 1989, 322, 49.

89AP(322)63 V. J. Ram, L. Mishra and D. S. Kushwaha, Arch. Pharm.(Weinheim), 989, 322, 63.

89AP(322)369 S. Massa, et al., Arch. Pharm. (Weinheim), 1989, 322, 369.

81CB(114)3253 H. Quast and L. Bieber, Chem. Ber, 1981, 114, 3253.

89CB(122)133 W. Adam, et al., Chem. Ber., 1989, 122, 133.

89CB(122)911 K. Banert, Chem. Ber., 1989, 122, 911.

89CB(122)185 R. Gleiter, T. Veszprémi and H. Quast, Chem. Ber., 1989, 122, 985.

89CB(122)1175 K. Banert, Chem. Ber., 1989, 122, 1175.

89CC126 M. DiVaira, F. Mani and P. Stoppioni, J. Chem. Soc., Chem. Commun., 1989, 126.

89CI(L)227 C. V. R. Sastry, V. S. H. Krishnan, G. R. A. S. Narayan and K. Vernama, Chem. Ind. (London), 1989, 227.

89CI(L)282 R. H. Khan and R. C. Rastogi, Chem. Ind. (London), 1989, 282.

89CJC(67)198 J. L. E-Valderrama, J. H. G.-Tapia, J. R-Ortiz, M. J. Rosales, R. A. Toscano and J. V-Martinez, Can. J. Chem., 1989, 67, 198.

89CJC(67)1144 A. R. Katritzky and J. N. Lam, Can. J. Chem., 1989, 67, 1144.

89CJC(67)1363 P. G. Potvin, R. Gau, C. M. Unrau and N. K. Gehr, Can. J. Chem., 1989, 67, 1363.

89CPB(37)322 M. Uchida, M. Komatsu, S. Morita, T. Kanbe and K. Nakagawa , Chem. Pharm, Bull., 1989, 37, 322.

89CPB(37)1009 S. Narita, T. Tasaki and T. Kitagawa, Chem. Pharm. Bull, 1989, 37, 1009.

89CZ(113)11 H. Wamhoff and W. Wambach, Chem.-Ztg., 1989, 113, 11.

88EUP273309 F. P. Miller, et al., European Patent Application, 1988, 273309.

88EUP276065 N. G. Bollinger, et al., European Patent Application, 1988, 276065.

88EUP288305 M. A. R. Rustad, European Patent Application, 1988, 288305.

89EUP302633 S. Shimizu and H. Fakano, European Patent Application, 1989, 302633.

89EUP317165 M. A. R. Rustad, European Patent Application, 1989, 317165.

89EUP318092	M. H. Fisher and M. J. Wyvratt, European Patent Application, 1989, 318092.
89EUP320750	European Patent Application, 1989, 320750.
89EUP320764	European Patent Application, 1989, 320764.
89EUP320765	European Patent Application, 1989, 320765.
89EUP324377	D. J. Carini, European Patent Application, 1989, 324377.
89H(28)245	J. Elguero, P. Goya and A. Martinez, Heterocycles, 1989, 29.245.
89H(28)103	A. Maquestiau, R. Flammang and F-B. B. Abdelouashale, Heterocycles, 1989, 28, 103.
89H(28)133	T. Aoyamo, S. Katsuta and T. Shioiri, Heterocycles, 1989, 28, 133.
89IC(28)1128	W. M. E. Koonen-van Oudenniel, R. A. G. de Graaff, J. G. Haasnoot, R. Prins and J. Reedigh, Inorg. Chem., 1989, 28, 1128.
89IC(28)1413	R. Hage, J. G. Haasnoot, D. J. Stufkens, T. L. Snoek, J. G. Vos and J. Reedijk, Inorg. Chem., 1989, 28, 1413.
89IC(28)2688	M. Inamo, T. Sumi, N. Nakagawa, S. Funahashi and M. Tanaka, Inorg. Chem., 1989, 28, 2688.
89IC(28)4068	O. Yamauchi, A. Odani, T. Kohzuma, H. Masuda, K. Toriumi and K. Saito, Inorg. Chem., 1989, 28, 4068.
89IJC(28B)83	U. S. Pathak, M. B. Devani, C. J. Shishoo and S. A. Shah, Indian Journal of Chemistry, Section B, 1989, 28B, 83.
89JA(111)4532	W. B. Tolman, R. L. Rardin and S. J. Lippard, J. Am. Chem. Soc., 1989, 111, 4532.
89JA(111)4897	S. Abuzar and H. Kohn, J. Amer. Chem. Soc., 1989, 111, 4897.
89JA(111)7348	F. Tomás, Jet al, J. Amer. Chem. Soc., 1989, 111, 7348.
89JAP25779	T. Kumagi, et al., Japanese Patent Application, 1989, 25779.
89JCR(S)16	L. Bruché, L. Garanti and G. Zecchi, J. Chem. Research (S), 1989, 16.
89JCR(S)206	P. Barraclough, D. Collard, S. Smith, S. J. Vine and C. J. Wharton, J. Chem. Research (S), 1989, 206.
78JCS(PI)378	M. G. Barlow, R. N. Haszeldine and J. A. Pickett, J. Chem. Soc., Perkin Trans. I, 1978, 378.
89JCS(PI)838	D. H. Davies, J. Hall and E. H. Smith, J. Chem. Soc. Perkin Trans. I, 1989, 838.
89JCS(PI)925	J. G. Buchanan, M. Harrison and R. H. Wightman, J. Chem. Soc. Perkin Trans. I, 1989, 925.
89JCS(PI)619	A. B. Theocharis, A. J. Maroulis, C. P. H-Maroulis and N. E. Alexandrou, J. Chem. Soc. Perkin Trans. I, 1989, 619.
89JHC(26)3	J. R. Beck, J. Aikins, M. P. Lynch, J. R. Rizzo and E. V. P. Tao, J. Heterocycl. Chem., 1989, 26, 3.
89JHC(26)177	S. M. El-Khawass and N. S. Habib, J. Heterocycl. Chem., 1989, 26, 177.
89JHC(26)225	D. B. Reitz and M. J. Finkes, J. Heterocycl. Chem., 1989, 26, 225.
89JHC(26)241	C. Baldoli, L. Lattuada, E. Licandro, S. Maiorana and A. Papagni, J. Heterocycl. Chem., 1989, 26, 241.
89JHC(26)327,763	Y. Miyamoto and C. Yamayahi, J. Heterocycl. Chem., 1989, 26, 327, 763.
89JHC(26)339	S. Tono-oka and I. Azuma, J. Heterocycl. Chem., 1989, 26, 339.
89JHC(26)713	N. P. Peet, J. Heterocycl. Chem., 1989, 26, 713.
89JMC(32)11	K. L. Kees, R. S. Cheeseman, D. H. Prozialeck and K. E. Steiner, J. Med. Chem., 1989, 32, 11.
89JMC(32)575	S. M. Campbell, J. Med. Chem., 1989, 32, 575.
89JMC(32)1360	P. L. Unangst, et al., J. Med. Chem., 1989, 32, 1360.
89JMC(32)1447	G. D. Kini, R. K. Robins and T. L. Avery, J. Med. Chem., 1989, 32, 1447.
89JMC(32)2038	N. Balasubramanian, P. J. Brown, J. D. Catt, W. T. Hau, R. A. Parker, S. Y. Set and J. J. Wright, J. Med. Chem., 989, 32, 2038.
89JMC(32)2116	J. Drummond, G. Johnson, D. G. Nickell, D. F. Ortwine, R. F. Bruns and B. Welbaum, J. Med. Chem., 1989, 32, 2116.
89JOM(361)C5	M. F. Mahon, K. C. Molloy and P. C. Waterfield, J. Organometallic Chem., 1989, 361, C5.

89JPC(93)3926 M. Alcamí, O. Mó and M. Yáñez, J. Phys. Chem., 1989, 93, 3929.
66JOC(31)781 H. C. Brown, H. J. Gisler and M. T. Chena, J. Org. Chem., 1966, 31, 781.
89JOC(54)428 H. H. Lee, B. F. Cain, W. A. Denny, J. S. Buckleton and G. R. Clark, J. Org. Chem., 1989, 54, 428.
89JOC(54)731 B. A. Astleford, G. L. Goe, J. G. Keay and E. F. V. Scriven, J. Org. Chem., 1989, 54, 731.
89JOC(54)1185 G. Morel, E. Marchand, A. Foucaud and L. Toupet, J. Org. Chem., 1989, 54, 1185.
89JOC(54)1256 S. C. Zimmerman, K. D. Cramer and A. A. Galan, J. Org. Chem., 1989, 54, 1256.
89JOC(54)1391 P. Navarro, M. I. R. Franco, C. F. Foces, F. Cano and A. Samat, J. Org. Chem., 1989, 54, 1391.
89JOC(54)1439 T. S. Manoharan and R. S. Brown, J. Org. Chem., 1989, 54, 1439.
89JOC(54)1760 D. B. Reitz and M. J. Finkes, J. Org. Chem., 1989, 54, 1760.
89JOC(54)3057 R. P. Thummel, V. Goulle and B. Chen, J. Org. Chem., 1989, 54, 3057.
89JOC(54)4356 R. W. Saalfran, K. Wirth and C-J. Lurz, J. Org. Chem., 1989, 54, 4356.
89JOC(54)5302 W. L. Mock, T. A. Irra, J. P. Wepsiec and M. Adhya, J. Org. Chem., 1989, 54, 5302.
89LA59 U. H. Granzer, I. Staatz and H. J. Roth, Liebigs Ann. Chem., 1989, 59.
89LA75 M. Veverka and J. Svetlik, Liebigs Ann. Chem., 1989, 75.
89LA83 W. Ried and S. Becker, Liebigs Ann. Chem., 1989, 83.
89LA415 G. Seitz, R. Hoferichter and R. Mohr, Liebigs Ann. Chem., 1989, 415.
89LA493 J. Altman and M. Wilcheck, Liebigs Ann. Chem., 1989, 493.
89LA689 D. Moderhack, K-H. Goos and L. Preu, Liebigs Ann. Chem., 1989, 689.
89MI(10)35 M. L. Williams and J. E. Gready, J. Comp. Chem., 1989, 10, 35.
89MI(24)73 G. Mouysset, M. Payard, J-C. Teulade, J. Couquelet and G. Grassy, Eur. J. Med. Chem., 1989, 24, 73.
89MI(8)261 A.-B. A. G. Ghattas and G. A. El-Saraf, Sulfur Letters, 1989, 8, 261.
89S66 A. Katritzķy, Z. Najzarek and Z. Dega-Szafran, Synthesis, 1989, 66.
89S269 G. Laus and W. Klötzer, Synthesis, 1989, 269.
89S399 L. Bruché, L. Garanti and G. Zecchi, Synthesis, 1989, 399.
89S714 F. Freeman, M. C. Keindl, H. N. Po, E. Brinkman and J. A. Masse, Synthesis, 1989, 714.
89T(45)259 I. R. Dunkin, C. J. Shields and H. Quast, Tetrahedron, 1989, 45, 259.
89T(45)637 S. Shimizu and M. Ogata, Tetrahedron, 1989, 45, 637.
89T(45)749 G. L'abbé and L. Beenaerts, Tetrahedron, 1989, 25, 749.
89T(45)1661 D. Janietz and W-D. Rudorf, Tetrahedron, 1989, 45, 1661.
89T(45)1823 P. Molina, M. Alajarin, C. Lpez-Leonardo, I. Madrid, C. Foces and F. H. Cano, Tetrahedron, 1989, 45, 1823.
89T(45)7041 B. Abarca, A. Asensio, G. Jones and D. J. Mofat, Tetrahedron, 1989, 45, 7041.
89TL(30)2049 R. J. Linderman and K. S. Kirollos, Tetrahedron Lett., 1989, 30, 2049.
89TL(30)6237 P. Molina, A. Arques, M. V. Vinader, Tetrahedron Lett., 1989, 30, 6237.
89TL(30)6469 F. A. L. Anet, R. Anet, M. P. Trova and L. A. Paquette, Tetrahedron Lett., 1989, 30, 6469.

CHAPTER 4.5

Three-Membered Ring Systems: With N & S (Se) Atoms

RIE TANAKA
Suntory Biomedical Research, Osaka, Japan

ICHIRO SHINKAI
Merck Sharp & Dohme Research Laboratories, Rahway, NJ, USA

4.5.1 ISOTHIAZOLES

The hypervalent sulfur 2,1-benzisothiazolo[2,3-b]-2,1-benzisothiazole system was reported. These compounds are of interest due to the hypervalency of the central sulfur atom and their display of symmetry. Treatment of diaminophenylmethane with N-sulfinylmethane-sulfonamide gave an isothiazole (**1**). Methylation of **1** with dimethyl sulfate gave two compounds, **2** (yellow) and **3** (red). Detailed NMR experiments support the proposed hypervalent fused structure for **3** <89JHC1095>(Scheme 1).

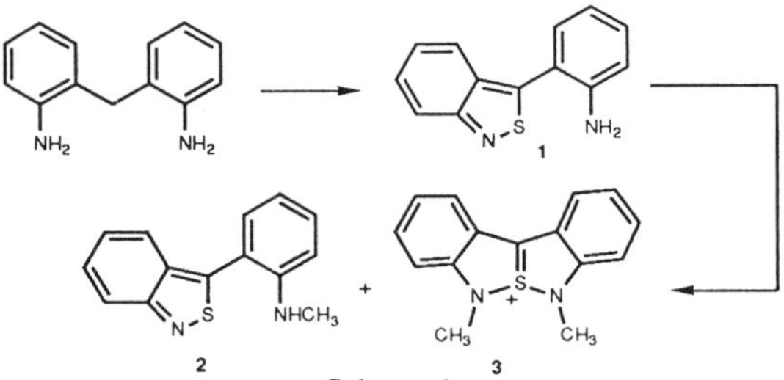

Scheme 1

The transformation of 5-substituted 3-phenyl-1,4,2-dithiazolium perchlorates (**4**) with malononitrile gave several products including isothiazole in moderate yield. The reaction has been explained by assuming three possible fission modes of the initial adducts, 5-dicyano-methylene-1,4,2-dithiazoles (**5**). Compounds **7** and **8** are formed in a simmilar manner from 1,3-dithiolium salts, whereas **4** gave **9** and **6**, depending on the substituent at the C-5 position and bases used, via novel reaction paths <89BCJ1086>(Scheme 2). A synthesis of 3,6-bis(acylamino)-3H,6H-isothiazolo[4,3-c]-1,2-isothiazole was achieved in 10-77% yield, depending on the acyl substituents employed by the reaction of 3,6-

bis(acylamino)-3H,6H-dithiolo[4,3-c]-1,2-dithiole with hydroxylamine <89ZC17>. Ring transformation of 3-methylbenzothiopyran-4-thione with diphenyl-sulfilimine ($Ph_2S{=}NH$) to the 4-methylisothiazole derivative was reported <88JCR(S)349>.

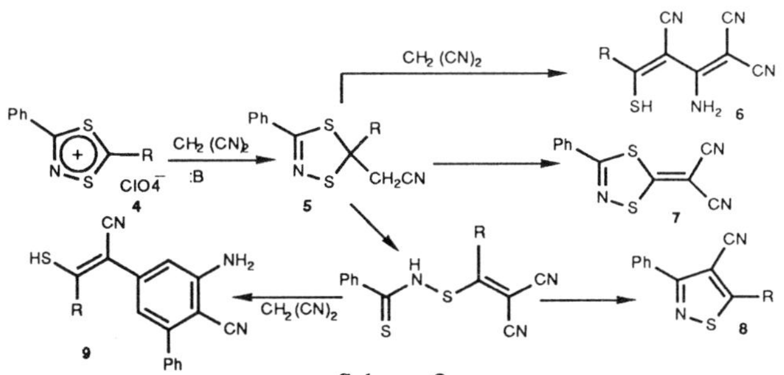

Scheme 2

Photochemical conversion of thieno[2,3-c]isothiazoles to thiophenes (**12**) was described. The reaction path involves the intermediacy of the open-chain valence bond isomer (**10**), formed via a homolytic cleavage of the N-S bond. Cyclization of **10** to **12** may proceed via the formation of the bicyclic thiirane (**11**) followed by the loss of elemental sulfur to give thiophenes (**12**).Infrared tracing of the

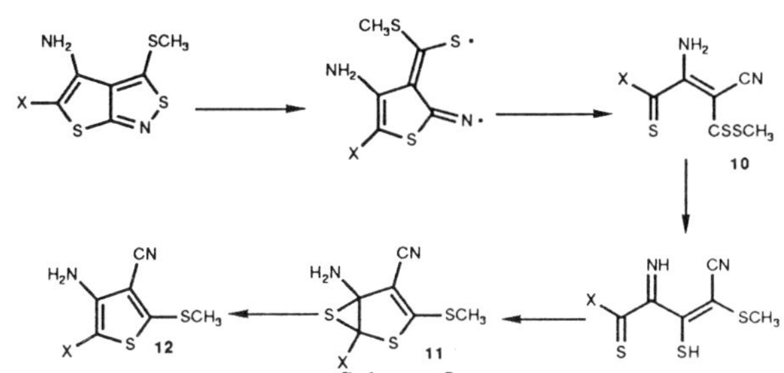

Scheme 3

reaction indicated the development of the nitrile band at 2200 cm^{-1} supporting the intermediacy of **10.** The ring instability of thieno[2,3-c]isothiazole was unexpected since isomeric thieno[2,3-d]isothiazoles containing the same substrates are stable upon irradiation <88H2539>(Scheme 3). Little work has been published on reactions involving reductive cleavage of the isothiazole ring except lithiation of the 5-position and desulfurization by Raney nickel. A new and simple method was reported for the reductive cleavage of isothiazolium salts by reaction with complex metal hydrides (sodium borohydride or lithium aluminum hydride) forming β-enaminothioketones (**13**) in high yields (82-93%). The isothiazolium ring acts as a masked β-enaminothioketones, which are useful building blocks for other heterocyclic ring systems <88SC1847>(Scheme 4).

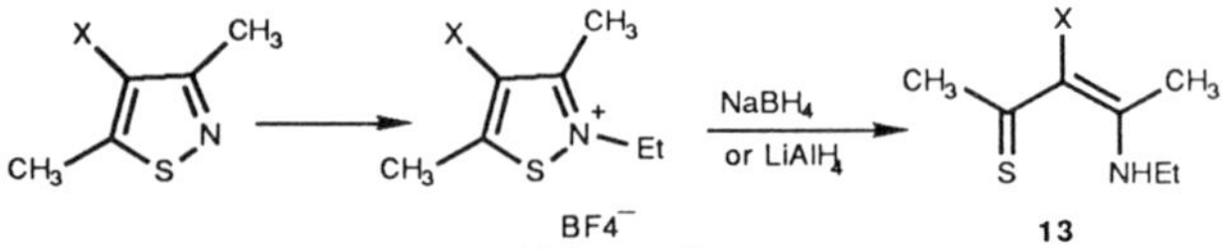

Scheme 4

Several synthetic methods for isothiazoles have appeared during 1989: a facile one-pot procedure for the synthesis of 1,2-benzisothiazole 1,1-dioxides (**14**) <89S394>, a one pot procedure for the synthesis of 4-unsubstituted isothiazoles (**15**) starting from α-acetylenic aldehydes or ketones, hydroxylamine-*O*-sulfonic acid and sodium hydrosulfide in a buffered aqueous solution <89H97>, formation of 5-(carbamoylthio)isothiazoles (**16**) by the reaction of the ammonium salts of 5-(2H)-isothiazolethione with isocyanates <88BCJ4443>.

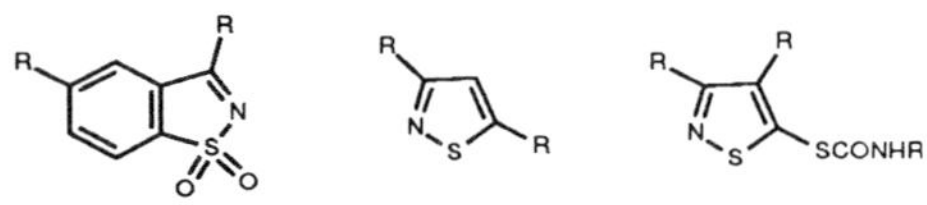

14 **15** **16**

The reaction of α-oxo ketene dithioacetals with hydroxylamine gave oximes, which further reacted with thionyl chloride and pyridine to afford isothiazoles <89JOC1089>(Scheme 5).

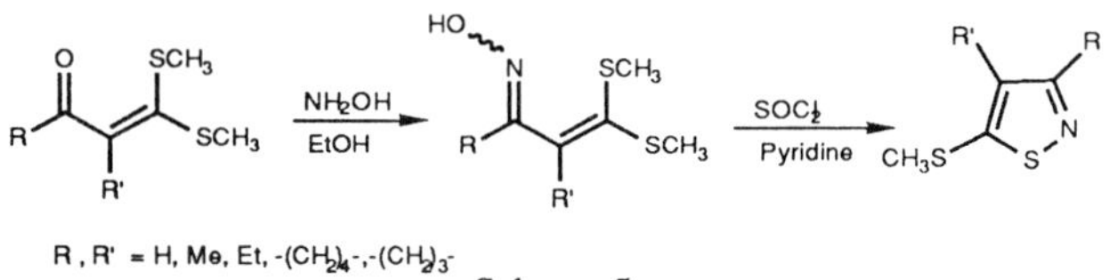

Scheme 5

An unusual 1,3-cycloaddition of 2,1-benzisothiazole with benzyne is known. Similarly, the 3-*N*-pyrrolidinylthieno[3,2-c]isothiazole (**17**) reacted with maleic anhydrides in a 1,3-cycloaddition mode giving thio-amides (**18**) <89JCR(S)29>(Scheme 6).

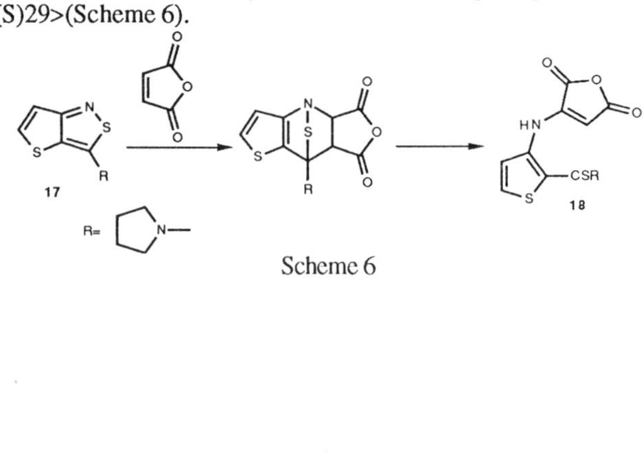

Scheme 6

There is no general procedure for the preparation of 4-acyl- and 4-alkenylisothiazoles, and it is not possible to obtain them by Friedel-Crafts reaction or related processes. A convenient synthesis of both 4-acyl- and 4-alkenylisothiazoles from 3,5-dimethyl-4-isothiazolylcarbonitrile (**19**) and carboxyaldehyde (**20**) with organoaluminum reagents and/or Grignard reagents in good yields was reported <88JHC235>(Scheme 7).

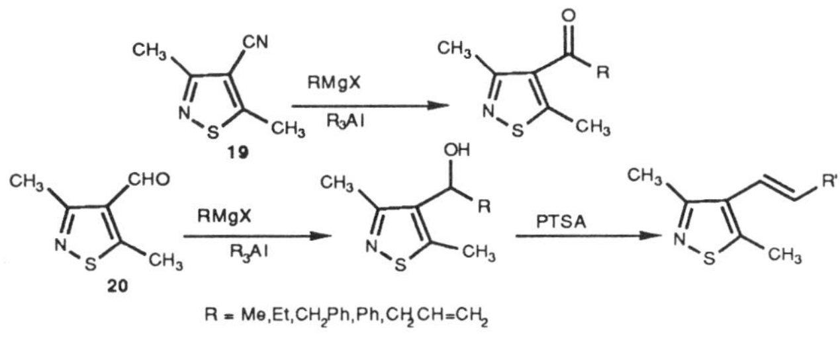

Scheme 7

4.5.2 THIAZOLES

Infrared and raman spectra of thiazoles were reinvestigated and the generally accepted hypothesis of a planar structure has been cofirmed.. The equilibrium configuration of the 2-aminothiazole molecule is not accurately known. Vibrational spectral work suggest a planar structure in which the lone-pair electrons of the nitrogen atom participate in the π-system of the aromatic ring. Therefore, the molecule is assumed to be planar <89CCC28>. MO calculations (3-21G) of thiazole-2-carboxyaldehyde show the formyl group and the heterocyclic ring as being coplanar and the S,O-cis conformer as being the more stable isomer. The difference in energy content between the two conformers is considerably higher in the thiazole derivatives <89JST269>. Spin-coupled theory was applied to the thiazole system <89JCS(P2)263>. The rate and equilibrium constants of the diazotization of 2-aminothiazole were measured. The kinetic data showed that the ring protonated 2-aminothiazole is attacked by the NO^+ ion. The reaction does not go to completion,but leads to an equilibrium <89HCA800>. Unexpected ring transformation byproducts, 1,2,4-triazoles (**21**), were formed in low yields during the cyclodehydration of 2-(2-aroylhydrazino)-thiazoles (**22**) with phosphorus oxychloride in refluxing xylene. The formation of nitrile imines (**23**) has been proposed <89LA75> (Scheme 8).

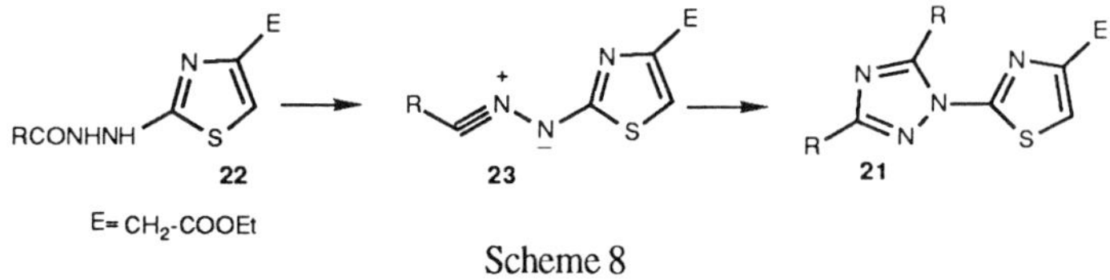

Scheme 8

Highly regio- and stereospecific intramolecular cycloadditions of anhydro-4-hydroxythiazolium hydroxide were well known. The intramolecular version of this mesoionic cycloaddition was applied by the attachment of a dipolarophilic side chain to the 2-, 3- and 5-positions of the nucleus. Anhydro - 2-[2-(allyloxy)phenyl]-4-hydroxythiazolium hydroxide gave a 6:1 mixture of exo-(**25**) : endo-(**26**)-cycloadducts. On the other hand, 3- and 5- allyloxy analogs gave single cycloadducts, (**27**) and (**28**), in regio- and stereospecific fashion. The rate of cycloaddition was influenced by the electronic nature of the substituents, by the spacers, and by the steric bulk of the dipolarophile. Alkynic side chains also undergo facile cycloaddition, leading to a number of ring fused heterocycles <89JOC1077>.

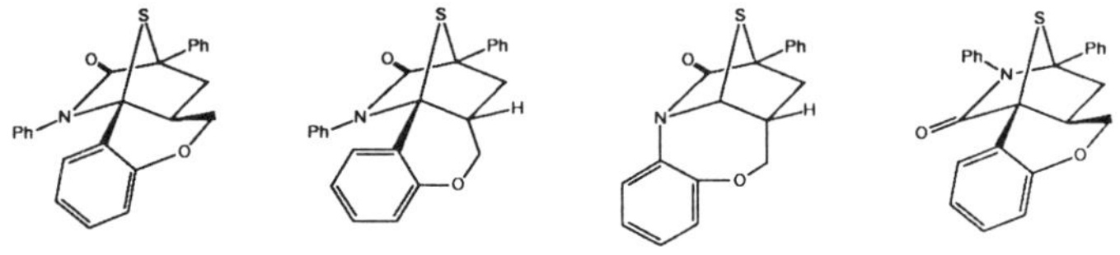

25 26 27 28

Treatment of aldehydes and ketones with o-(methylamino)benzothiazole gave 3-methylbenzothiazoline, which serves as an efficient method of protecting the carbonyl group under both acidic and basic conditions. 3-Methylbenzothiazolium salts, obtained by oxidation, have been found to be effective formyl cation synthons <89BCJ1215>. Stereoselective and efficient methodology using 2-trimethylsilylthiazole (**29**) as a formyl anion synthon for the synthesis of long-chain chiral polyhydroxylated aldehydes was reported (Scheme 9). The strategy involves a linear one-carbon chain elongation and the conversion of the thiazole ring to the aldehyde.

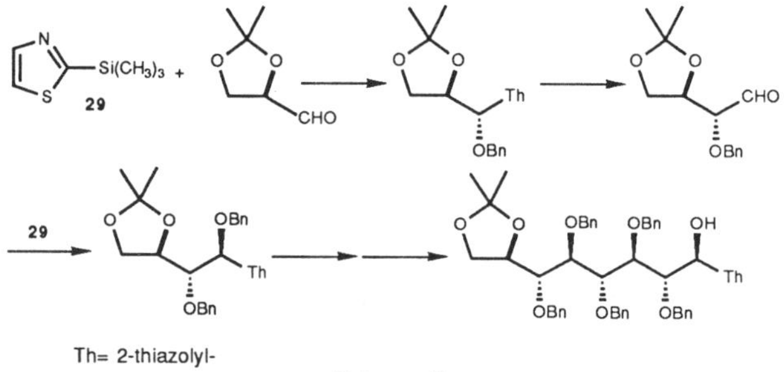

Scheme 9

An interesting feature of this methodology is the iterative repetition of the sequences over several cycles with a high degree of acyclic stereoselectivity (ds >95%) and good chemical yields. The C7, C9 and C10 thiazole sugars were prepared by this methodology <89JOC693>. Futhermore, anti-1,2-diols can be converted in good yields to syn-1,2-diols via oxidation-reduction sequences employing K-selectride® as a reducing reagent <89JOC702>. The α-bromoketones react with the ring nitrogen of 2-aminothiazoles (**30**) to give 2-imino-2,3-dihydrothiazole (**31**) rather than 2-acylaminothiazole. The intermediates (**31**) cyclize to imidazo[2,1-b]thiazoles (**32**) <89JCS(PI)643> (Scheme 10).

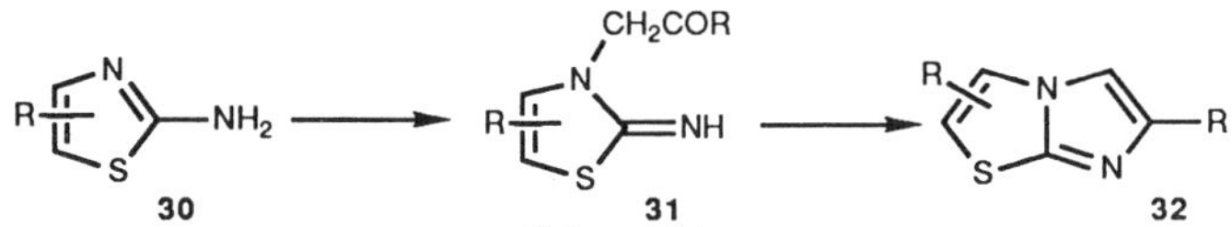

30 31 32

Scheme 10

Condensation of 2-hydrazinobenzothiazoles with acetylacetaldehyde dimethyl acetal gave the isomeric 2-(3-methylpyrazol-1-yl)benzothiazoles and 2-(5-methylpyrazol-1-yl)benzothiazoles, depending on the reaction conditions <89IJC(B)27>. Formation of azepine derivatives via the reaction of 2-methylbenzothiazole with methyl propiolate was reported. Convenient synthesis of nor-alcohols starting from carboxylic acids via reaction of carbon centered radicals with antimony trisphenylsulfide followed by the reaction with air. was discovered. The corresponding organobismuth intermediate could actually be isolated <89T2615>(Scheme 11).

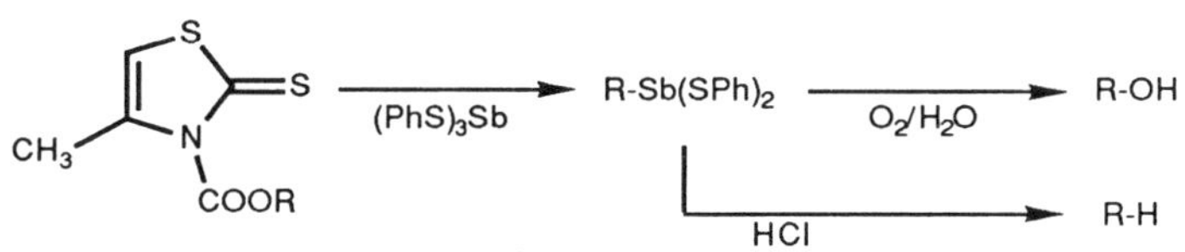

Scheme 11

A rearrangement product, 2-benzamido-4-phenylthiazole , was detected by the reaction of 2-bromo-1,3-diphenylpropane-1,3-dione with thiourea in the presence of triethylamine. The common intermediate, 2-amino-4-hydroxy-4,5-dihydrothiazoline eliminates water to form 2-amino-thiazole under normal acid catalyzed Hantzsch conditions. On the other hand, in the presence of base, a retro-aldol reaction is observed leading to a open chain keto-intermediate followed by the cyclization of unsubstituted amine to form a new 2-acylamino-thiazole<89JCS(PI)1169>(Scheme 12).

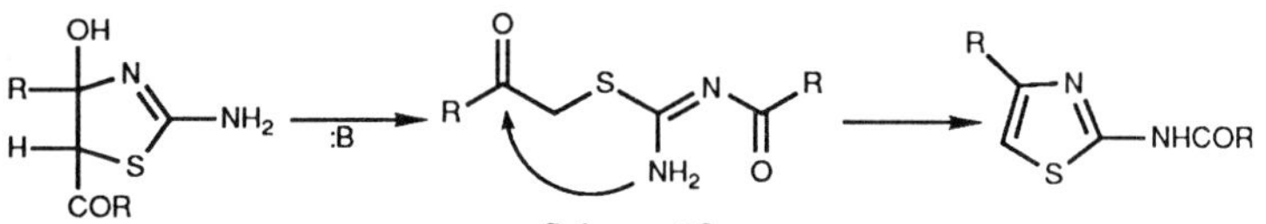

Scheme 12

Several pharmacologically interesting thiazoles were synthesized and their biological importance was discussed <89JMC2486>. 5-Thiazolylalkyl-substituted tetrahydro-naphthalene carboxylic acid derivatives showed potent thromboxane A2 (TXA2) synthetase inhibitory activities in vitro and had long duration of action in rats by oral or iv administration <89JMC1326>. Analgesic and anti-inflammatory activities were reported for thiazolo[3,2-a]thieno[2,3-d]pyrimidines <89MI55000>. Similarly, anticholinergic antisecretory activity of pyridylthiazole and 2-methylimidazo [1,2-a] pyridine derivatives are claimed <89MI55001,89MI55006>.The one pot synthesis of a 2-aminobenzothiazole derivative ,which is useful to synthesize fused imidazo[2,1-b]thiazoles was reported <89IJC81>. Condensation of 2-alkylthio-4-aminothiazoles with formamide gave a thiazolo[4,5-d]pyrimidine derivative <89LA409>. Various bithiazoles were prepared and evaluated for their antibacterial activities <89MI55003,89MI55004,89MI55005>. Polymer-bound thiazolium salts can be easily prepared from cross-linked chloromethylated XAD-2 resin and 4-methyl-5-hydroxyethylthiazole. The decrease in catalytic activity (during the benzoin condensation) upon immobilizing the thiazole moiety is partly a result of the increase in steric hindrance by the polymeric support. Pretreatment of the polymeric thiazolium salt with a cocatalyst ,such as triethylamine, catalyzed benzoin condensation in the absence of a cocatalyst, which strongly supports the hypothesized conversion of the inactive thiazolium salt into its active ylide form by a cocatalyst <89MI55002> The carbonylation reaction of binuclear complexes [{Rh(u-benzothiazole-2-thiolate)(diolefin)}$_2$] with carbon monoxide gave tetracarbonyl complexes with replacement of diolefin <89JC(D)25>.

4.5.3 THIADIAZOLES

4.5.3.1 1,2,3-Thiadiazoles

1,3-dipolar addition of stable thiocarbonyl compounds with ethyl diazoacetate followed by 1,2-elimination of the azole afforded 4,5-disubstituted 1,2,3-thiadiazoles in excellent yield under mild conditions. Though two possible isomers, a 1,2,3- & 1,3,4-thiadiazole, are predicted,the reaction appears to be regiospecific to give 1,2,3-thiadiazole. The structure was confirmed by ^{13}C NMR. The rearrangement of 1-aryl-5-mercapto-1,2,3-triazoles has also been used to prepare these compounds <88MI55007>(Scheme 13).

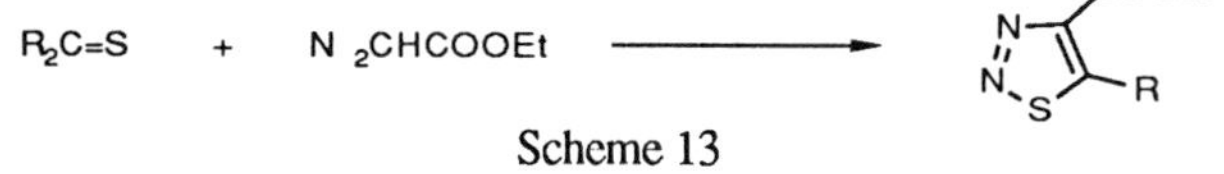

Scheme 13

Syntheses of some new thiadiazoles containing the 4-quinazolone moiety were reported. Treatment of semicarbazone, which was obtained from the condensation of 2-methyl-3-(p-acetylphenyl)quinazol-4-one with semicarbazide hydrochloride, with thionyl chloride afforded 1,2,3-thiadiazoles <88MI55002>. The novel synthesis of 5-amino-1,2,3-thiadiazoles via phthalimido ethylidenecarbazate was prepared<89KGS112>. Reactions of 1,2,3-thiadiazoles with $Fe_2(CO)_9$ in the presence of ethanol gave thiolato-hydrazonato complexes. The initial step was suggested to be a transition-metal initiated ring cleavage of the thiadiazole ring similar to that of alpha-diazocarbonyl compounds <89MI55008>.

4.5.3.2 1,2,4-Thiadiazoles

Reversible bond switching of 5-(1-aminoethylideneamino)-3-methyl-1,2,4-thiadiazole (**33**) was reported. It occurred via a symmetrical hypervalent sulfurane (**34**) intermediate. ^{15}N labeling experiments and kinetic studies on equilibration were carried out under neutral and acidic conditions to confirm the existence of ring transformation equilibrium and position of protonation under acidic condition. To present unambiguous evidence for bond switching, 5-(1-aminoethylideneamino)-3-phenyl-1,2,4-thiadiazole (**36**) was synthesized. It was made from 5-amino-3-phenyl-1,2,4-thiadiazole (**37**) and acetonitrile,and was identical to the compound previously prepared from 5-amino-3-methyl-1,2,4-thiadiazole (**38**) and benzonitrile. Treatment of the imidate,prepared from **38** and triethylorthoacetate, with $^{15}NH_3$ gave ^{15}N labeled **33**. Two peaks with almost equal height were observed in the proton decoupled ^{15}N NMR spectrum of ^{15}N labeled **33**. The rate of bond switching was accelerated by the addition of trifluoroacetic acid. Furthermore, a ^{15}N NMR study confirmed that protonation took place at the amino nitrogen of the amidino group. A kinetic study showed that the activation free energy from **33** to **35** is low because of the existance of the intermediate thiapentalenium salt containing a hypervalent bond. The central sulfur atom is estimated to move back and forth as a pendulum in the range of 0.38 Å nine times a second at 0°C<89BCJ479>(Scheme 14).

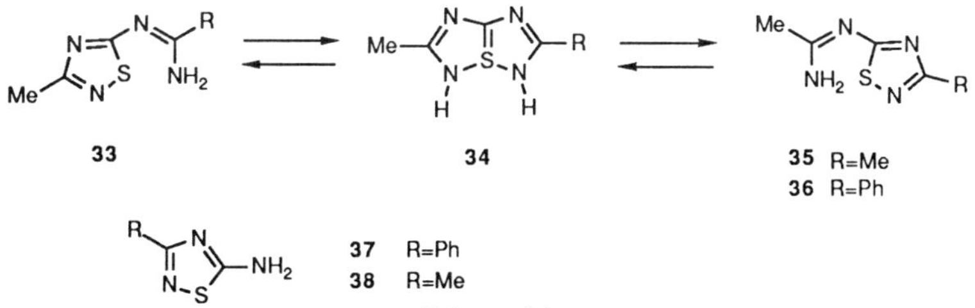

Scheme 14

Reaction of 3-amino-5-methylisoxazole(**39**) with arylcarboxyisothiocyanate gave a thiourea, which further reacted with PCl_5 or$POCl_3$ to afford E and Z isomers of 1,2,4-thiadiazoles(**41**). Nucleophilic attack by sulfur on the nitrogen of isoxazole with concomitant fission of N-O bond led to the formation of (**40**), which further replaced by Cl in its enol form <88MI55006>. Treatment of pyridine carboxamide oximes with excess of phenyl isothiocyanate gave 3-pyridyl-5-phenylamino-1,2,4-thiadiazoles in 30-40% yield. <88MI55004>. Several 1,2,4-thiadiazoles were found to be active as anticonvulsant, analgesics and nitrification inhibitors<89MI55009;88MI55001>.

4.5.3.3. 1.2.5-Thiadiazoles

Complexes of 4-methylbenzo-2,1,3-thiadiazole with Cu(I)-halogen (Cl, Br, I) were prepared, and a crystal structure of the iodide adduct (CuLI) was shown <89MI55012>. The thermal stability of the complexes CuL_2Cl_2, and $CuLCl_2$ (L = substituted benzo-2,1,3-thiadiazole) were studied <89MI55010>. Reduction of nitrosubstituted benzo-2,1,3-thiadiazoles by immobilized E. Coli cells ,as compared with native cells, afforded the corresponding amine in high yield<89MI55011>. 3-Substituted imidazo[1,2-a]pyridines were prepared. that showed antiulcer activity. 4-Amino-1,2,5-thiadiazole functionalities were converted from the sulfoxide, which derived from amines with methoxythiadiazole 1-oxide in 28-90% yield. A two-step process which involved extrusion of the sulfoxide afforded diimidamides, which were subsequently treated with thiobisphthalimide to thiadiazole since its inability to directly reduce by conventional methods<89 JMC 2204>.

4.5.3.4 1,3,4-Thiadiazoles

A new method for the preparation of 1,3,4-thiadiazole was reported. The intramolecular cyclization of 1-acylbithiourea (**42**) in the presence of base gave 1,2,4-triazoles (**43**) or 1,3,4-thiadiazoles (**44**) depending upon the substituents on the 1-acylbithoureas employed. Treatment of **42** with p-toluenesulfonyl chloride or with brominating reagents in a protic solvent also gave 1,3,4-thiadiazoles in good yields <88JHC1071>.

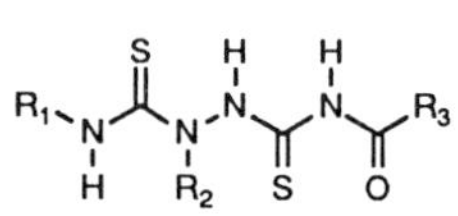

42

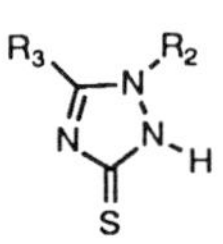

43

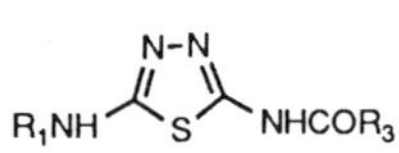

44

Alkoxycarbonyl isothiocyanates are bifunctional reagents used widely for the preparation of many heterocycles. The synthesis begins with a nucleophilic addition and is followed by an intramolecular cyclization to form a ring system. The reaction of ethoxycarbonyl isothiocyanate with hydrazine gave bithiourea (**45**) in excellent yield. Treatment of **45** with ethanolic hydrochloric acid afforded 1,3,4-thiadiazole (**46**). Interestingly, treatment of **45** with methyl iodide also gave **46** via a monomethyl-thioester intermediate. On the other hand, the reaction of **45** with base gave moderate yields of 1,2,4-triazoles (**47**). Adducts obtained from the reaction of aminoguanidines with two equivalents of ethoxycarbonyl isothiocyanate cyclized to give 1,3,4-thiadiazole in good yield. The reaction of ethoxycarbonyl hydrazide with ethoxycarbonylisothiocyanate gave 1,4-bis(ethoxycarbonyl)-3-thiosemicarbazides (**48**). Ring closure of **48** with polyphosphoric acid afforded 1,3,4-thiadiazole in moderate yield; on the other hand, base treatment gave 3-mercapto-5-hydroxy-1,2,4-triazole <89JHC355> (Scheme 15).

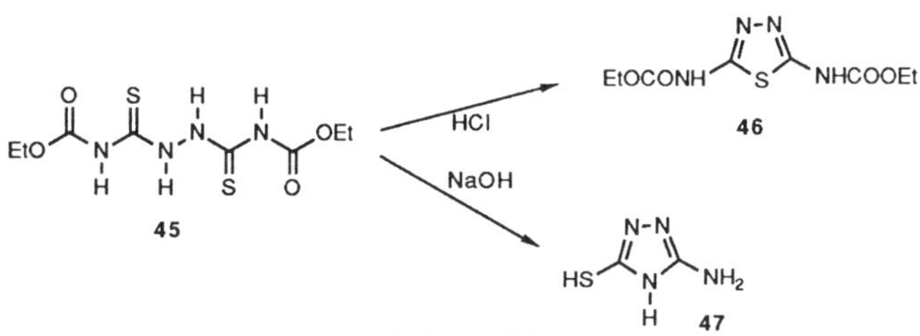

Scheme 15

Monothioacylation of carbonohydrazide (**49**,X=O) with ethoxythiocarbonylthioacetic acid gave 1,3,4-oxadiazole derivative (**50**) in moderate yield. Dithioacylation of thiocarbonohydrazide (**49**,X=S) gave 1,3,4-thiadiazole (**51**) by the spontaneous cyclization of the initial dithioacyl intermediate <89JCR(S)48>.

49 X=O,S **50** X=O **51** X=S

4-Aryl-1-[(2-benzothiazolylthio)acetyl]-3-thiosemicarbazides obtained from the reaction of (2-benzothiazolylthio)acetic acid hydrazides with aryl isothiocyanates cyclized to 1,3,4-thiadiazoles upon treatment with conc. sulfuric acid, whereas base treatment gave 1,2,4-triazole derivatives <88MI55008>. The formation of 2,6,7-trisubstituted-6,7-dihydro-1,3,4-thiadiazolo[3,2-a]-s-triazine-5(H)-thiones (**52**) via 4+2 cycloaddition of 2-benzylideneamino-1,3,4-thiadiazoles with aryl isothiocyanates were reported The thiones showed antifungal activivty.<89IJC(B)78>. Several 2-aryloxymethyl analogs were synthesized similarly and screened for their antibacterial, herbicidal, and antifungal activities. Reaction of the 2-amino-5-aryloxymethyl-1,3,4-thiadiazoles with chloroacetyl chloride gave chloroacetyl derivatives that were cyclized to 1,3,4-thiadiazolo[3,2-a]imidazol-6-ones upon treatment with dry pyridine at room temperature. Most of the compounds showed very little biological activity <89IJC(B)358>. In addition, 2-aryloxymethyl-1,3,4-thiadiazolo[2,3-b]quinazolin-4-ones and 2-aryloxymethyl-5-substituted-1,3,4-thiadiazolo[3,2-a]-1,3,5-triazine-7-thiones were prepared as the antifungal agents <89IJC(B)200, 88IJC(B)542, 88JIC571>. The synthesis and conformational analysis of medium-sized S-bridged metacyclophanes have been of interest in recent years. Reaction of the dipotassium salt of 2,5-dimercapto-1,3,4-thiadiazole with 1,3-dibromopropane under high-dilution heterocyclization conditions gave mesoionic bicyclic 1,3,4-thiadiazole-2-thione (**53**), which was formerly assigned as a dimeric 1,10-dithia[4,4](3,5)-1,3,4-thiadiazolinophane-6,15-dithione . The structure of **53** was established by X-ray. In the original report, the molecular formula was deduced from the mass spectral studies, however, **53** dimerizes thermally under MS conditions. Further structural proof of monomeric **53** was obtained from the conversion of **53** to the S-methyl-1,3,4-thiazolium salt by treatment with methyl iodide <89JOC2024>. The main products from the reaction of 3-aryl-5-benzoyl-2-imino-2,3-dihydro-1,3,4-thiadiazoles (**54**) with dimethyl acetylene-dicarboxylate were the thiazoles (**55**) in good yields. Careful mechanistic studies on the effect of reaction solvent and substituents led to the conclusion that the vinyl by-products and the thiazole are formed competitively through different intermediates, hypervalent sulfurane (**56**) and zwitterion (**57**) (Scheme 16). 2-Iminothiazolines gave only simple adducts,and no thiazole was obtained. On the other hand, 1,2,4-thiadiazolines gave only ring-transformed thiazoles in excellent yield via sulfurane intermediates. These results can be explained by the stability of two intermediates involved <89BCJ211>.

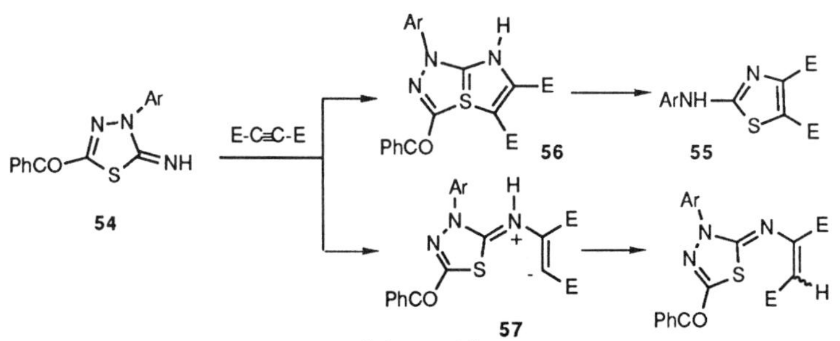

Scheme 16

Several new classes of fused mesoionic compounds, 5-phenylthiazolo[2,3-b]-1,3,4-thiadiazolium-2-arylaminides, were prepared from 3-amino-2-benzylthio-4-phenylthiazolium halides <88S729>. The reaction of 2,5-dimethyl-1,3,4-thiadiazole with acid chloride was investigated. In the case of aromatic acid chloride, bis-acylated products were formed; on the other hand, the reaction with acetyl chloride gave the bis-thiadiazolidine derivative <89CZ125>.

Cyclization of thiosemicarbazone in the presence of acetyl chloride in pyridine gave bisacyl-dihydro-1,3,4-thiadiazoles in good yields <89ZC166>. Treatment of the 4-methyl-3-thiosemicarbazone of 3-acetylpropionic acid with acetic anhydride gave the bicyclic dihydro-1,3,4-thiadiazole derivative via the formation of the mixed anhydride, followed by a novel intramolecular bis-annulation with the elimination of an acetate ion. On the other hand, bis-annulation was not observed when the ethyl ester was reacted with acetic anhydride <89H133>. New syntheses of fused 1,3,4-thiadiazoles using iminophosphoranes appeared. The reaction of 1-amino-2-methylthio-4-phenylimidazole with triphenylphosphine dibromide gave the aza-phosphorane, which reacted with aroyl chlorides to give imidazo[2,1-b]-1,3,4-thiadiazoles in moderate to good yields. This reaction clearly involves the formation of an imidoyl chloride intermediate, followed by cyclization and elimination of methyl chloride. Similarly, 4-amino-3-thioxo-tetrahydro-1,2,4-triazine was reacted with triphenylphosphine dibromide to give fused thiadiazolo[2,3-c]-1,2,4-triazinium chloride derivatives <88H1935> (Scheme 17).

Scheme 17

4.5.4 Selenazoles, Selenadiazoles

A few 1,3,4-selenadiazole derivatives have been shown to possess significant antibacterial activity. Arenediazonium chlorides reacted with 2-(selenocyanatoacetyl)thiophene (**58**) to give 3-aryl-5-thienoyl-2-imino-2,3-dihydro-1,3,4-selenadiazoles (**59**). Treatment of **59** with sodium nitrite gave the nitrosoimino derivatives which decomposed to the selenadiazolones (**60**) upon refluxing in xylene. Similarly, the reaction of 58 with diazotized anthranilic acid and its methyl ester afforded the corresponding 1,3,4-selenadiazole[2,3-b]quinazoline <88MI55005>(Scheme 18).

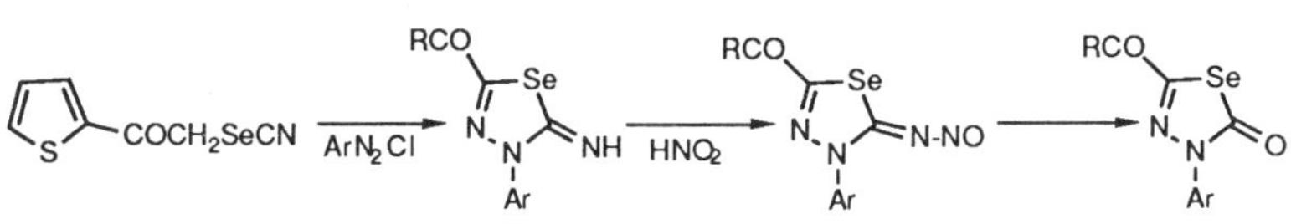

58 **59** **60**

Scheme 18

The reaction of methyl 4-bromomethyl-2-phenyl-5-selenazolyl ketone with N,N-diethylselenopropionamide gave 2-phenyl-6-methylselenolo[3,4-d]selenazole. In the NMR spectrum of selenolo[3,4-d]selenazole, the proton that is geminal to the selenium atom appears as a strong singlet and a weak doublet centered around the singlet. This doublet is assigned to the splitting caused by the presence of the selenium isotope ^{77}Se with a natural abundance of 7.5% <89JHC709>. Oxidation of the semicarbazone derivative with selenium dioxide in glacial acetic acid gave 1,2,3-selenadiazole <88MI55002>. 5-Arylazo-4-aryl-2-aminoselenazoles and 2-imino-2,3-dihydro-1,3,4-selenadiazoles were prepared by the reaction of C-aroyl-N-arylformaldehyde hydrazones with potassium selenocyanate or with selenourea, respectively. Treatment of the selenadiazole derivatives with sodium nitrite in acetic acid gave the N-nitrosoimino derivatives which were converted to the corresponding 2-oxo-2,3-dihydro-1,3,4-selenadiazoles by heating in xylene <88OPP505>. The behavior of isoselenazoles towards electrophilic reagents was investigated. Isoselenazoles are weak bases and they are stable in acidic media. Bromination of isoselenazoles at the 4-position was achieved in good yield in an ionic reaction with NBS in acetic acid at room temperature. The nitration of isoselenazoles gave exclusively 4-nitro derivatives. When the 3-methyl-5-phenyl-isoselenazole was submitted to nitration, the phenyl ring reveals itself more reactive than the heterocyclic ring. The parent isoselenazole ringshows poor stability for bromination and nitration. Thus, only electrophilic substitution in the 4-position of isoselenazoles fully parallels the behavior of isoxazoles and isothiazoles <88H2431>. The reaction of 1,2,3-selenadiazoles with $Fe_2(CO)_9$ in the presence of ethanol gave a new bis(selenoketoketene) complex <89MI55008>. The reaction of 1,2,3-selenadiazoles with transition metal complexes gave a mixture of mononuclear and binuclear complexes <89MI55007>. Ring transformation of isoselenazolo[4,3-d]pyrimidines gave isoselenazolo[4,3-e]-1,4-diazepines, which showed no significant antitumor activity <88CPB2902>. Substituted 2,1,3-benzoselenadiazoles were prepared from the reaction of substituted o-phenylenediamines with selenous acid and their pesticidal activities were examined <88MI55003>.

References

88BCJ4443	N. Fukada, T. Mori, M. Muraoka, T. Yamamoto and T.Takeshima; Bull.Chem.Soc. Jpn., 1988, **61**, 4443.
88CPB2902	T. Ueda, Y. Kato, J. Sakakibara and M. Murata; Chem. Pharm. Bull., 1988, **36**, 2902.
88H1935	P. Molina, A. Lorenzo, M. J. Vilaplana, E. Aller and J. Planes; Heterocycles, 1988, **27**, 1935.
88H2431	F. Lucchesini, N. Picci, M. Pocci, A. De Munno and V. Bertini; Heterocycles, 1988, **27**, 2431.
88H2539	A. Corsaro, F. Guerrera, M. Sarva and M. Siracusa; Heterocycles, 1988, **27**, 2539.
88IJC(B)542	V. Adhikari and V. Badiger; Indian J. Chem., Sect. B, 1988, **27B**, 542.

88JCR(S)349	K. Buggle and B. Fallon; J. Chem. Research (S), 1988, **349**, 2764.
88JHC235	A. Alberola, F. Alonso, P. Cuadrado and M. Sanudo; J. Heterocyclic Chem. 1988, **25**, 235.
88JHC1071	T. Okawara, Y. Tateyama, T. Yamasaki and M. Furukawa; J. Heterocyclic Chem. 1988, **25**, 1071.
88JHC1095	D. McKinnon and K. Duncan; J. Heterocyclic Chem. 1988, **25**, 1095.
888JIC571	R. Kumar, S. Giri and Nizamuddin; J. Indian Chem. Soc., 1988, **65**, 571.
88MI55001	S. Pandeya and A. Khan; Indian J. Physiol. Pharmac., 1988, **32**, 164.
88MI55002	B. Bauomy, M. Al-Ashinawl and M. El-Sadek; Egypt. J. Pharm. Sci., 1988, **29**, 269.
88MI55003	J. Kochansky, C. Cohen, W. Lusby, J. Svoboda, J. Feldmesser and F. Wright; J. Agric. Entomol. 1988, **5**, 131.
88MI55004	M. Kocevar, S. Polanc, M. Sollner and B Vercek; J. Serb. Chem. Soc., 1988, **53**, 75.
88MI55005	A. Abdelhamid and F. Abdel-Galil; Sulfur Letters, 1988, **8**, 11.
88MI55006	G. Sridevi, P. Rao and K. Reddy; Sulfur Letters, 1988, **8**, 101.
88MI55007	D. Harpp, J. MacDonald and M. Ryan; Sulfur Letters, 1988, **7**, 155.
88MI55008	G. Murthy, V. Reddyand K. Mogilaiah; Sulfur Letters, 1988, **7**, 171.
88OPP505	H. Hassaneen, A. Farag, M. Algharib and A. Shawali; Org. Preparations and Procedures Int., 1988, **20**, 505.
88S729	P. Molina, A. Arques, M. Velasco and J. Villalgordo; Synthesis, 1988, 729.
88SC1847	P. Cuadrado, A. Gonzalez and F. Pulido; Synth. Commun., 1988, **18**, 1847.
89BCJ211	Y. Yamamoto, T. Tsuchiya, M. Ochiumi, S. Arai, N. Inamoto and K. Akiba; Bull. Chem. Soc. Jpn., 1989, **62**, 211.
89BCJ479	Y. Yamamoto and K. Akiba; Bull. Chem. Soc. Jpn., 1989, **62**, 479.
89BCJ1086	K. Yonemoto, I. Shibuya and K. Honda; Bull. Chem. Soc. Jpn., 1989, **62**, 1086.
89BCJ1215	H. Chikashita, S. Komazawa, N. Ishimoto, K. Inoue and K. Itoh; Bull. Chem. Soc. Jpn., 1989, **62**, 1215.
89CCC28	J. Arenas, J. Perez-Pena and M. Gonzalez-Davila; Collect. Czech. Chem. Commun., 1989, **54**, 28.
89CZ125	Von W. Kantlehner, E. Hang, W. Kinzy, H. Hagen, P. Speh and H. Brauner; Chemiker-Zeitung, 1989, **113**, 125.
89H97	F. Lucchesini, N. Picci, M. Pocci, A. DeMunno and V. Bertini; Heterocycles, 1989, **29**, 97.
89H133	E. Tao, J. Brennan, J. Swartzendruber and J. Deeter; Heterocycles, 1989, **29**, 133.
89HCA800	H. Diener, B. Gulec, P. Skrabal and H. Zollinger; Helvetica Chimica Acta, 1989, **72**, 800.

89IJC(B)27 S. Singh, S. Sehgal and L. Tarar; Indian J. of Chem., 1989, **28B**, 27.

89IJC(B)78 L. Yadav, K. Shukla and H. Singh; Indian J. of Chem., 1989, **28B**, 78.

89IJC(28)81 V. Ahluwalia and J. Rao; Indian J. of Chem., 1989, **28B**, 81.

89IJC(13)200 N. Tiwari, B. Chaturvedi and Nizamuddin; Indian J. of Chem., 1989, **28B**, 200.

89IJC(B)358 B. Chaturvedi, N. Tiwari and Nizamuddin; Indian J. of Chem., 1989, **28B**, 358.

89JCR(S)29 J. Barker, P. Huddleston and P. Needs; J. Chem. Research (S), 1989, 29.

89JCR(S)48 F. Kurzer and R. Esmail; J. Chem. Research (S), 1989, 48.

89JCR(S)115 R. Letcher, K. Cheung and D. Sin; J. Chem. Research (S), 1989, 115.

89JCS(D)25 M. Ciriano, F. Viguri, J. Perez-Torrente, F. Lahoz, L. Oro, A. Tiripicchio and M. Tiripicchio-Camellini; J. Chem. Soc. Dalton Trans., 1989, 25.

89JCS(P1)643 G. Meakins, S. Mush, C. Robertson and L. Woodhouse; J. Chem. Soc. Perkin Trans. 1, 1989, 643.

89JCS(P1)1169 J. Chakrbarti and C. Smith; J. Chem. Soc. Perkin Trans. 1, 1989, 1169.

89JCS(P2)263 D. Cooper, S. Wright, J. Gerratt and M. Raimondi; J. Chem. Soc. Perkin Trans. II, 1989, 263.

89JHC355 F. Kurzer and J. Secker; J. Heterocyclic Chem. 1989, **26**, 355.

89JHC709 A. Shafiee, A. Shafaati, and B. Habibi-Khameneh; J. Heterocyclic Chem., 1989, **26**, 709.

89JMC1326 M. Kanao, Y. Watanabe, Y. Kimura, J. Saegusa, K. Yamamoto, H.Kanno, N. Kanaya, H. Kubo, S. Ashida and F.Ishikawa; J. Med. Chem., 1989, **32**, 1326.

89JMC2204 J. Starrett, Jr., T. Montzka, A. Crosswell and R.Cavanagh; J. Med. Chem. 1989, **32**, 2204.

89JMC2486 O.W. Woltersdorf, Jr., et al.; J. Med. Chem., 1989, **32**, 2486.

89JOC693 A. Dondoni, G. Fantin, M. Fogagnolo, A. Medici and P. Pedrini; J. Org. Chem., 1989, **54**, 693.

89JOC702 A. Dondoni, G. Fantin, M. Fogagnolo, A. Medici and P. Pedrini; J. Org. Chem., 1989, **54**, 702.

89JOC1077 K. Potts, M. Dery and w. Juzukonis; J. Org. Chem., 1989, **54**, 1077.

89JOC1088 R. Kieter and H. Chang; J. Org. Chem., 1989, **54**, 1088.

89JOC2024 F. Bottino, U. Chiacchio, F. Fronczek and S. Pappalardo; J. Org. Chem., 1989, **54**, 2024.

89JST269 G. Barbieri, R. Benassi and F. Taddei; J. Mol. Structure (Theochem), 1989, **184**, 269.

89JST325 G. Buemi and C. Gandolfo; J. of Mol. Structure (Theochem), 1989, **187**, 325.

89KGS112 R. Belen'kaya, A. Zimichev and O. Yukova; Khim. Geterotsikl. Soedin, 1989, 112.

89LA75	M. Veverka and J. Svetlik; Liebigs Ann. Chem., 1989, 75.
89LA409	D. Wobig; Liebigs Ann. Chem., 1989, 409.
89MI55000	F. Russo, A. Santagati, M. Santagati, A. Caruso, M. Leone, A. Felice and M. Amico-Roxas; Eur. J. Med. Chem., 1989, **24**, 91.
89MI55001	J. Kosary, E. Kasztreiner and F. Andrasi; Pharmazie, 1989, **44**, 191.
89MI55002	H. Van Den Berg, G. Challa and U. Pandit; J. of Mol. Catal., 1989, **51**, 13.
89MI55003	J. Parvate, V. Bhagwat and S. Redkar; Indian Drugs, 1989, **26**, 172.
89MI55004	J. Parvate, V. Bhagwat, M. Doshi and H. Mondkar; Indian Drugs, 1989, **26**, 222.
89MI55005	P. Vicini and P. Mazza; IL Farmaco, 1989, **44(5)**, 511.
89MI55006	A. Santilli, A. Scotese, R. Morris, S. Nielsen and D. Strike; Eur. J. Med. Chem., 1989, **24**, 87.
89MI55007	C. Morley; Organometallics, 1989, **8**, 800.
89MI55008	A. Mayr, K. Pannell, B. Carrasco-Flores and F. Cervantes-Lee; Organometallics, 1989, **8**, 2961.
89MI55009	M. Germann-Bauer and A. Amberger; Fertilier Research, 1989, **19**, 13.
89MI55010	E. Bezzubets, E. D'yachenko and V. Ivanov; Zh. Neorg. Khim, 1989, **34**, 1895.
89MI55011	T. Davidenko, and I. Romanovskaya; Khim. Farm. Zh., 1989, **23**, 473.
89MI55012	A.. Antsyshkina,O. Tokarskaya,V. Tsveniashvili, V. Ostrikova, L. Ukhin,M. Porai-Koshits,A. Garnovskii, Koord. Khim.,1989,**15**,214.
89OPP479	A. Elghandour, M. Ramizt, M. Ibrahim and M. Elmoghayar; Org. Preparations and Procedures Int., 1989, **21(4)**, 479.
89S394	D. Hellwinkel and R. Karle; Synthesis, 1989, 394.
89T2626	D. Barton, D. Bridon and S. Zard; Tetrahedron, 1989, **45**, 2615.
89ZC17	H. Kibbel ad c. Knebusch; Z. Chem., 1989, **19**, 17.
89ZC166	B. Schulze, J. Hibig and M. Muhlstadt; Z. Chem., 1989, **29**, 166.

CHAPTER 4.6

Five-Membered Ring Systems: With O & S (Se, Te) Atoms

R ALAN AITKEN
University of St Andrews, UK

4.6.1 1,3-Dioxoles and Dioxolanes

Several new methods for the synthesis of dioxoles and dioxolanes have been reported. Reaction of *o*-quinones such as (**1**) with nitroalkanes, such as $EtNO_2$, in basic solution gives the dioxoles (**2**) [88JAP63301883]. The anions of active methylene compounds react with 2-chloroethyl chloroformate to give methylene dioxolanes (**3**) [88S981]. A new efficient procedure for the

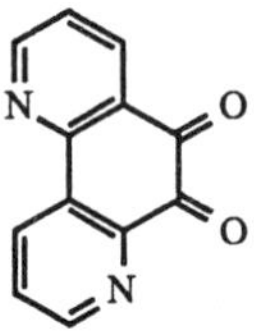

(1)

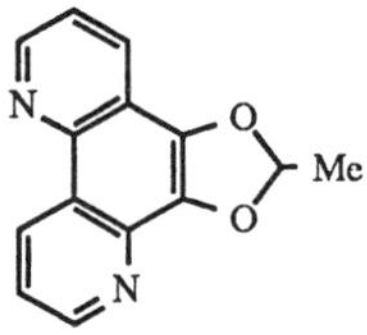

(2)

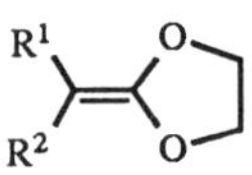

(3)

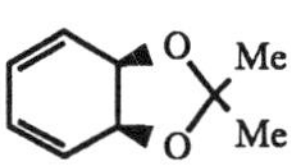

(4)

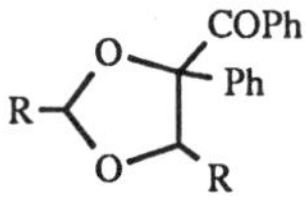

(5)

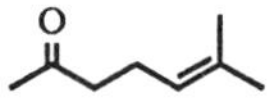

(6)

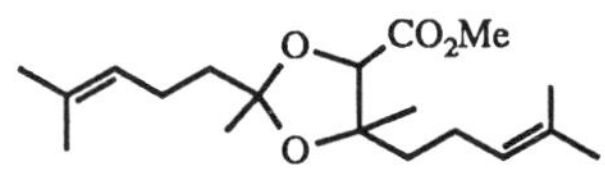

(7)

preparation of acetonide (**4**) from the diol has been described [89OPP255]. Treatment of benzil with aldehydes, RCHO, in the presence of $TiCl_3$ gives (**5**) [88SC2281]. Copper bronze catalysed decomposition of methyl diazoacetate in the presence of (**6**) gives 30% of the dioxolane (**7**) in addition to the expected cyclopropane [88CI(L)631]. The formation of spiro-ortho esters (**8**) and (**9**) by treatment of epoxides with γ-butyrolactone [88SUP1397449] and ε-caprolactone

[88JAP6329587] respectively is reported and the hydrolysis of (**8**) and the corresponding orthocarbonates has been studied [89BCJ945]. Photochemical dehydrobromination of (**10**) by NaOPr^i proceeds by *ipso* substitution to give (**11**) [89CC1154] while electrochemical or $Pb(OAc)_4$ oxidation of (**12**) gives (**13**) [88JCR(S)241]. Dechlorination of dioxolane (**14**) with $LiAlH_4/TiCl_4$ affords the corresponding dioxole [89GEP3842986].

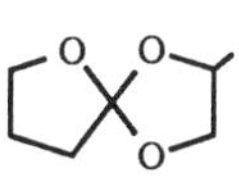
(8)

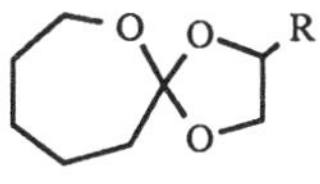
(9)

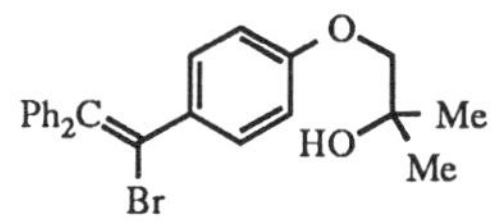
(10)

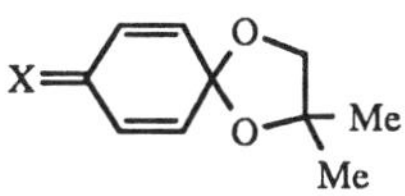
(11) X= Ph_2C=C
(13) X= ArN

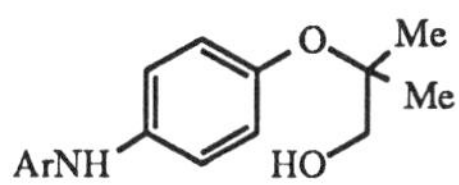
(12)

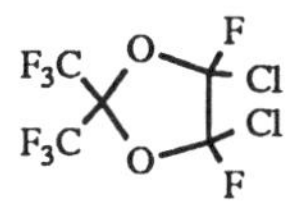
(14)

1,3-Dioxolan-2-ones have been prepared by reaction of 1,2-diols with $ClSO_2NCO$ [88SC2295] and 2-haloalcohols with $NaHCO_3$ in DMSO [89GEP3723782]. The ring-expansion of oxiranes with CO_2 to give dioxolanones is effectively catalysed by ytterbium naphthalenide [89MI392] and an example bearing a polyether side chain has been reported [88Z0R1623]. Caesium carbonate and 3Å molecular sieves catalyse the mild carboxylation of 2,3-epoxyalcohols to C-2 inverted dioxolanones [88TL6389]. Treatment of 3-hydroxy-4-pyridone (**15**) with phosgene or carbonyl diimidazole in the presence of DBU gives (**16**) [89JHC401]. The mild carboxylation of propargyl alcohols (**17**) to give (**18**) is catalysed by $Bu^n{}_3P$ [89TL3981] while reaction of (**19**) with aqueous LiOH gives 48% (**20**), increased to 68% in the presence of acetone

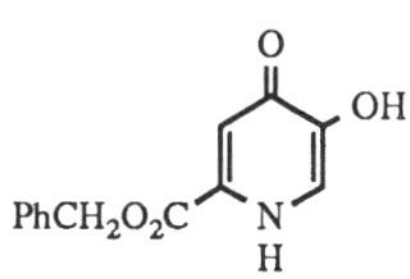
(15)

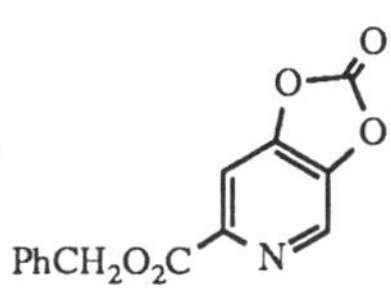
(16)

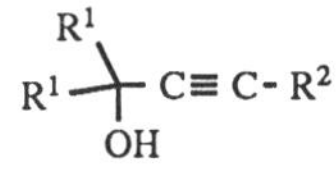
(17) R^2= H
(19) R^1= Me, R^2= CN

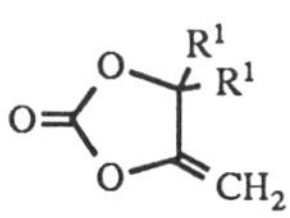
(18)

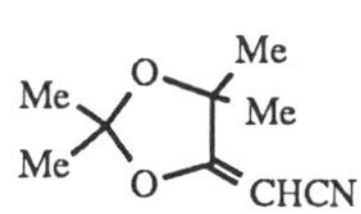
(20)

[89IZV1421]. Diphosgene, $Cl_3COCOCl$ is effective in cyclising α-hydroxy acids to the corresponding 1,3-dioxolane-2,4-diones [89H(29)975].

The enantiomers of chiral 1,3-dioxolan-4-ones can be separated on a commercially available chiral HPLC stationary phase [89LA103]. Reaction of hexafluoropropylene oxide with DMF gives (**21**) and with hexafluoroacetone (**22**) [88IZV392]. Treatment of (**23**) and with Na_2CO_3 gives mainly (**24**) [89IZV938] which dimerises by a head-head 2+2 cycloaddition on heating to 150°C [89IZV201].

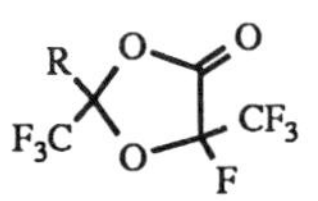

(21) R= COF
(22) R= CF_3

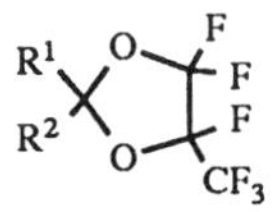

(23) R^1= COF, R^2= CF_3
(24) R^1,R^2= F_2C=

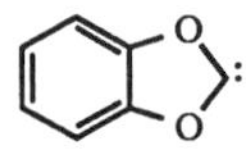

(25)

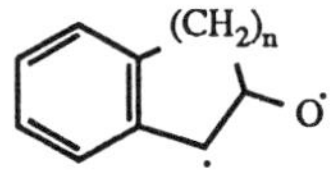

(29)

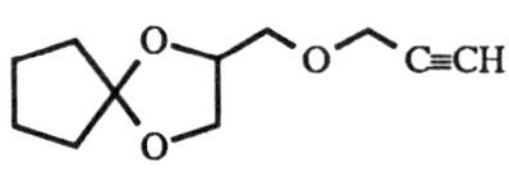

(26)

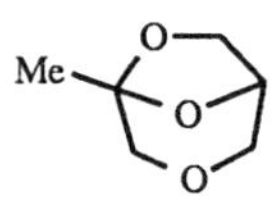

(27)

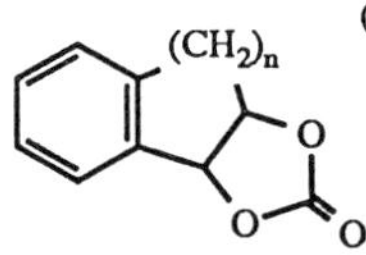

(28) n= 1,2

Trichloro- or trifluoroacetic acid converts 2-ethoxy-1,3-benzodioxole to a carbene-like intermediate (**25**) which dimerises and can be trapped by Ph_3P to form an ylide [88KGS852]. The conversion of (**26**) to the tricyclic ortho ester (**27**) with HgO/H_2SO_4 has been reported [88MI34]. Photochemical decarboxylation of (**28**) gives products derived from the 1,3-diradical (**29**) [88JHC1781]. The synthesis and Diels-Alder reactions of a series of 1,3-dioxol-2-ones have been described [89H(28)93]. The highly stereoselective Diels-Alder reaction of (**30**) with cyclopentadiene gives mainly (**31**) [89CB328]. Good diastereoselectivity is observed in the reactions of (**32**) with dienophiles

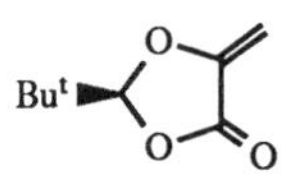

(30)

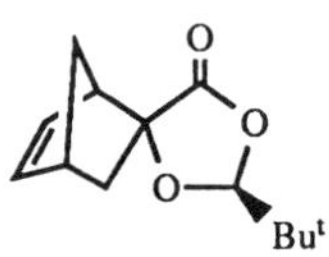

(31)

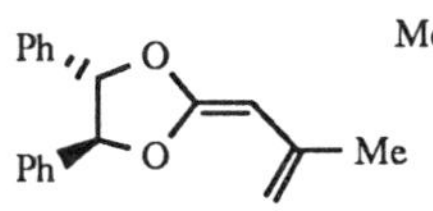

(32)

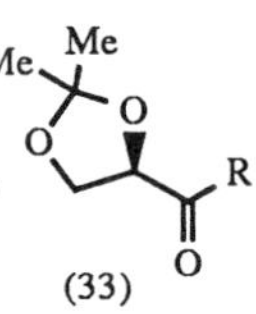

(33)

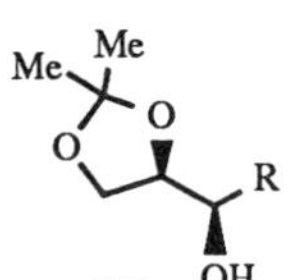

(34)

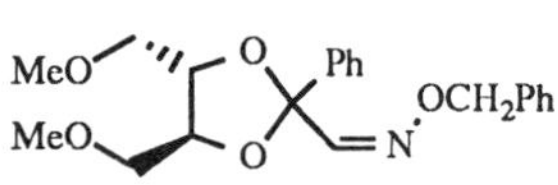

(35)

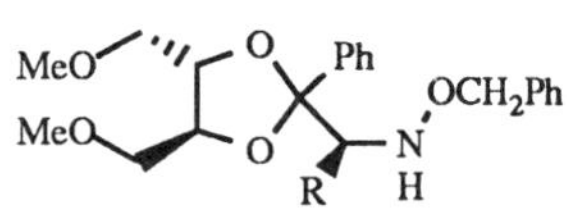

(36)

[89JA4515], of (**33**) with $LiBu^s{}_3BH$ to give (**34**) [89BCJ2121], and of chiral oximes (**35**) with organocerium reagents to give (**36**) [89CPB602].

4.6.2 1,3-Dithioles and Dithiolanes

There has been a continued high level of interest in the preparation and electrical properties of tetrathiafulvalenes. The syntheses of examples bearing long alkyl chains [88CC1391][89SM(31)379] as well as alkylated [88SM(27)89] and tetrachloro and tetrabromo derivatives [89S207] have been reported. Some of the more significant TTF derived systems to be examined are (**37**) [88BCJ3455], (**38**) [88TL6177], (**39**) and (**40**) [88SM(27)B651], (**41**)

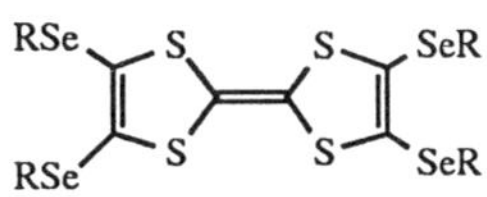

(37)

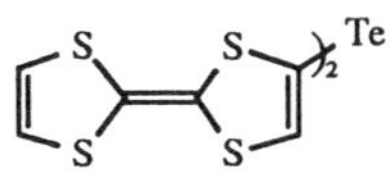

(38)

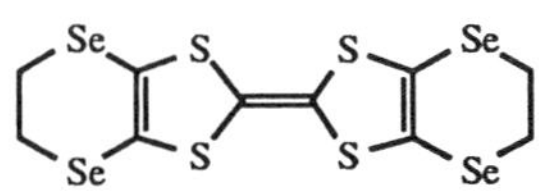

(39)

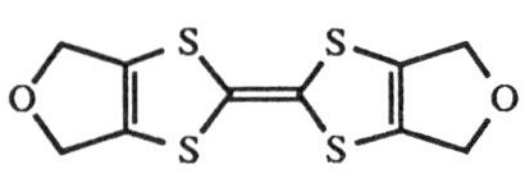

(40)

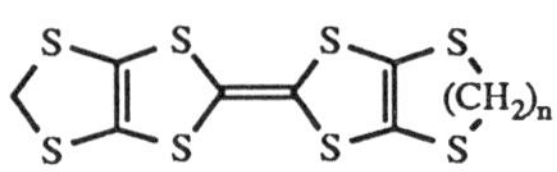

(41) n=2,3

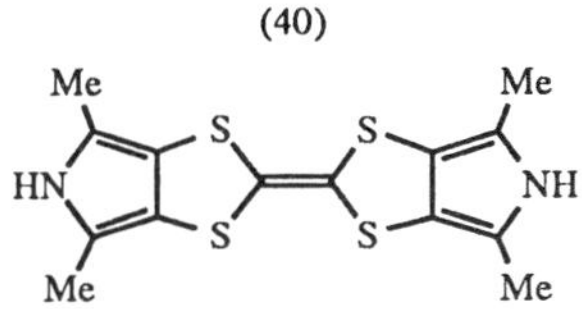

(42)

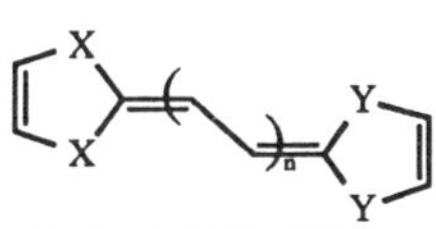

(43) n=1 X,Y= S,Se
(44) n=2 X=Y=S

[88CS365], (**42**) [88JA7903], (**43**) and (**44**) [89CMA535]. The electrochemical behaviour of the dication salts (**45**) derived from (**46**) has been studied [88MI47]. The preparation and properties of *p*-quinodimethane analogues of TTF such as (**47**) and (**48**) have been described [88SM(27)B557] [89AG(E)1052] as well as those of (**49**) [89CMA421] and (**50**) [88PS221]. An unusual example of TTF formation is the reaction of 3-thiacyclooctyne with CS_2 to form (**51**) which then dimerises [89LA435]. Since the cycloalkyne is made from a 1,2,3-selenadiazole a minor side reaction results in the formation of (**52**) in 4% yield.

The reactions of $(CN)_2C{=}C(SNa)_2$ with methyl bromopropiolate to give (**53**) [89ZOR1512] and with $Cl_2C{=}CHCOMe$ to give (**54**) [88ZOR1773] have been described. Reaction of ethanedithiol with $BF_3{\cdot}Me_2S$ gives (**55**) which is

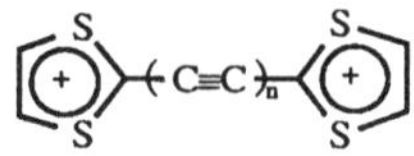
(45) n= 1,2

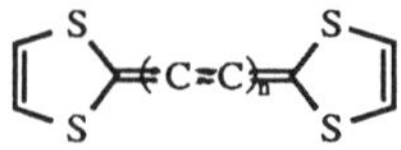
(46) n= 1,2

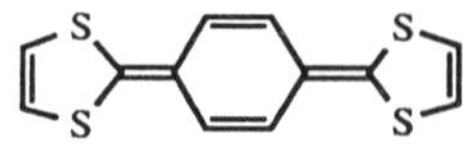
(47)

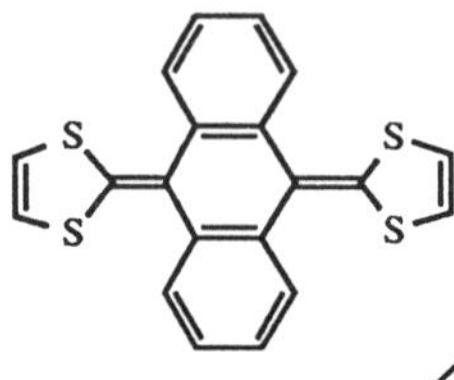
(48)

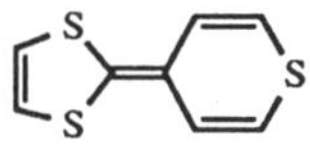
(49)

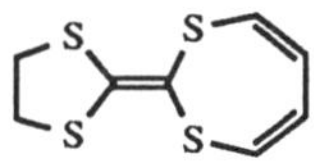
(50)

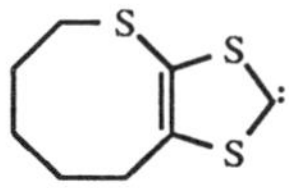
(51)

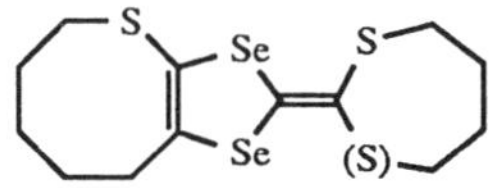
(52)

a useful reagent for the one-step conversion of acids, RCO_2H, to dithiolanes (**56**) [89MI219]. Treatment of 1,2-dibromides with sodium trithiocarbonate provides a convenient route to 1,3-dithiolane-2-thiones [89BCJ339]. Dimethylazines RC(Me)=N–N=C(Me)R react with $ClSCF_2CF_2SCl$ to give azo(methylenedithiolanes) (**57**) [89ZN(B)543]. Convenient new syntheses of 1,3-dithiole-2-thione and -2-selenone have been reported [89JCS(P1)1068]. A detailed study of the high pressure reaction of epoxides with excess CS_2 to give

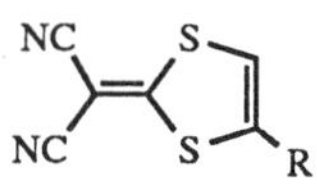
(53) R= CO_2Me
(54) R= COMe

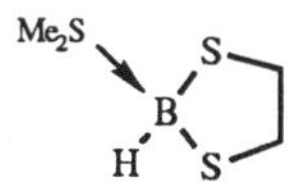
(55)

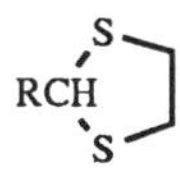
(56)

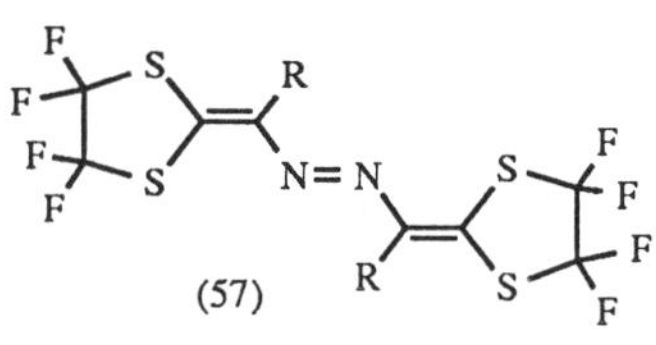
(57)

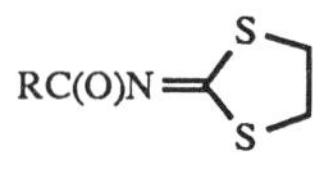
(58)

1,3-dithiolane-2-thiones, 1,3-oxathiolane-2-thiones and 1,3-oxathiolan-2-ones as well as thiiranes has appeared [88BCJ921]. When one equiv. CS_2 is used the main product is the 1,3-dithiolan-2-one which is not formed with excess CS_2 [89BCJ474] [89JAP01230574]. The effect of various additives on the formation of 2-imino-1,3-dithiolanes *vs.* thiazolidine-2-thiones in the reaction of simple thiiranes with MeNCS has been examined [89NKK63]. With PhNCS the iminodithiolane is the main product. The formation of iminodithiolanes (**58**) by sequential reaction of amides with NaH, CS_2, and $BrCH_2CH_2Br$ has been reported [89ZC16]. The benzo-*tris*-dithiole derivatives (**59**) have been prepared [88ZC284].

The mass spectra of substituted 1,3-dithiolane S-oxides have been studied [88OMS841] as well as the separation of their enantiomers by chiral stationary phase HPLC [88MI199]. A new procedure for the deprotection of 2,2-disubstituted 1,3-dithiolanes to the corresponding ketones involves treatment

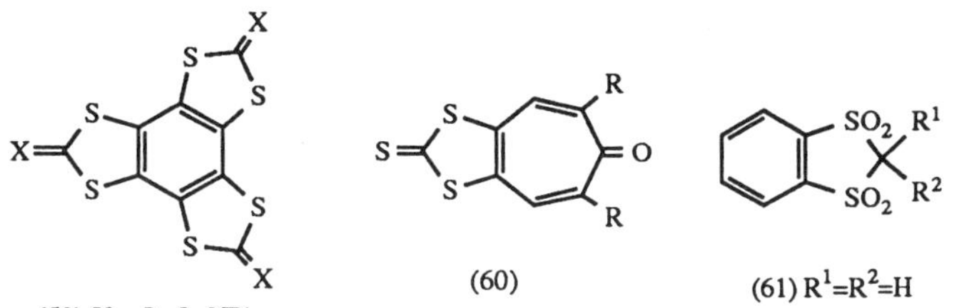

(59) X= O, S, NPh (60) (61) R^1=R^2=H

with $PhOP(O)Cl_2$ and NaI in DMF [88TL5471]. Condensation of 1,3-dithiole-2-thione-4,5-dicarboxaldehyde with ketones gives (**60**) [88AP757]. The 1,3-benzodithiole tetroxide (**61**) is a more efficient synthetic equivalent of CH_2^{2-} than $(PhSO_2)CH_2$ [88T6855]. It can be sequentially alkylated even with bulky alkyl groups and cleaved to give $R^1CH_2R^2$ by treatment with Mg/MeOH.

4.6.3 1,3-Oxathioles and Oxathiolanes

The 4-hydroxy-2-pyridones (**62**) react with ClSCOCl to afford fused 1,3-oxathiol-2-ones (**63**) [88EGP258988]. A novel equilibrium between thioketene ylide (**64**) and 1,3-oxathiole (**65**), formed by decomposition of dialkyl diazomalonates in the presence of di-t-butylthioketene, has been described

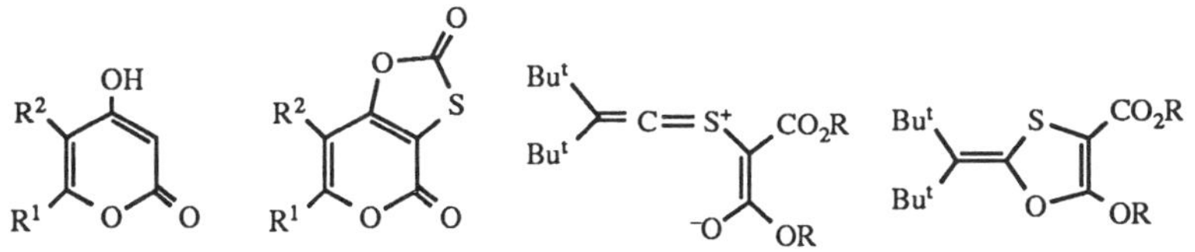

(62) (63) (64) (65)

[89TL1249]. Ketones are conveniently converted to 5,5-disubstituted 1,3-oxathiolanes by treatment with $Me_3SiCH_2SCH_2Cl$ in the presence of F^-

[89EP307000]. Reaction of (**19**) with KSCN gives oxathiolanone (**66**) [88ZOR2454] while with NH_4SCN oxathiolane (**67**) is formed [89ZOR221] [89SUP1468901].

4.6.4 1,2-Dioxolanes

The cyclisation of alk-3-enyl hydroperoxides upon halogenation occurs stereoselectively with Br_2 in pyridine and non-stereoselectively with NBS and NIS to give 1,2-dioxolanes (**68**) [89CC173]. The cycloaddition of α,ß-unsaturated imines such as (**69**) with $^1O_2$ gives products (**70**) [89H(28)445].

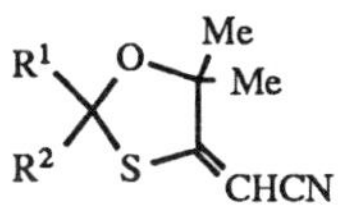

(66) R^1,R^2 = O
(67) R^1= CH_2CN, R^2= CMe_2SCN

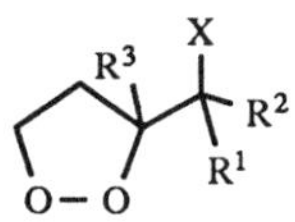

(68)

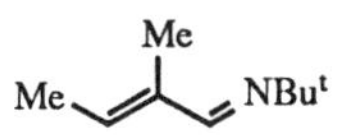

(69)

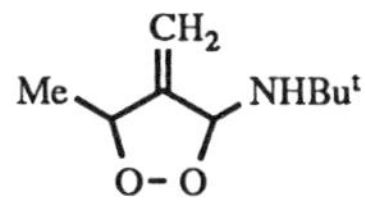

(70)

4.6.5 1,2-Dithioles and Dithiolanes

A convenient new method for the formation of 1,2-dithiolanes involves treatment of 1,3-ditosylates with $(NaS)_2C{=}C(CN)CONH_2$ followed by cleavage with ethylenediamine [88ACS(B)629]. A simple but low yielding preparation of 1,2-dithiol-3-one is achieved by treatment of propargyl alcohol with MeSSMe or MeSSSMe [88KGS999]. The convenient preparation of 1,2-dithiole-3-thiones (**71**) has been reported [89JOC2165] and this has led to the first synthesis of (**72**) [89CC1716]. The equilibrium between (**74**) and (**75**), formed

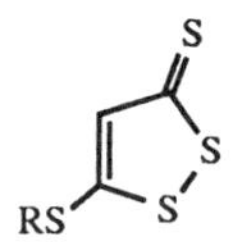

(71)

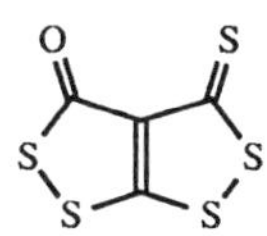

(72)

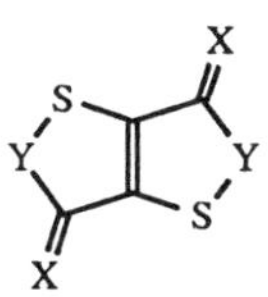

(73) X=Y=S
(74) X=S, Y=ArN
(75) X=ArN, Y=S

by treating (**73**) with aromatic amines has been examined as a function of solvent polarity and the presence of added electron donors and acceptors [89T125]. Oxidation of methylene dithiomalonamides (**76**) with $NOBF_4$ gives (**77**) [89CC1346].

4.6.6 1,2-Oxathiolanes

A convenient one-pot synthesis of oxathiolane S-oxides (**78**) from the corresponding 3-mercaptoalcohols involves treatment with Cl_2 followed by

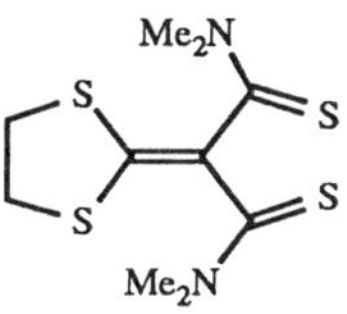

(76)

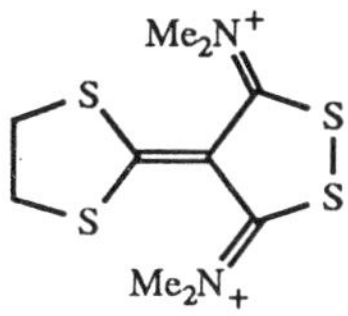

(77)

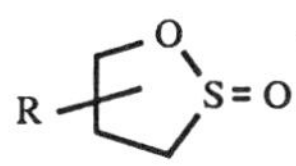

(78)

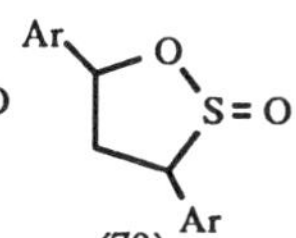

(79)

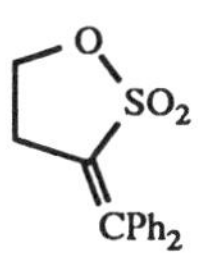

(80)

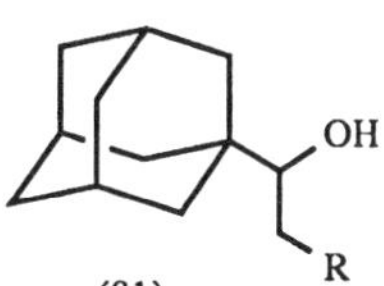

(81)

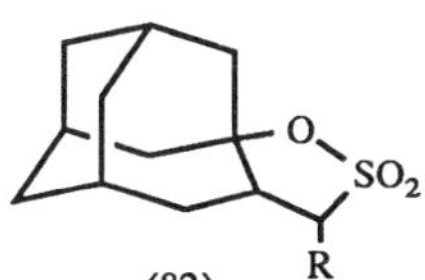

(82)

H_2O [89TL2763]. Reaction of 1,2-diarylcyclopropanes with SO_2 gives (**79**) [88ZOR1937] while (diphenylmethylene)cyclopropane reacts with SO_3 to give (**80**) [89JOC1680]. The formation of sultone (**82**) from (**81**) by treatment with H_2SO_4 in acetic or trifluoroacetic anhydride involves an adamantane-homoadamantane rearrangement [88ZOR738]. The regio-selectivity of nucleophilic ring-opening of cyclic sulphinic-carboxylic anhydrides (1,2-oxathiol-5-one S-oxides) has been examined [88IZV1633].

4.6.7. References

88ACS(B)629 L. Teuber and C. Christophersen, *Acta Chem. Scand.*, 1988, **B42**, 629.

88AP757 G. Seitz and P. Imming, *Arch. Pharm. (Weinheim, Ger.)*, 1988, **321**, 757.

88BCJ921 Y. Taguchi, K. Yanagiya, I. Shibuya, and Y. Suhara, *Bull. Chem. Soc. Jpn.*, 1988, **61**, 921.

88BCJ3455 P. Wang, T. Mori, C. Nakano, Y. Maruyama, H. Inokuchi, N. Iwasawa, H. Yamochi, H. Urayama, and G. Saito, *Bull. Chem. Soc. Jpn.*, 1988, **61**, 3455.

88CC1391 A. S. Dhindsa, M. R. Bryce, J. P. Lloyd, M. C. Petty, K. Kobayashi, and H. Tukada, *J. Chem. Soc., Chem. Commun.*, 1988, 1391.

88CI(L)631	G. S. Joshi, G. H. Kulkarni, and E. A. Shapiro, *Chem. Ind. (London)*, 1988, 631.
88CS365	G. C. Papavassiliou, G. A. Mousdis, S. Y. Yiannopoulos, and J. S. Zambounis, *Chem. Scr.*, 1988, **28**, 365.
88EGP258988	B. Hirsch, N. Höfgen, and L. Litauszki, *E. Ger. Pat.*, 258 988 (1988) [*Chem. Abstr.*, 1989, **110**, 173214].
88IZV392	V. S. Yuminov, S. V. Kartsov, V. L. Maksimov, and A. V. Fokin, *Izv. Akad. Nauk SSSR, Ser. Khim.*, 1988, 392.
88IZV1633	T. P. Vasil'eva, V. M. Bystrova, and O. V. Kil'disheva, *Izv. Akad. Nauk SSSR, Ser. Khim.*, 1988, 1633.
88JA7903	W. Chen, M. P. Cava, M. A. Takassi, and R. M. Metzger, *J. Am. Chem. Soc.*, 1988, **110**, 7903.
88JAP63295587	T. Endo, T. Yokozawa, and M. Sato, *Jap. Pat.*, 63 295 587 (1988) [*Chem. Abstr.*, 1989, **111**, 77990].
88JAP63301883	Y. Oshiro and S. Ito, *Jap. Pat.*, 63 301 883 (1988) [*Chem. Abstr.*, 1989, **111**, 39351].
88JCR(S)241	I. G. C. Coutts and M. R. Southcott, *J. Chem. Res. (S)*, 1988, 241.
88JHC1781	R. C. White, P. Drew, and R. Moorman, *J. Heterocycl. Chem.*, 1988, **25**, 1781.
88KGS852	O. G. Safiev, D. V. Nazarov, V. V. Zorin, and D. L. Rukhmankulov, *Khim. Geterosikl. Soedin.*, 1988, 852 [*Chem. Abstr.*, 1989, **110**, 212656].
88KGS999	L. A. Ostroukhova, L. P. Turchaninova, N. A. Korchevin, E. N. Deryagina, and M. G. Voronkov, *Khim. Geterosikl. Soedin.*, 1988, 999 [*Chem. Abstr.*, 1989, **111**, 97123].
88MI34	B. G. Babaev, U. B. Imashev, E. A. Mamedov, and S. F. Karaev, *Dokl. Akad. Nauk Az. SSR*, 1988, **44**, 34 [*Chem. Abstr.*, 1989, **110**, 114723].
88MI47	H. Awaji, T. Sugimoto, and Z. Yoshida, *J. Phys. Org. Chem.*, 1988, **1**, 47 [*Chem. Abstr.*, 1989, **110**, 75365.].
88MI199	W. H. Pirkle and B. C. Hamper, *J. Chromatogr.*, 1988, **450**, 199.
88OMS841	O. Bortolini, P. Traldi, and G. Licini, *Org. Mass Spectrom.*, 1988, **23**, 841.
88PS221	C. J. H. Long, M. Naylor, C. Moreland, J. Bordner, and R. D. Bereman, *Phosphorus Sulfur*, 1988, **39**, 221.
88S981	R. Neidlein and D. Kikelj, *Synthesis*, 1988, 981.
88SC2281	A. Clerici and O. Porta, *Synth. Commun.*, 1988, **18**, 2281.
88SC2295	S. P. Joseph and D. N. Dhar, *Synth. Commun.*, 1988, **18**, 2295.
88SM(27)B9	K. Lerstrup, I. Johannsen, and M. Joergensen, *Synth. Met.*, 1988, **27**, B9.
88SM(27)B557	M. R. Bryce and A. J. Moore, *Synth. Met.*, 1988, **27**, B557.
88SM(27)B651	S. Y. Hsu and L. Y. Chiang, *Synth. Met.*, 1988, **27**, B651.

88T6855 E. P. Kündig and A. F. Cunningham, *Tetrahedron*, 1988, **44**, 6855.

88TL5471 H.-J. Lu and V. Wiszniewski, *Tetrahedron Lett.*, 1988, **29**, 5471.

88TL6177 J. Y. Becker, J. Bernstein, S. Bittner, J. A. R. P. Sarma, and L. Shahal, *Tetrahedron Lett.*, 1988, **29**, 6177.

88TL6389 A. G. Myers and K. L. Widdowson, *Tetrahedron Lett.*, 1988, **29**, 6389.

88ZC284 A. M. Richter, N. Beye, and E. Fanghänel, *Z. Chem.*, 1988, **28**,.284.

88ZOR738 V. V. Kovalev and E. A. Shokova, *Zh. Org. Khim.*, 1988, **24**, 738.

88ZOR1623 N. A. Nedolya, T. F. Tatarova, E. P. Vyalykh, and B. A. Trofimov, *Zh. Org. Khim.*, 1988, **24**, 1623.

88ZOR1773 V. N. Drozd and E. N. Komarova, *Zh. Org. Khim.*, 1988, **24**, 1773.

88ZOR1937 O. B. Bondarenko, A. V. Buevich, T. I. Voevodskaya, L. G. Saginova, and Yu. S. Shabarov, *Zh. Org. Khim.*, 1988, **24**, 1937.

88ZOR2454 A. G. Mal'kina, Yu. M. Skvartsov, E. I. Moshchevitina, and B. A. Trofimov, *Zh. Org. Khim.*, 1988, **24**, 2454.

89AG(E)1052 Y. Yamashita, Y. Kobayashi, and T. Miyasha, *Angew. Chem. Int. Ed. Engl.*, 1989, **28**, 1052.

89BCJ339 A. Sugawara, T. Sato, and R. Sato, *Bull. Chem. Soc. Jpn.*, 1989, **62**, 339.

89BCJ474 Y. Taguchi, M. Yasumoto, I. Shibuya, and Y. Suhara, *Bull. Chem. Soc. Jpn.*, 1989, **62**, 474.

89BCJ945 H. Tagoshi and T. Endo, *Bull. Chem. Soc. Jpn.*, 1989, **62**, 945.

89BCJ2121 H. Chikashita, T. Nikaya, H. Uemura, and K. Itoh, *Bull. Chem. Soc. Jpn.*, 1989, **62**, 2121.

89CB327 J. Mattay, J. Mertes, and G. Moas, *Chem. Ber.*, 1989, **122**, 327.

89CC173 A. J. Bloodworth and R. J. Curtis, *J. Chem. Soc., Chem. Commun.*, 1989, 173.

89CC1154 T. Kitamura, I. Nakamura, T. Kabashima, S. Kobayashi, and H. Taniguchi, *J. Chem. Soc., Chem. Commun.*, 1989, 1154.

89CC1346 R. Gompper, K. Polborn, and W. Strassberger, *J. Chem. Soc., Chem. Commun.*, 1989, 1346.

89CC1716 F. Closs, G. Srdanov, and F. Wudl, *J. Chem. Soc., Chem. Commun.*, 1989, 1716.

89CMA421 D. J. Sandman, G. P. Ceasar, A. P. Fisher, E. Schramm, D. D. Titus, and A. D. Baker, *Chem. Mater.*, 1989, **1**, 421.

89CMA535 T. Sugimoto, H. Awaji, I. Sugimoto, Y. Misaki, T. Kawase, S. Yoneda, Z. Yoshida, T. Kobayashi, and H. Anzai, *Chem. Mater.*, 1989, **1**, 535.

89CPB602	H. Fujioka, M. Fuji, Y. Okaishi, T. Yoshida, H. Annoura, Y. Kita, and Y. Tamura, *Chem. Pharm. Bull.*, 1989, **37**. 602.
89EUP307000	A. Hosomi, *Eur. Pat.*, 307 000 (1989) [*Chem. Abstr.* 1989, **111**, 134130].
89GEP3723782	G. Grahe and A. Lachowicz, *Ger. Pat.*, 3 723 782 (1989) [*Chem. Abstr.*, 1989, **111**, 134129].
89GEP3842986	M. H. Hung, *Ger. Pat.*, 3 842 986 (1989) [*Chem. Abstr.*, 1990, **112**, 55839].
89H(28)93	M. E. Jung, R. B. Blum, B. J. Gaede, and M. R. Gisler, *Heterocycles*, 1989, **28**, 93.
89H(28)445	T. Akasaka, K. Takeuchi, Y. Misawa, and W. Ando, *Heterocycles*, 1989, **28**, 445.
89H(29)975	K. Toyooka, Y. Takeuchi, and S. Kubota, *Heterocycles*, 1989, **29**, 975.
89IZV201	V. S. Yuminov, S. V. Kartsov, V. L. Maksimov, O. A. Popova, and K. M. Makarov, *Izv. Akad. Nauk SSSR, Ser. Khim.*, 1989, 201.
89IZV938	V. S. Yuminov, S. V. Kartsov, V. L. Maksimov, M. D. Bergamova, and A. V. Fokin, *Izv. Akad. Nauk SSSR, Ser. Khim.*, 1989, 938.
89IZV1421	L. V. Andriankova, N. D. Abramova, A. G. Mal'kina, and Yu. M. Skvortsov, *Izv. Akad. Nauk SSSR, Ser. Khim.*, 1989, 1421.
89JA4515	J. P. Konopelski and M. A. Boehler, *J. Am. Chem. Soc.*, 1989, **111**, 4515.
89JAP01230574	Y. Taguchi, I. Shibuya, and Y. Suhara, *Jap. Pat.*, 01 230 574 (1989) [*Chem. Abstr.*, 1990, **112**, 98512].
89JCS(P1)1068	F. S. Guziec, J. M. Russo, F. F. Torres, G. C. Long, and M. R. Tellez, *J. Chem. Soc., Perkin Trans. 1*, 1989, 1068.
89JHC401	J. Singh, T. P. Kissick, and R. H. Müller, *J. Heterocycl. Chem.*, 1989, **26**, 401.
89JOC1680	B. H. Bakker, H. Cerfontain, and H. P. M. Tomassen, *J. Org. Chem.*, 1989, **54**, 1680.
89JOC2165	F. L. Lu, K. M. Keshavarz, G. Srdanov, R. H. Jackson, and F. Wudl, *J. Org. Chem.*, 1989, **54**, 2165.
89LA103	U. Gerlach, T. Haubenreich, S. Hünig, and N. Klaunzer, *Liebigs Ann. Chem.*, 1989, 103.
89LA435	E. Stravridou, H. Schuhmacher, and H. Meier, *Liebigs Ann. Chem.*, 1989, 435.
89MI219	S. Kim, S. Lim, and S. S. Kim, *Bull. Korean Chem. Soc.*, 1989, **10**, 219 [*Chem. Abstr.*, 1990, **112**, 98425.].
89MI392	E. A. Federova, O. A. Vasina, Yu. F. Rad'kov, G. S. Kalinina, and M. N. Bochkarev, *Metalloorg. Khim.*, 1989, **2**, 392 [*Chem. Abstr.*, 1989, **111**, 232627].
89NKK63	Y. Taguchi, I. Shibuya, and Y. Suhara, *Nippon Kagaku Kaishi*, 1989, 63 [*Chem. Abstr.*, 1989, **110**, 212660].
89OPP255	M. G. Banwell, *Org. Prep. Proced. Int.*, 1989, **21**, 255.
89S207	M. Joergensen and K. Bechgaard, *Synthesis*, 1989, 207.

89SM(31)379	A. S. Dhindsa, M. R. Bryce, M. C. Petty, K. Kobayashi, and H. Tukada, *Synth. Met.*, 1989, **31**, 379.
89SUP1397449	N. V. Zhidkova and V. M. Kobryanskii, *USSR Pat.*, 1 397 449 (1989) [*Chem. Abstr.*, 1989, **110**, 8197].
89SUP1468901	B. A. Trafimov, Yu. M. Skvortsov, E. I. Moshchevitina, A. G. Mal'kina, and V. K. Bel'skii, *USSR Pat.*, 1 468 901 (1989) [*Chem. Abstr.*, 1989, **111**, 134131].
89T125	E. Fanghänel, B. Kordts, and A. M. Richter, *Tetrahedron*, 1989, **45**, 125.
89TL1249	N. Tokitoh, T. Suzuki, A. Itami, M. Goto, and W. Ando, *Tetrahedron Lett.*, 1989, **30**, 1249.
89TL2763	J. F. King and R. Rathore, *Tetrahedron Lett.*, 1989, **30**, 2763.
89TL3981	J. Fournier, C. Bruneau, and P. H. Dixneuf, *Tetrahedron Lett.*, 1989, **30**, 3981.
89ZC16	M. Richter and W. Richter, *Z. Chem.*, 1989, **29**,.16.
89ZN(B)543	H. W. Roesky, U. Otten, R. Herbst, and M. Noltemeyer, *Z. Naturforsch., Teil B*, 1989, **44**, 543.
89ZOR221	B. A. Trofimov, Yu. M. Skvartsov, E. I. Moshchevitina, A. G. Mal'kina, and V. K. Bel'skii, *Zh. Org. Khim.*, 1989, **25**, 221.
89ZOR1512	E. N. Komarova, D. S. Yufit, Yu. T. Struchkov, and V. N. Drozd, *Zh. Org. Khim.*, 1989, **25**, 1512.

CHAPTER 4.7
Five-Membered Ring Systems: With O and N Atoms

G V BOYD
The Hebrew University, Jerusalem, Israel

This part deals with five-membered heterocyclic systems containing oxygen and one or more nitrogen atoms; 1,2,4-dioxazolidines are also mentioned. The order in each section is that of increasing saturation, so that the fully conjugated 'aromatic' compounds are mentioned first, dihydro compounds and their oxo derivatives follow and the section concludes with tetrahydro compounds.

ISOXAZOLES

Treatment of the oxime of glyoxylic acid with chlorine, followed by propargyl alcohol and potassium carbonate, yields the isoxazole (1) (88SC 1171). The azirine (2) is converted into the benzamidoisoxazole (3) by the action of phenyl isocyanate (88BCJ2881). The reaction of 2,3-diarylcyclopropanones with nitrosonium tetrafluoroborate affords mixtures of isoxazoles (4) and (5) (88CL 233).

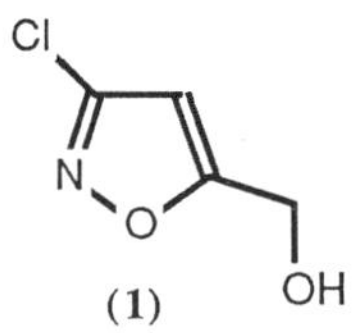

(1)

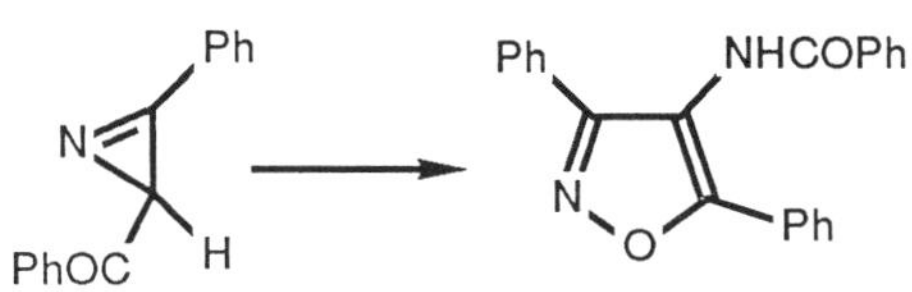

(2) (3)

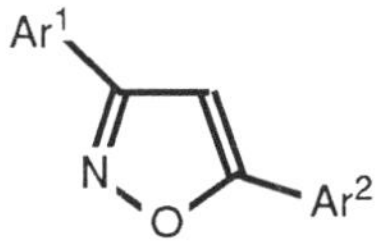

(4)

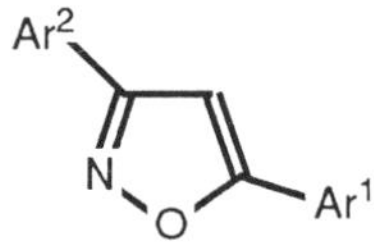

(5)

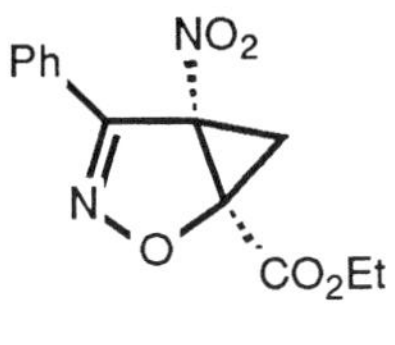

(6)

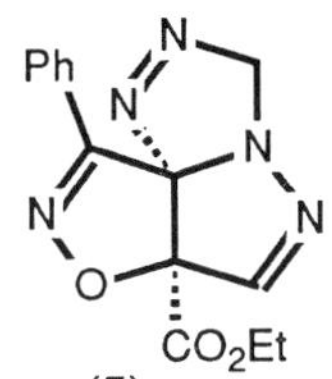

(7)

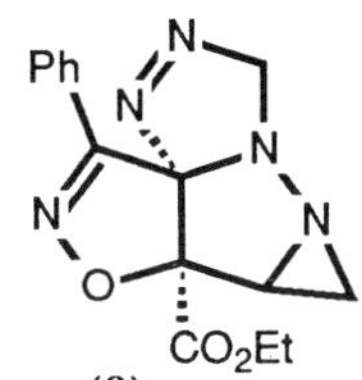

(8)

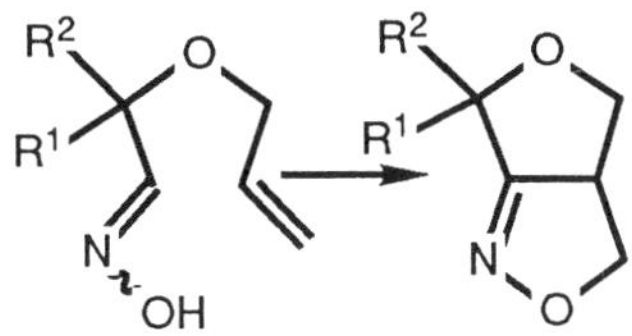

(9) (10)

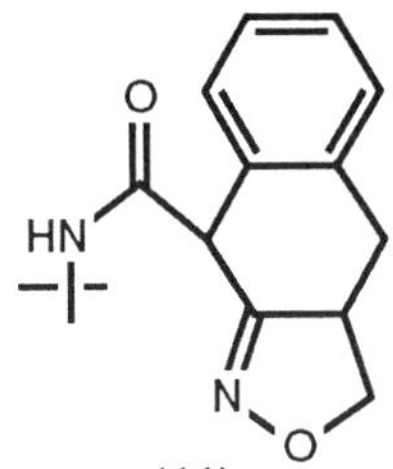

(11)

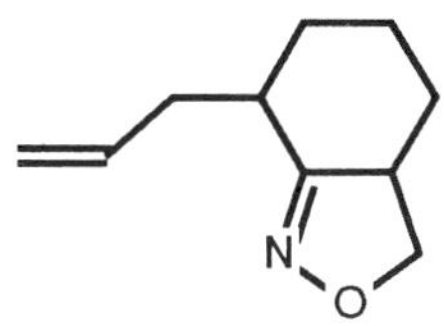

(12)

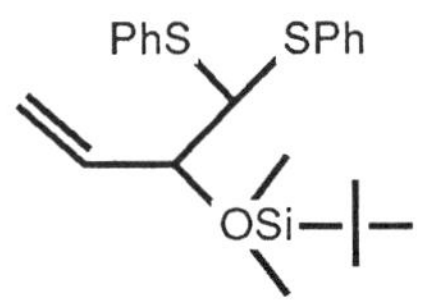

(13)

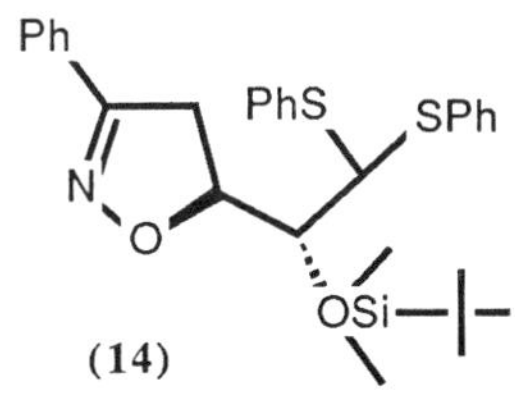

(14)

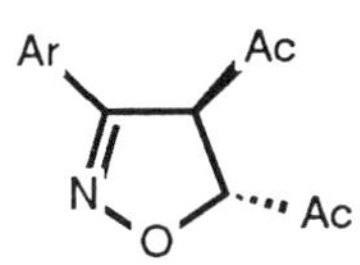

(15)

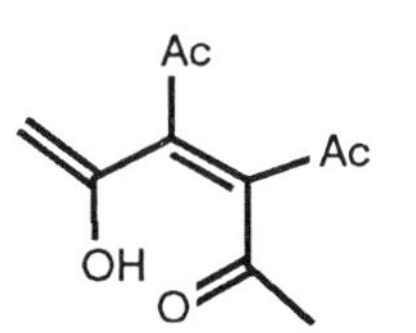

(16)

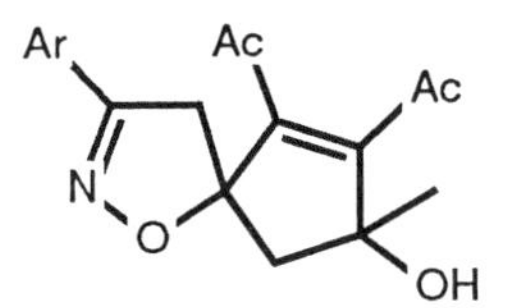

(17)

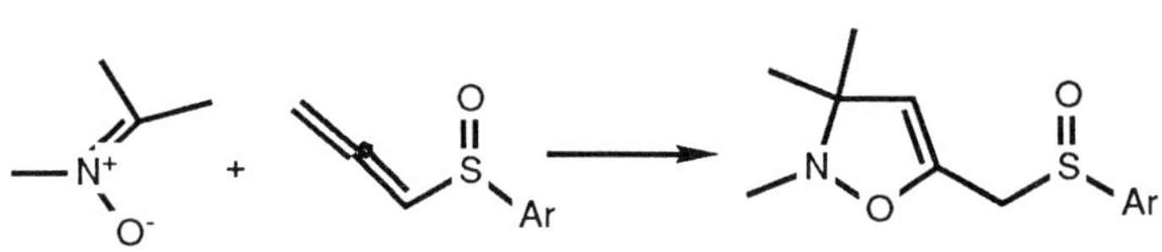

(18) (19) (20)

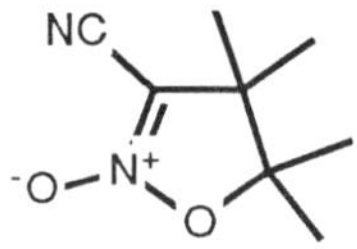
(21)

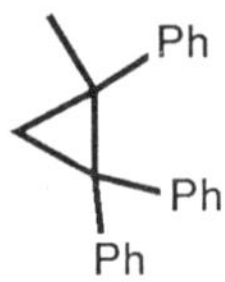
(22)

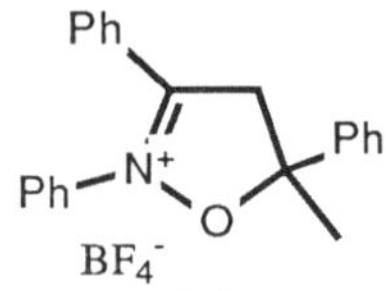
(23)

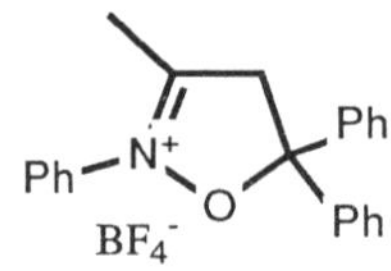
(24)

Isoxazolines

A mixture of products (6) - (8) is formed when ethyl 4-nitro-3-phenylisoxazole-5-carboxylate is treated with an excess of diazomethane (89JOC706). The stereochemistry of the isoxazolines formed by the 1,3-dipolar cycloaddition of nitrile oxides to chiral enitols and pentenolides has been reviewed (88MI58). Aldoximes react with a variety of olefins in the presence of Chloramine T to give 2-isoxazolines via transient nitrile oxides (89S57). α-Bromo-O-trimethylsilyl)oximes react with allyl alcohol in the presence of cesium fluoride to yield ethers (9; R^1 = Me or Ph, R^2 = H or Me), which on treatment with sodium hypochlorite afford the intramolecular cycloadducts (10) (88TL469, 89JOC5277). Other intramolecular nitrile oxide cycloadditions are the formation of the tricyclic isoxazoline (11) from o-allyl-β-nitrostyrene and t-butyl isocyanide (89JCS (P1) 979) and the reaction of 1-nitrohepta-1,6-diene with allyltrimethylstannane in the presence of a Lewis acid to afford compound (12) (89JCS(P1)289). The chiral allyl ether (13) adds benzonitrile oxide to give mainly the isomer (14) (88T4645). Whereas both cis- and trans-3-hexene-2,5-dione react with nitrile oxides to form the trans-isoxazolines (15), 3,4-diacetyl-3-hexene-2,5-dione yields the spiro compounds (17) by initial addition to the enol (16) (88JHC1621). A tandem cycloaddition - [2,3] sigmatropic rearrangement in the reaction of the nitrone (18) with the allenyl sulfoxide (19; Ar = 2,4-$(O_2N)_2C_6H_3$) leads to the isoxazoline (20) (89TL663). The isoxazoline N-oxide(21) results from tetramethylethylene and dibromonitroacetonitrile (88H2059). A mixture of rearranged isoxazolinium salts (23) and (24) is formed by the action of nitrosonium tetrafluoroborate on the cyclopropane (22) (89CL457).

The photochemistry of isoxazolines has been reviewed (88MI12). Racemic 5-acetylisoxazolines (25; R = Me, Ph, COOEt, etc.) are reduced by baker's yeast to mixtures of (26) and (27) in good yield and high enantiomeric excess (88TL6167). The isoxazoline (28) is cleaved by m-chloroperbenzoic acid to the hydroxy ketone PhCH(OH)CH_2COMe (88TL6703). Flash-vacuum pyrolysis of the spiro isoxazoline (29) gives the fused heterocycle (30) (89JCS(P1)1253).

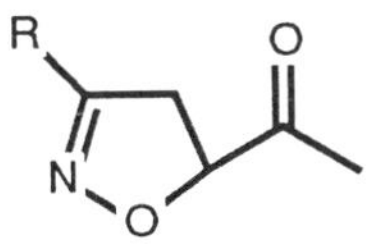

(25)

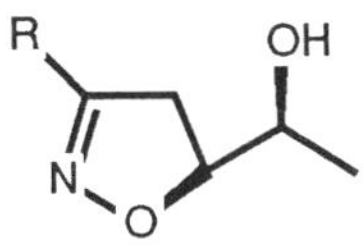

(26)

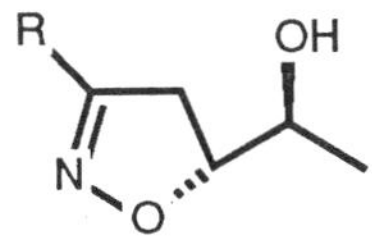

(27)

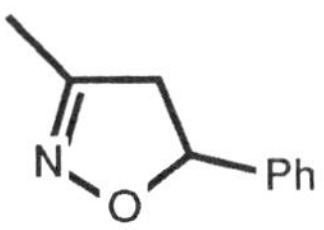

(28)

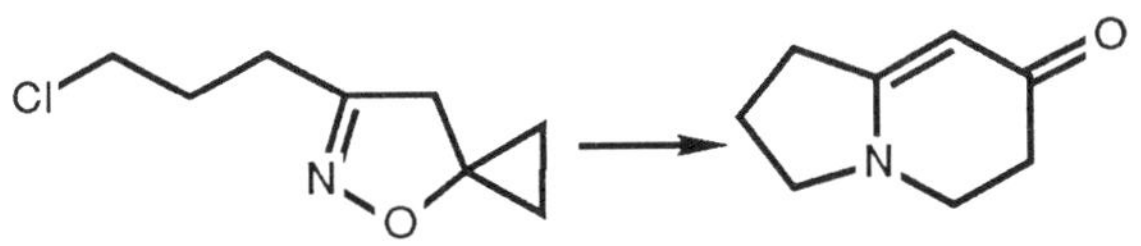

(29) **(30)**

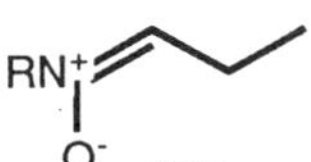

(31)

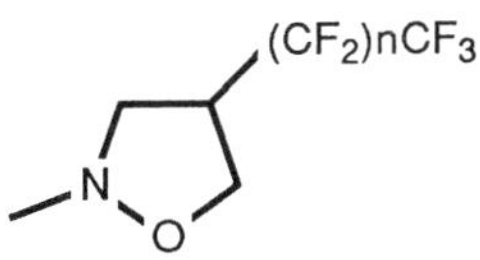

(32)

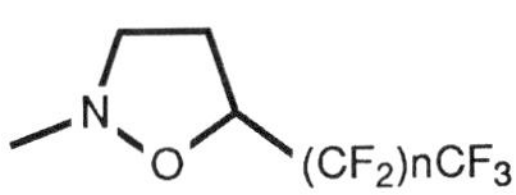

(33)

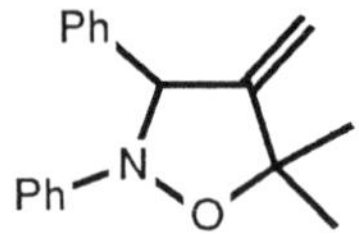

(34)

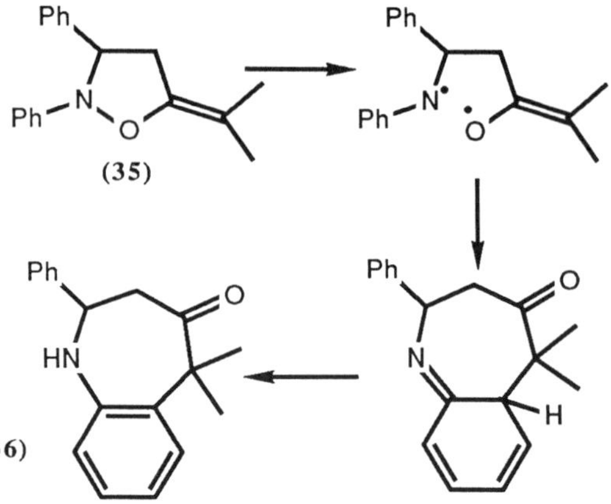

(35) **(36)**

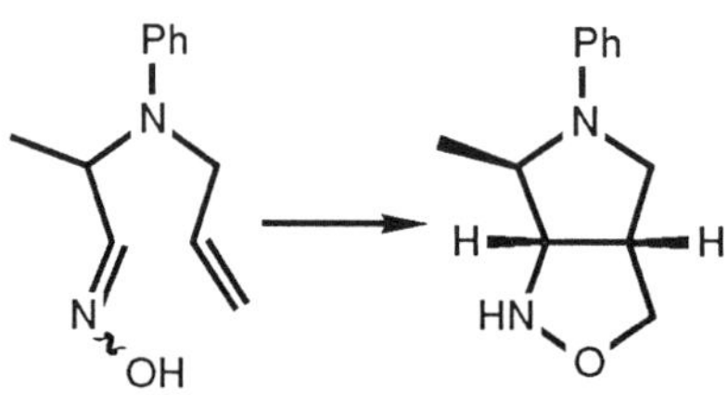

(37) **(38)**

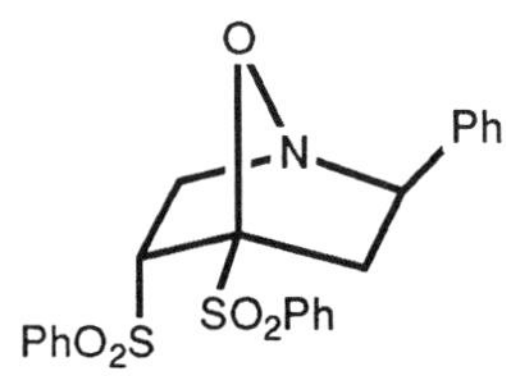

(39)

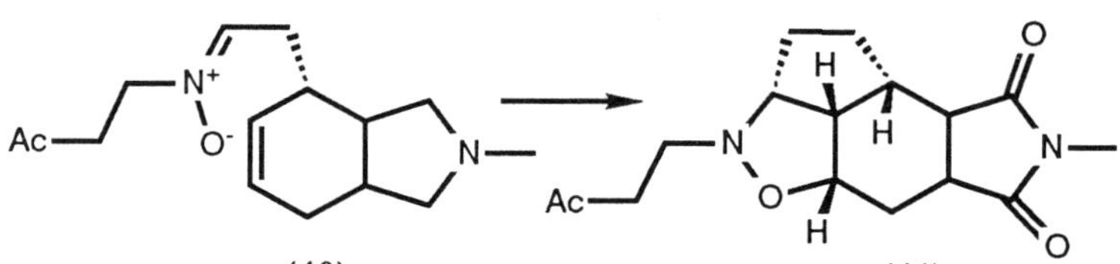

(40) **(41)**

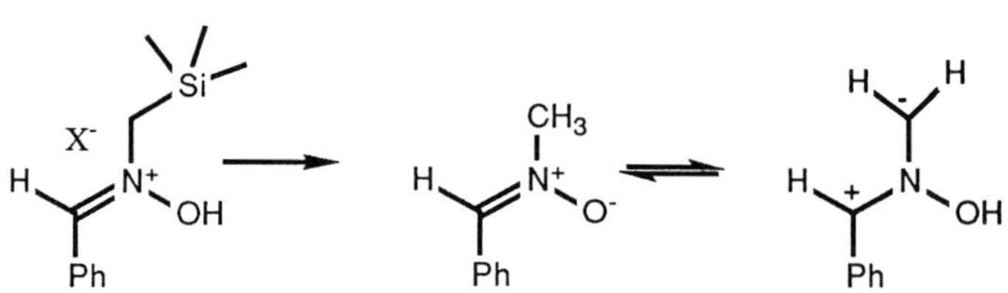

(42) **(43)** **(44)**

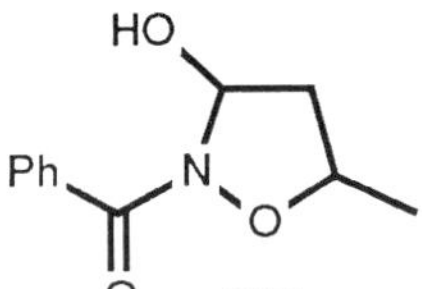

(45)

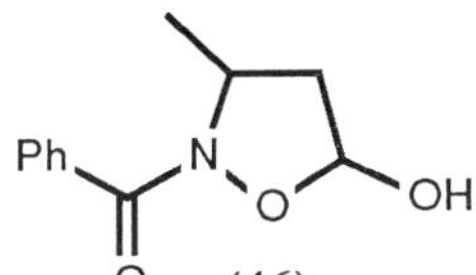

(46)

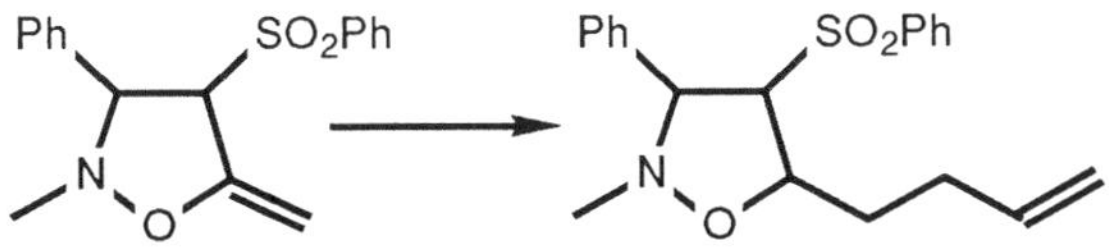

(47) (48)

Isoxazolidines

Ultrasound accelerates the cycloaddition of nitrones to styrene (88CC1445). The regioselectivity of 1,3-cycloaddition reactions of the nitrones (31) is reversed on changing the group R from methyl to t-butyl (88TL307). N-Methylnitrone, generated from formaldehyde and N-methylhydroxylamine, adds to the fluoro-olefins $H_2C{=}CH(CF_2)_n CF_3$ (n = 3, 5 or 7) to yield mixtures of isoxazolidines (32) and 33), in which the latter predominate (88MI63). The reaction of C,N-diphenylnitrone with the allene $Me_2C{=}C{=}CH_2$ gives the cycloadduct (34), together with the benzazepinone (36), which is thought to arise from the isomeric adduct (35) by way of a biradical as shown (89JOC2862). Heating the oxime (37) yields the pyrrolidinoisoxazolidine (38) with the stereospecific introduction of three chiral centres (88TL5313). The intramolecular cycloaddition of a transient nitrone, formed from benzaldoxime and 2,3-di(phenylsulfonyl)-buta-1,3-diene, leads to the bridged isoxazolidine (39) (88TL2417). Another intramolecular nitrone cycloaddition is the formation of compound (41) from the precursor (40)(89TL609). Desilylation of the triflate (42; X = CF_3SO_3) with cesium fluoride gives both the nitrone (43) and the tautomeric azomethine ylid (44); these are trapped by reactive olefins to yield isoxazolidines and N-hydroxypyrrolidines respectively (89JOC4430). Crotonaldehyde and benzhydroxamic acid, adsorbed on silica gel yield the isoxazolidine (45), whereas adsorption on triethylaminoethylcellulose leads to the isomer (46) (89TL117).

The methyleneisoxazolidine (47) is attacked by allyl bromide at the exocyclic carbon atom to give the isoxazoline (48) (88JCS(P1)2639).

OXAZOLES AND BENZOXAZOLES

A new general synthesis of oxazoles is exemplified by the oxidation of deoxybenzoin by iron (II) perchlorate in acetonitrile to afford 2-methyl-4,5-diphenyloxazole (49) (89CPB606). Ethyl 2-diazo-3,3,3-trifluoropropanoate reacts with aliphatic or aromatic nitriles RCN in the presence of rhodium (II) acetate to yield 5-ethoxyoxazoles (50) (89CC60). In a new synthesis of 4-amino-oxazoles (51), α-oxonitriles R^1COCN are treated with aldehydes R^2CHO and ammonium acetate in acetic acid (88JHC1413, 89OPP141). The action of triphenylphosphine on the enol acetate of ω-azidoacetophenone, $AcOCPh{=}CHN_3$, affords the oxazole (52) by a Staudinger reaction, followed by an intramolecular aza-Wittig reaction (89JOC431). In the presence of boron trifluoride, diazoacetophenone and chloroacetonitrile yield 2-chloromethyl-5-phenyloxazole (53) (89BCJ618). Cyclodesulfurization of o-hydroxythioanilides (54) with potassium superoxide produces benzoxazoles (55) in high yield (89H213).

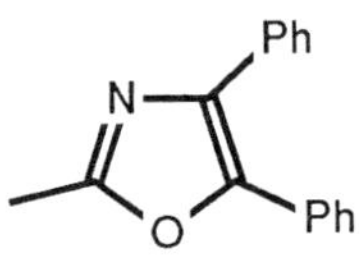

(49)

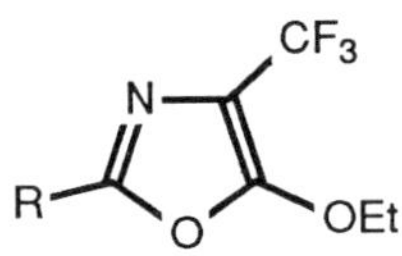

(50)

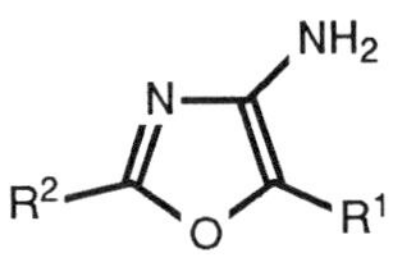

(51)

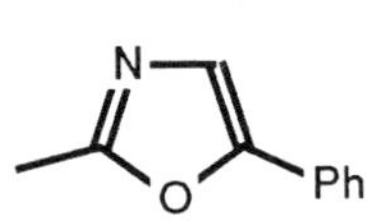

(52)

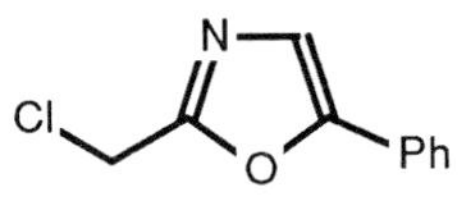

(53)

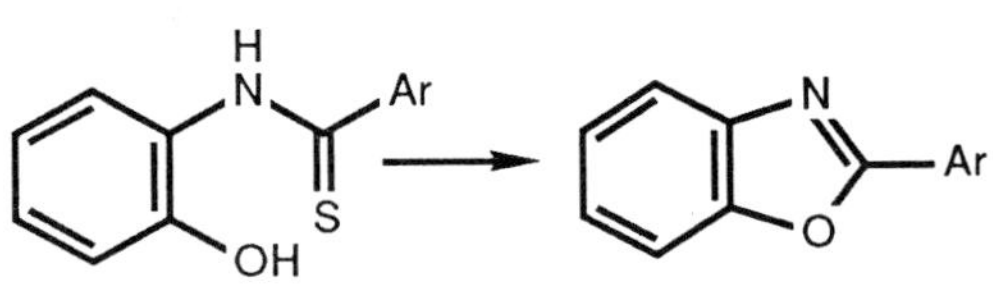

(54) **(55)**

Simple oxazoles form isolable endo-peroxides (56; R^1, R^2, R^3 = H, Me, or Ph) with singlet oxygen (88TL1003). Lithiation of 4-methyloxazole gives the 2-lithio derivative, which reacts with numerous aryl and heteroaryl halides RX under tetrakis(triphenylphosphine)palladium catalysis to yield the coupling products (57)(87S693). The action of 4-phenyl-1,2,4-triazoline-3,5-dione on various oxazoles leads to anomalous adducts, which arise by fission of the oxazole ring; thus 2,5-diphenyloxazole gives compound (58) (88CL1551). Acetylation-desilylation of 5-trimethylsilyloxy)-2-phenyloxazole (59) generates the N-acetylmünchnone (60), which is in equilibrium with the benzoylmünchnone (62) via the valence tautomer (61); in the presence of dimethyl acetylenedicarboxylate, the isomeric pyrroles (63) and (64) (E = COOMe) are isolated (88TL2027).

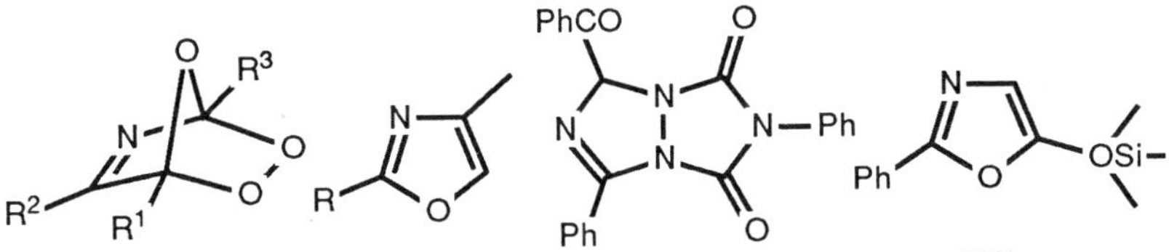

(56) (57) (58) (59)

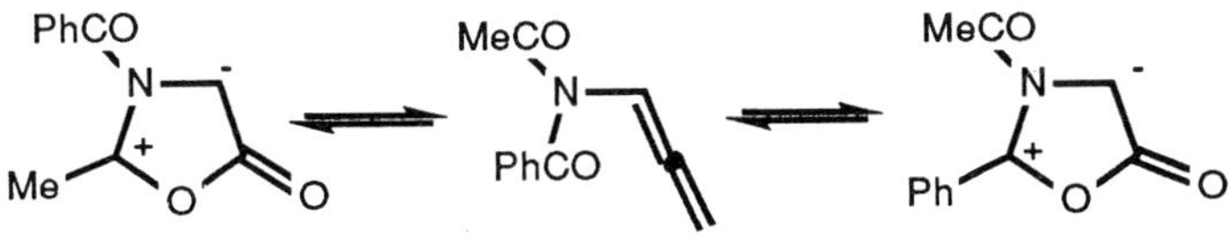

(62) (61) (60)

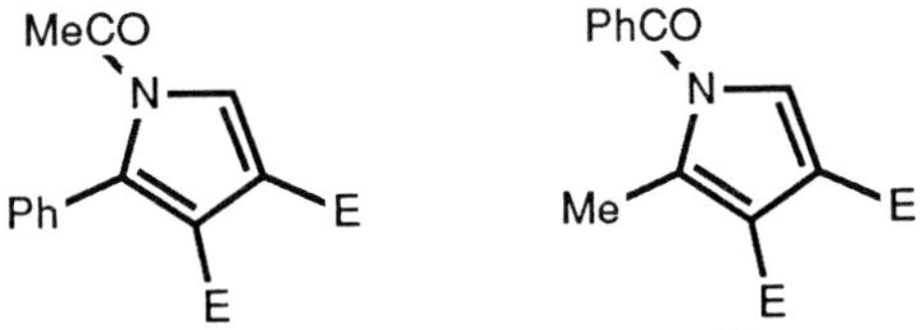

(63) (64)

Oxazolines

Silicon tetrafluoride promotes the reaction of phenyloxirane with acetonitrile to yield 2-methyl-4-phenyl-2-oxazoline (65) (88H2527). Alkenes react with acetonitrile and phenyltellurium trifluoroacetate in the presence of boron trifluoride etherate to give 2-methyl-2-oxazolines; thus with cyclopentene, compound (66) is obtained (88TL1049). An asymmetric synthesis of 2-oxazolines (67) which proceeds with up to 98.6% enantiomeric excess, is the reaction of aliphatic or aromatic aldehydes RCHO with NN-dimethylisocyanoacetamide in the presence of catalytic amounts of a chiral ferrocenylphosphine - gold complex (88TL6321).

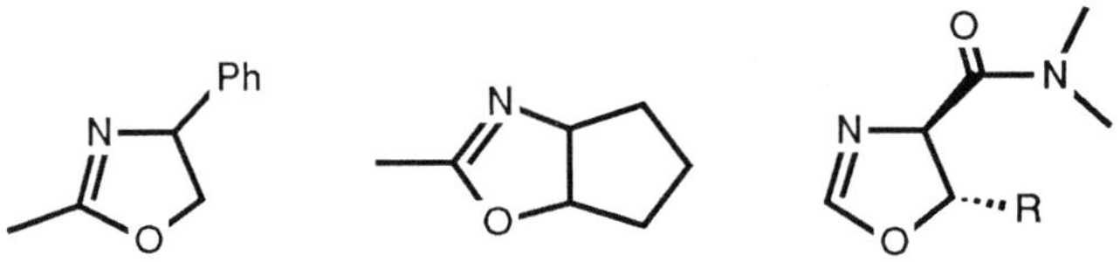

(65) (66) (67)

Iodolactonization of the γ,δ-unsaturated alkyl-oxazoline(68) affords the pyrrolidinone (69) (89JOC4411). Deprotonation of the optically active oxazolinium toluene-p-sulfonate (70) with lithium isopropoxide, followed by a thermal aza-Claisen rearrangement, gave (71), accompanied by a little of the epimer (72) (88S362). Optically active 2-oxazolines, e.g. (73), have been used as co-catalysts with $[Rh(COD)Cl]_2$ for the enantioselective hydrosilylation of prochiral ketones with diphenylsilane; thus acetophenone gave, after hydrolysis, phenylethanol in 83.4% ee (89CB499). Thermolysis of the benzylthio-oxazoline (74) gives carbon dioxide and the nitrile ylid (75), which can be trapped by suitable dipolarophiles (88HCA1177).

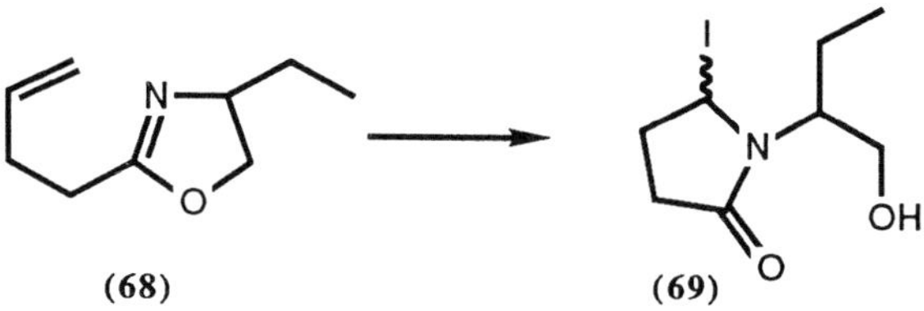

(68) (69)

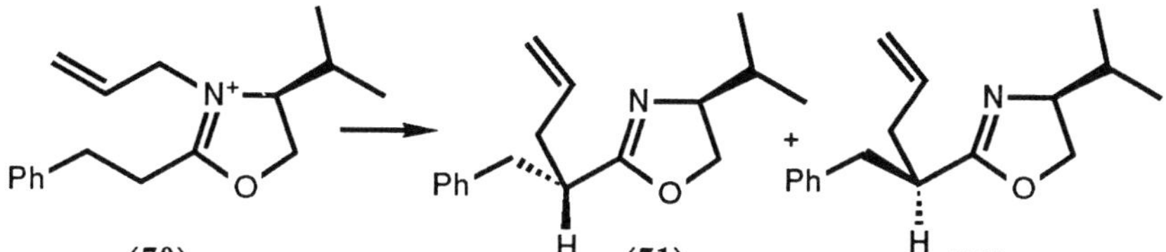

(70) (71) (72)

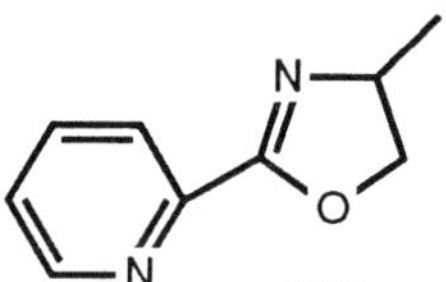

(73)

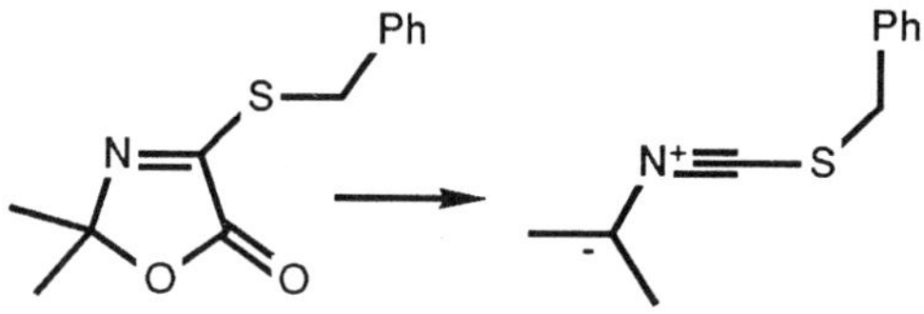

(74) (75)

Oxazolidines

The reaction of sarcosine with aromatic aldehydes results in the oxazolidines (76) (88T519). The optically active alcohol (R)-(77) condenses with 2-(hydroxymethylene)cyclohexanone to give the oxazolidine (78) which can be used as a chiral template. Thus it can be transformed successively into the alcohol (79) and then via the dithiane (80) into the chiral hydroxy aldehyde (81) (89AG(E)62). The carbonyl epoxide rearrangement can be extended to δ,ε-epoxyimines, which undergo intramolecular nucleophilic displacement by attack of the imino nitrogen atom on the oxirane ring. For example treatment of the ketone (82) with ammonia gave the oxa-azabicyclo [3.2.1] octane (83) (88TL4973). The silver-catalysed cyclization of the O-acetylenic isourea (84) results in the imino-oxazolidine (85) (88CC1175). Heating the optically active pyrrolidino-oxazolidine (86) with toluene-p-sulfonic acid generates the acyliminium zwitterion (87), which cyclises to the optically active tricyclic product (88) as shown (89JA1905).

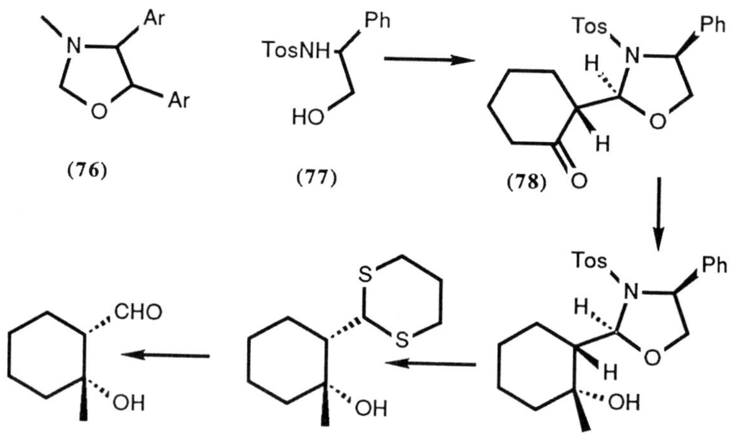

(76) (77) (78) (79) (80) (81)

(82) (83)

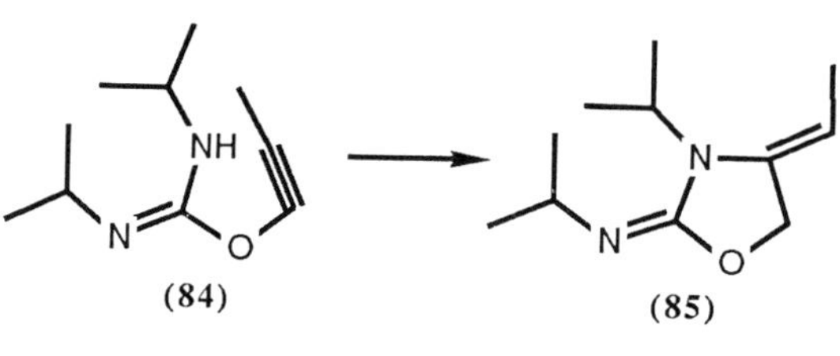

(84) (85)

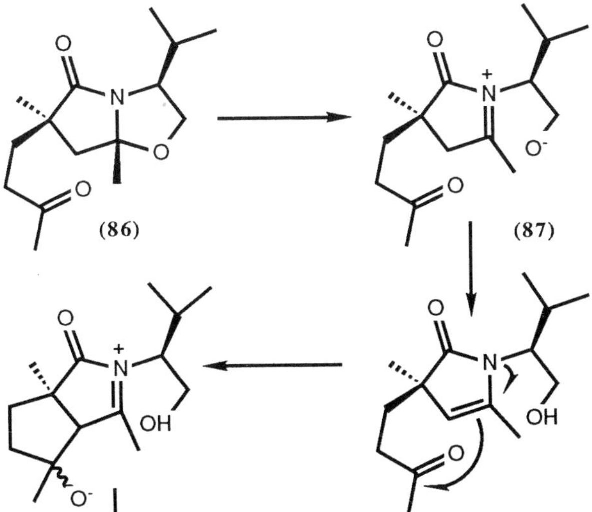

(86) **(87)**

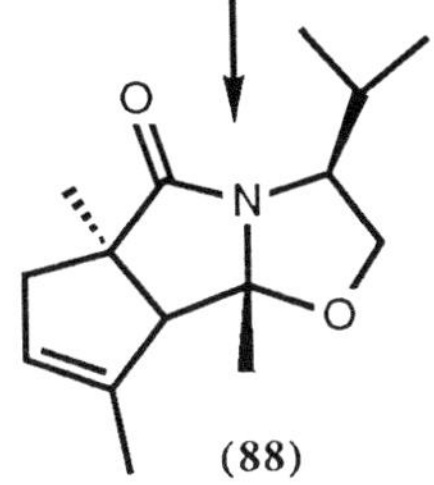

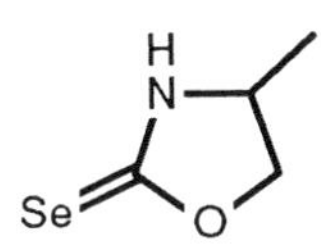

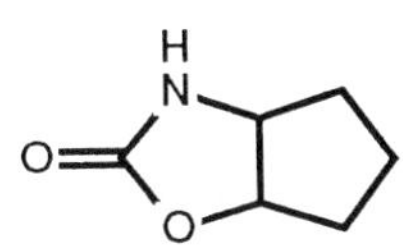

(88) **(89)** **(90)**

Oxazolidin-2-ones are formed from oxiranes and isocyanates in the presence of tributylphosphine oxide - lithium bromide (88JHC1431) or tetraphenylstibonium iodide (88JHC1351); with p-tolyl isothiocyanate, oxazolidine-2-thiones are obtained (88JHC 1435). Mercury (II) chloride promoted the reaction of (S)-alaninol with carbon diselenide to give the optically active oxazolidine-2-selenone (89) (89H269). The action of ethyl carbamate and PhTe (O) $OOCCF_3$ on cyclopentene under boron trifluoride etherate catalysis leads to the cyclopentaoxazolidinone (90) (89JOC4398). N-3-Butenyl-4-(phenylthio)oxazolidinones (91; R = Me, Et, allyl or benzyl) undergo free-radical cyclization in the presence of tributyltin hydride and AIBN to yield the cyclohexanooxazolidinones

(92) with good diastereoselectivity (88H1437); the method has been extended to the tandem cyclization (93)→(94) (88CPB2934). The isomünchnone (96), produced from the diazo compound (95) under rhodium acetate - catalysis, gives the intramolecular cycloadduct (97) with

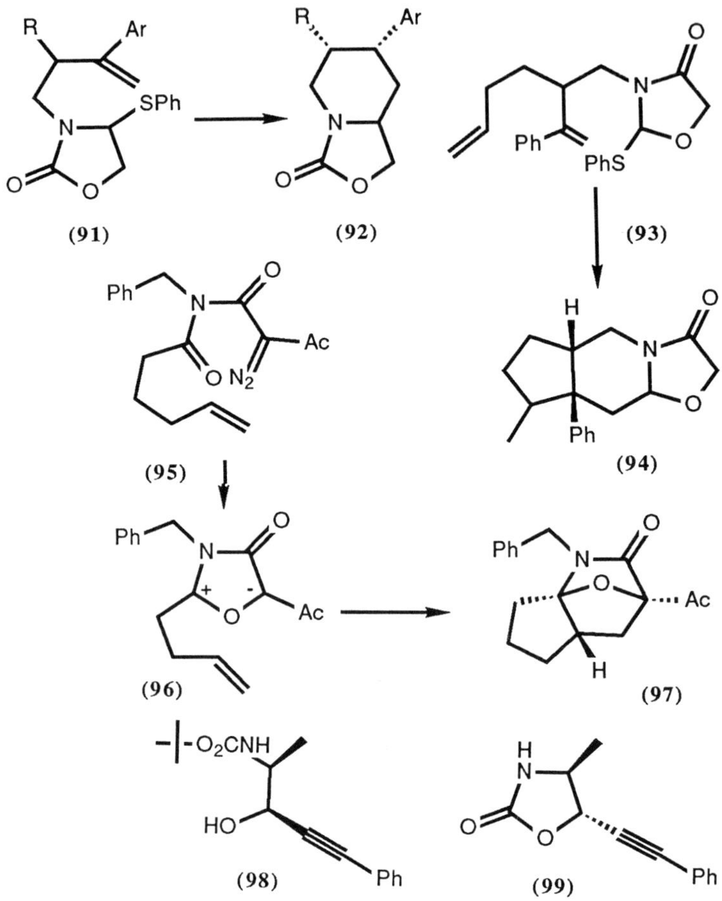

(91) (92) (93) (94) (95) (96) (97) (98) (99)

high stereospecificity (88TL1677). Diastereoconversion of (R,S)-(98) was accomplished by treatment with thionyl chloride, which gave the oxazolidinone (99) with inversion at one chiral centre; reaction with t-butyl chloroformate, followed by hydrolysis, then yielded (S,S)-(98) (88CPB3296).

SYSTEMS CONTAINING THREE HETEROATOMS

1,2,4-Dioxazolidines

Ozonolysis of vinyl ethers generates carbonyl oxides, e.g. (100), which add to aldimines $R^1CH=NR^2$ (R^1, R^2 = alkyl or Ph) to form 1,2,4-dioxazolidines (101) (88CC1550).

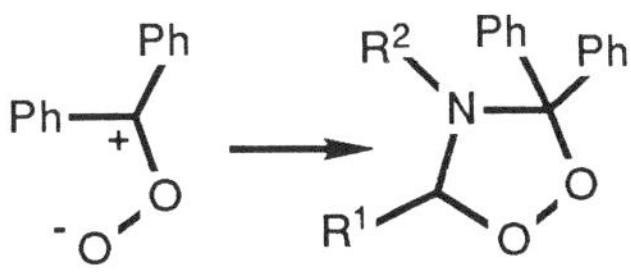

(100) **(101)**

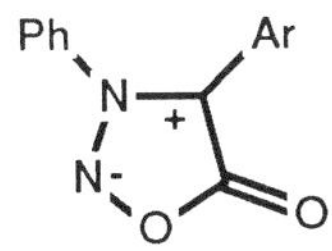

(102)

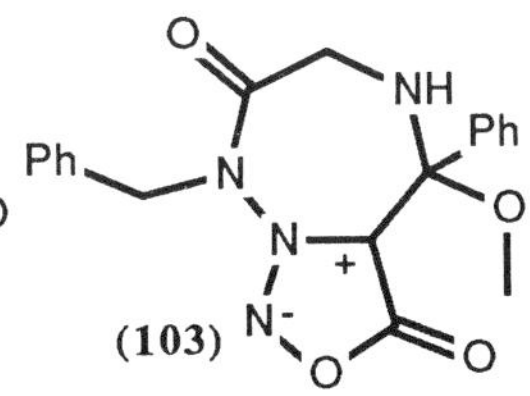

(103)

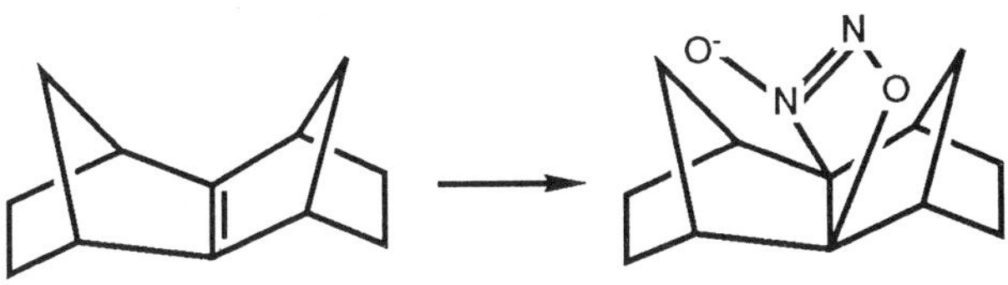

(104) **(105)**

1,2,3-Oxadiazoles

N-Phenylsydnone on sequential treatment with butyllithium and copper (I) bromide yields a stable copper derivative, which couples with aryl iodides in the presence of palladium tetrakis(triphenylphosphine) to afford 4-aryl-3-phnylsydnones (102) (88JOM34). The synthesis of the sydnotriazepine (103) has been reported (89JHC453). The first 4,5-dihydro-1,2,3-oxadiazole 3-oxide, compound (105), has been obtained from syn-sesquinorbornene (104) and nitrosonium hexafluorophosphate (88JOC5930).

1,2,4-Oxadiazoles

3-Amino-1,2,4-oxadiazoles (106) rearrange to 2-amino-1,3,4-oxadiazoles (107) on irradiation (88JCS(P1)1313, 88JHC931). The 1,2,4-oxadiazolin-5-ones (108) are produced from nitrile oxides and chlorocarbonyl isocyanate (88S994). The reaction of C-arylnitrones with aryl isocyanates furnished 1,2,4-oxadiazolidinones (109) rather than 1,3,4-oxadiazolidinones, as was thought previously (87JCS(P1)2593).

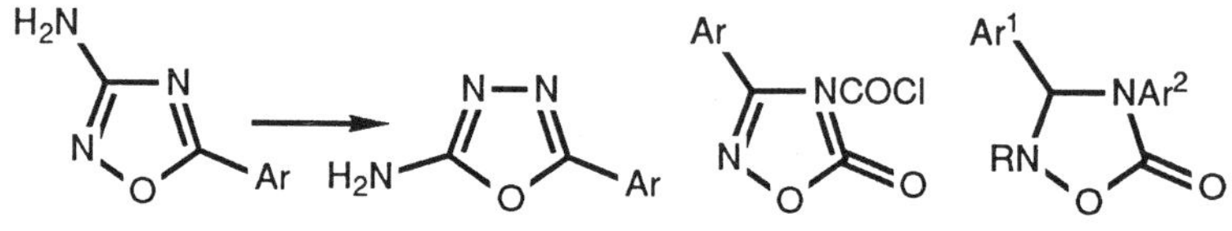

(106) (107) (108) (109)

1,2,5-Oxadiazoles

The oxime (110) undergoes a thermal degenerate rearrangement to (111) as revealed by mass spectrometry (88KGS 1701). Treatment of the azidofurazan (112) with sodium nitrite in acetic acid yields the tetrazole derivative (113) (88KGS1666). Two successive 1,3-dipolar cycloadditions of N-p-nitrophenylmaleimide to the furazan N-oxide (114) lead to the tetracyclic product (115; Ar = p-$O_2NC_6H_4$) (88TL3331). The di(phenylsulfonyl)benzofurazan N-oxide (116) undergoes a complex ring transformation on treatment with sodium azide to give the tetrazoloazepine (117) (88JCS(P1)145).

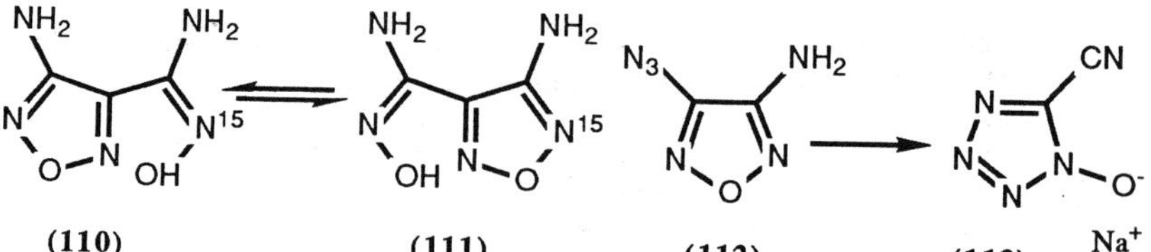

(110) (111) (112) (113)

1,3,4-Oxadiazoles

A derivative (119) of oxadiazoloazepine is formed by the action of perfluorobut-2-yne on 2,5-dimethyl-1,3,4-oxadiazole (118) (89IZV744). Photolysis of the oxadiazoline (120) yields 2-diazopropane and methyl acetate (89JOC1842).

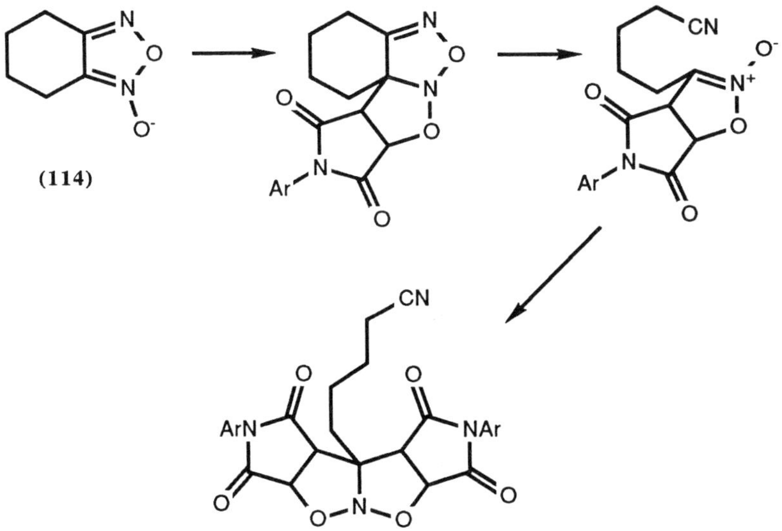

(114) **(115)**

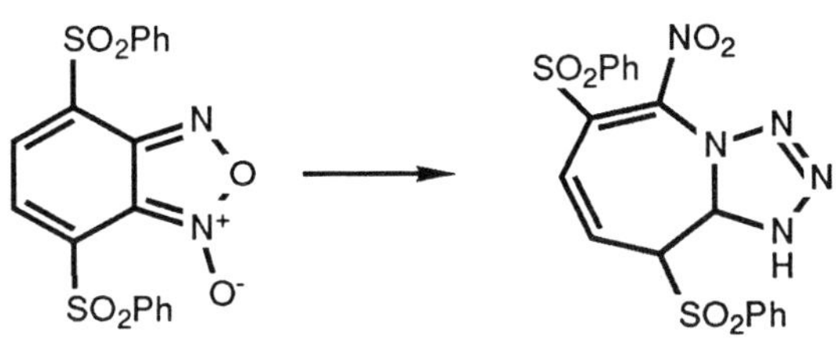

(116) **(117)**

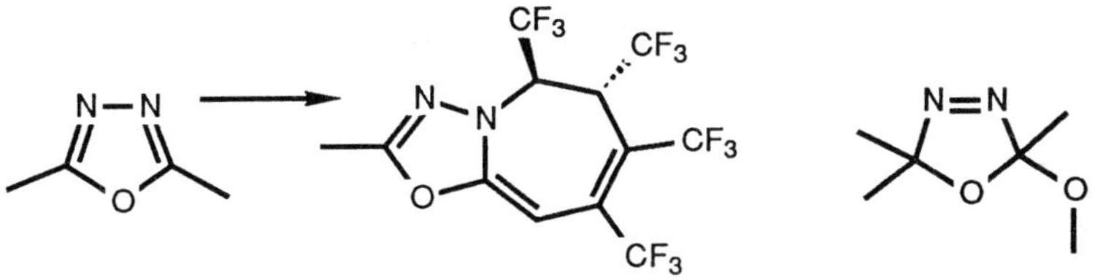

(118) **(119)** **(120)**

REFERENCES

87JCS(P1)2593 A.M.T.Bell, J.Bridges, R.Cross, C.P. Falshaw, B.F.Taylor, G.A.Taylor, I.G. Whittaker and M.J.Begley, J. Chem. Soc., Perkin Trans. 1, 1987, 2593

87S693 A.Dondoni, G.Fantin, M.Fogagnolo, A. Medici and P.Pedrini, Synthesis, 1987, 693

88BCJ2881 S.Zen, K.Harada, H.Nakamuro and Y. Iitaka, Bull. Chem. Soc. Jpn., 1988, 61, 2881

88CL233 N.Ichinose, K.Mizuno, T.Tamai and Y. Otsuji, Chem. Lett., 1988, 233.

88CL1551 T.Ibata, Y.Isogami and H.Tamura, Chem. Lett., 1988, 1551

88CC1175 W.Clegg, S.P. Collingwood, B.T.Golding and S.M.Hodgson, J. Chem. Soc., Chem. Commun., 1988, 1175.

88CC1445 D.R.Borthakur and J.S.Sandhu., J. Chem. Soc., Chem. Commun., 1988, 1444.

88CC1550 M.Mori, M.Nojima, S.Kusabayashi and K. J.McCullough, J. Chem. Soc., Chem. Commun., 1988, 1550.

88CPB2934 S.Kano, Y.Yuasa, T.Yokomatsu, K.Asami and S.Shibuya, Chem. Pharm. Bull., 1988 36, 2934.

88CPB3296 S.Kano, T.Yokomatsu, H.Iwasawa and S. Shibuya, Chem. Pharm. Bull., 1988, 36, 3296.

88H1437 S.Kano, Y.Yuasa, K.Asami and S.Shibuya, Heterocycles, 1988, 27, 1437.

88H2059 J.H.Boyer and T.Manimaran, Heterocycles, 1988, 27, 2059.

88H2527 M.Shimizu and H.Yoshioka, Heterocycles, 1988, 27, 2527

88HCA1177 P.Wipf, R.Prewo, J.H.Bieri, G.Germain and H.Heimgartner, Helv. Chim. Acta, 1988, 71, 1177.

88JCS(P1)145 J.G.Belton and R.S.McElhinney, J. Chem. Soc., Perkin Trans. 1, 1988, 145.

88JCS(P1)1313 S.Buscemi, M.G.Cicero, N.Vivona and T. Caronna, J. Chem. Soc., Perkin Trans 1, 1988, 1313.

88JCS(P1)2639 A.Padwa, S.P.Carter, U.Chiaccio, D.N. Kline and J.Perumattam, J.Chem. Soc., Perkin Trans. 1, 1988, 2639.

88JHC931 S.Buscemi, M.G.Cicero, N.Vivona, J. Heterocycl. Chem., 1988, 25, 931.
88JHC1351 M.Fujiwara, A.Baba and H.Matsuda, J. Heterocycl. Chem., 1988, 25, 1351.
88JHC1413 R.Lakhan and R.L.Singh, J.Heterocycl. Chem. 1988, 25, 1413.
88JHC1431 J.W.McFarland, J.J.Beaulieu, L.N.Arrey and L.M.Frey, J. Heterocycl. Chem., 1988, 25,1431
88JHC1435 J.W.McFarland, J.J.Beaulieu, L.N.Arrey and L.M.Frey, J. Heterocycl. Chem., 1988, 25,1435
88JHC1621 G.Adembri, A.DiTommaso, L.R.Lampariello and M.Scotton, J. Heterocycl. Chem., 1988, 25, 1621.
88JOC5930 S.F.Nelsen, M.F.Teasley and M.Kaftory, J. Org. Chem., 1988, 53, 5930.
88JOM34 V.N.Kalinin and P.M.Se, J. Organomet. Chem. 1988, 352, C34.
88KGS1666 A.M.Churakov, S.L.Ioffe, V.S.Kuzmin, Y.A.Strelenko, Y.T.Struchkov and V.A. Tartakovski, Khim. Geterotsikl. Soedin, 1988,1666
88KGS1701 V.G.Andrianov, V.G.Semenikhina, A.V.Eremeev and A.P.Gaukhman, Khim. Geterotsikl. Soedin. 1988, 1701.
88MI12 L.Fisera, Stud. Org. Chem. (Amsterdam), 1988, 35(Chem. Heterocycl. Compd.), 12.
88MI58 V.Jaeger, R.Franz, W.Schwab, B.Haefele, D.Schroeter, D.Schaefer, W.Huemmer, E. Guntrum and B.Seidel, Stud. Org. Chem. (Amsterdam), 1988, 35(Chem. Heterocycl. Compd.) 58.
88MI63 J.Fayn and A.Cambon, J. Fluorine Chem., 1988, 40, 63.
88S362 M.J.Kurth and E.G.Brown, Synthesis, 1988, 362.
88S994 K.R.Rao, Y.V.D.Nageswar, A.Gangadhar and P.P.Sattur, Synthesis, 1988, 994.
88SC1171 D.Chiarino, M.Napoletano and A.Sala, Synth. Commun., 1988, 18, 1171.
88T519 F.Orsini, F.Pelizzoni, M.Forte, R.Destro and P.Gariboldi, Tetrahedron, 1988, 44, 519.
88T4645 R.Annunziata, M.Cinquini, F.Cozzi and L. Raimondi, Tetrahedron, 1988, 44, 4645.
88TL307 H.G. Aurich, H.P.Kesselheim and M.Schmidt Tetrahedron Lett., 1988, 29, 307

88TL469 A.Padwa, U.Chiacchio, D.C.Dean, A.M. Schoffstall, A.Hassner and K.S.K.Murthy, Tetrahedron Lett., 1988, 29, 469.

88TL1003 K.Gollnick and S.Koegler, Tetrahedron Lett., 1988, 29, 1003.

88TL1049 N.X.Hu, Y.Aso, T.Otsubo and F.Ogura, Tetrahedron Lett., 1988, 29, 1049.

88TL1677 M.E.Maier and K.Evertz, Tetrahedron Lett., 1988, 29, 1677.

88TL2027 R.G.Wilde, Tetrahedron Lett., 1988, 29, 2027.

88TL2417 A.Padwa and B.H.Norman, Tetrahedron Lett., 1988, 29, 2417.

88TL3331 R.N.Butler, D.Cunningham, E.G.Marren and P.McArdle, Tetrahedron Lett., 1988, 29, 3331

88TL4973 H.H.Wasserman, M.Thyes, S.Wolff and V. Rusiecki, Tetrahedron Lett., 1988, 29, 4973.

88TL5313 A.Hassner, R.Maurya and E.Mesko, Tetrahedron Lett., 1988, 29, 5313.

88TL6167 C.Ticozzi and A.Zanarotti, Tetrahedron Lett., 1988, 29, 6167.

88TL6321 Y.Ito, M.Sawamura, M.Kobayashi and T.Hayashi, Tetrahedron Lett., 1988, 29, 6321.

88TL6703 P.U.Park and A.P.Kozikowski, Tetrahedron Lett., 1988, 29, 6703.

89AG(E) I.Hoppe, D.Hoppe, C.Wolff, E.Egert and R.Herbst, Angew. Chem., Int. Ed. Engl. 1989, 28, 62.

89BCJ618 T.Ibata and Y.Isogami, Bull. Chem. Soc. Jpn., 1989, 62, 618.

89CB499 H.Brunner and U.Obermann, Chem. Ber., 1989, 122, 499.

89CC607 G.Shi and Y.Xu, J.Chem. Soc., Chem. Commun., 1989, 607.

89CL457 N.Ichinose, K.Mizuno and Y.Otsuji, Chem. Lett., 1989, 457.

89CPB606 E.Kotani, S.Kobayashi, M.Adachi, T. Tsujioka, K.Nakamura and S.Tobinaga, Chem. Pharm. Bull., 1989, 37, 606.

89H213 Y.H.Kim, Y.I.Kim, H.S.Chang and D.C. Yoon, Heterocycles, 1989, 29, 213.

89H269 A.Kjaer and T.Skrydstrup, Heterocycles, 1989, 28, 269.

89IZV744	V.M.Koshelev, A.N.Chekhlov, N.V.Vasilev, A.F.Gontar and I.V.Martynov, Izv. Akad. Nauk SSSR, Ser. Khim, 1989, 744.
89JA1905	S.Bienz, C.Busacca and A.I.Meyers, J. Am. Chem. Soc., 1989, 111, 1905.
89JCS(P1)289	H.Uno, K.Goto, N.Watanabe and H.Suzuki, J.Chem. Soc., Perkin Trans 1, 1989, 289
89JCS(P1)979	J.Knight and P.J.Parsons, J. Chem. Soc. Perkin Trans. 1, 1989, 979.
89JCS(P1)1253	A.Goti, A.Brandi, G.Danza, A.Guarna, D.Donati and F.DeSarlo, J. Chem. Soc, Perkin Trans 1, 1989, 1253.
89JHC453	W.Fleischhacker and E.Urban, J. Heterocycl. Chem., 1989, 26, 453.
89JOC411	M.J.Kurth and S.H.Bloom, J. Org. Chem., 1989, 54, 411.
89JOC431	H.Takeuchi, S.Yanagida, T.Ozaki, S. Hagiwara and S.Eguchi, J. Org. Chem., 1989, 54, 431.
89JOC706	R.Nesi, D.Giomi, S.Papaleo, S.Bracci and P.Dapporto, J. Org. Chem., 1989, 54, 706.
89JOC1842	M.W.Majchrzak, M.Bekhazi and I.Tse-Sheepy, J. Org. Chem., 1989, 54, 1842.
89JOC2862	A.Padwa, W.H.Bullock, D.N.Kline and J. Perumattam, J.Org.Chem., 1989, 54,2862.
89JOC4398	N.X.Hu, Y.Aso, T.Otsubo and F.Ogura, J. Org. Chem., 1989, 54, 4398.
89JOC4430	A.Padwa, W.H.Dent, A.M.Schoffstall and P.E.Yeske, J.Org.Chem., 1989, 54, 4430.
89JOC5277	A.Hassner, K.S.K.Murthy, A.Padwa, U. Chiacchio, D.C.Dean and A.M.Schoffstall J. Org. Chem., 1989, 54, 5277.
89OPP141	R.Lakham and R.L.Singh, Org. Prep. Proced. Int., 1989, 21, 141.
89S57	A.Hassner and K.M.C.Rai, Synthesis, 1989, 57.
89TL117	I.A.Motorina, L.A.Sviridova, G.A. Golubeva and Y.G.Bundel, Tetrahedron Lett., 1989, 30, 117.
89TL609	G.Donegan, R.Grigg, F.Heaney, S.Surendrakamur and W.J.Warnock, Tetrahedron Lett., 1989, 30, 609.
89TL663	A.Padwa, B.H.Norman and J.Perumattam, Tetrahedron Lett., 1989, 30, 663.

CHAPTER 5.1

Six-Membered Ring Systems: Pyridine and Benzo Derivatives

J G KEAY & A R SHERMAN
Reilly Industries, Inc, Indianapolis, IN, USA

5.1.1 SYNTHESIS

5.1.1.1 Pyridines

Cobalt catalyzed cocyclizations of acetylene and nitriles are well-established, however, the relatively high reaction temperature and pressure pose limitations. Schulz et al. [89TL1229] have developed a photoassisted cyclization catalyzed by cobalt complexes that allows the reaction to take place under mild conditions (15-40°C). In this manner, good yields of 2-substituted pyridines were obtained.

An interesting ring expansion reaction to form 2-phenylpyridines is based on the pyrolysis of dichlorocyclopropanecarboxaldimine derivatives. These compounds thermally rearrange to substituted pyridine derivatives in the presence of tungsten (VI) oxide. In most instances, the non-chlorinated product **1** is the major product obtained [89JOC4275].

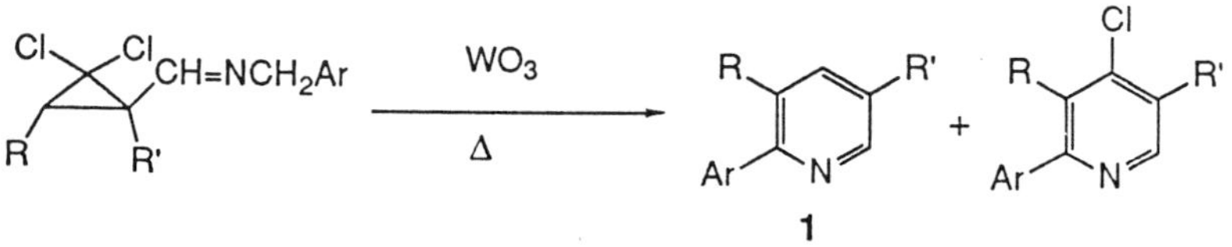

R = H, Ar; R' = Ph, Me, *p*-tolyl; Ar = Ph, *p*-tolyl, *o*-ClC_6H_4

SCHEME 1

The cyclization of δ-oxonitriles to chloropyridines has been explored. The δ-oxonitriles are obtained from α,α-dichloropropionitrile and α,β-unsaturated carbonyl compounds. Under the reaction conditions, the nitriles cyclize to dihydropyridones which are thermally converted to pyridones. Dichlorooxonitriles in DMF in the presence of HCl, cyclize to either pyridines or pyridones, depending upon the reaction time (**Scheme 2**). Longer reaction times lead to a predominance of the pyridone (2:1). However, exclusively pyridines can be obtained by the cyclization in DMF in the presence of 100-120% $POCl_3$ in yields of 70-80% [89CHE428].

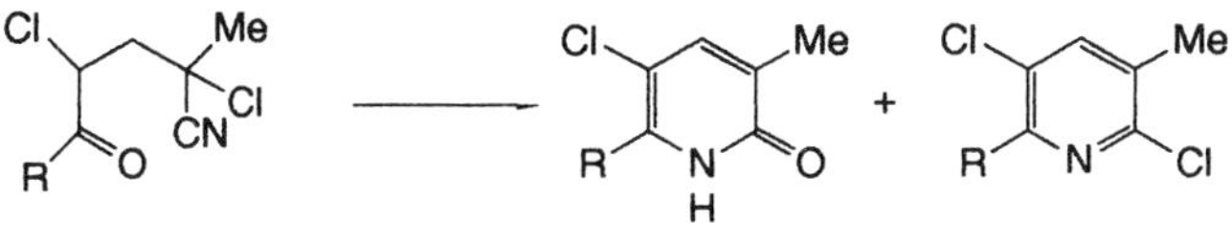

R = H, Me, Et, Pr

SCHEME 2

The inverse electron demand Diels-Alder reaction methodology as a route to pyridine ring construction has been widely studied. H. C. van der Plas and co-workers have been very active in this field, synthesizing pyridine rings from pyrimidines, triazinium salts, and pyrazines and their quaternary salts. Scheme 3 depicts the preparation of annelated pyridines employing this intramolecular reaction. When X = NH, the cycloaddition could not be accomplished [89T6511; 89T6891 and references therein]. Similarly, triazines afforded highly functionalized pyridines upon reaction with enamines and enaminones [89JOC1249].

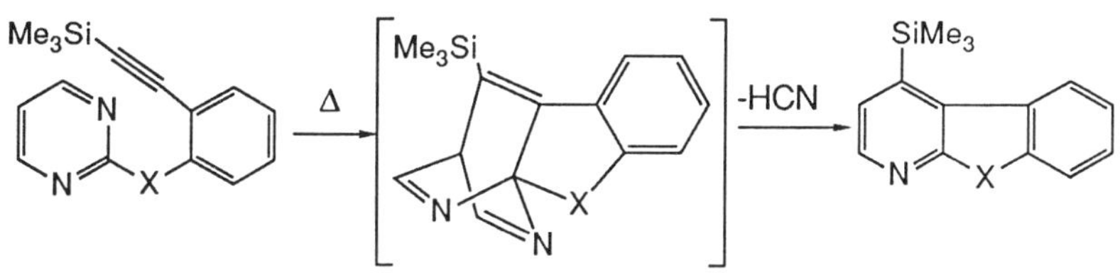

X = O, NAc, S, CH_2, CO

SCHEME 3

Intramolecular amidocarbonylation of alkenamides is a new route to pyridones. 3-Butenamide, in the presence of a rhodium catalyst, afforded three products as shown in Scheme 4. The ratio of **2** and **3** can be altered by varying the phosphine ligands, however, the conditions to selectively prepare **4** have not been found. By employing 20 equivalents of triphenylphosphine, **2** was formed with 92% selectivity. Similarly, with 10 equivalents of triphenylphosphite, **3** was obtained in 90% yield with 94% selectivity [89TL6283].

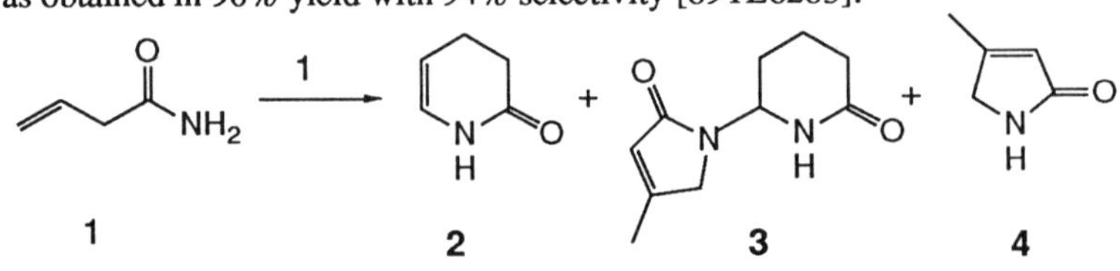

1 2 3 4

1.[Rh] /Phosphorus Ligand , H_2/CO

SCHEME 4

Tetrahydropyridines have been prepared from iminium ion cyclization reactions of α-cyanoamines in the presence of titanium tetrachloride. An interesting aza-Cope reaction, outlined in Scheme 5, occurred with iminium ion **5** that ultimately generated an allene or a tetrahydropyridine derivative, depending upon the reaction conditions [89TL4973].

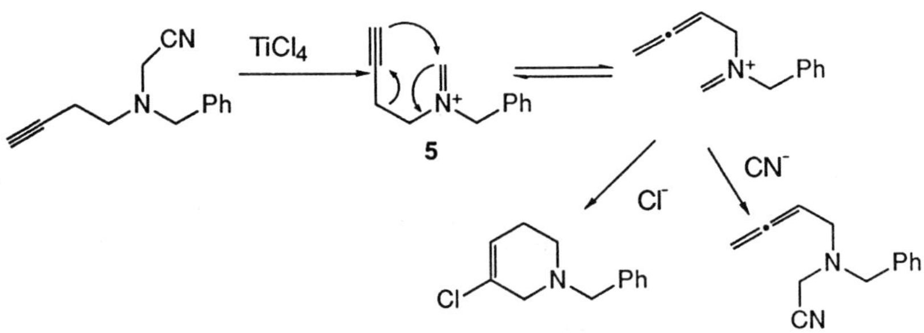

SCHEME 5

Diels-Alder methodology is frequently used to construct the tetrahydropyridine framework. Examples of both inter- and intramolecular reactions have been described previously. The reaction can be carried out with *N*-sulfonyl imines and alkyl-substituted dienes [89TL3037], imines and bis(phenylsulfonyl)substituted dienes [89TL3259], and α,β-unsaturated *N*-benzenesulfonyl imines and 1-aza-1,3-butadienes [89JA1517; 89TL2481]. Scheme 6 summarizes the various Diels-Alder approaches that have been considered for the preparation of tetrahydropyridines.

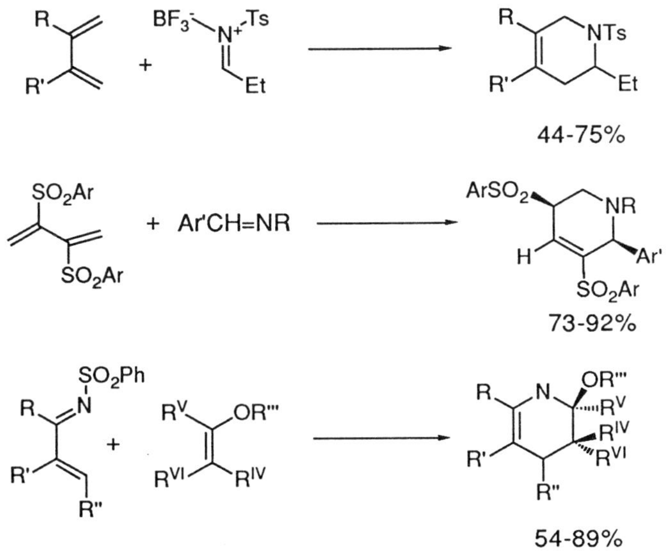

SCHEME 6

The synthesis of piperidines resulted from the reaction of azides with chloroboranes generated in situ by an intramolecular cyclization reaction (**Scheme 7**). The stereochemistry of the reaction can readily be controlled by the use of diastereomeric or enantiomeric pure boronic esters [89CC142].

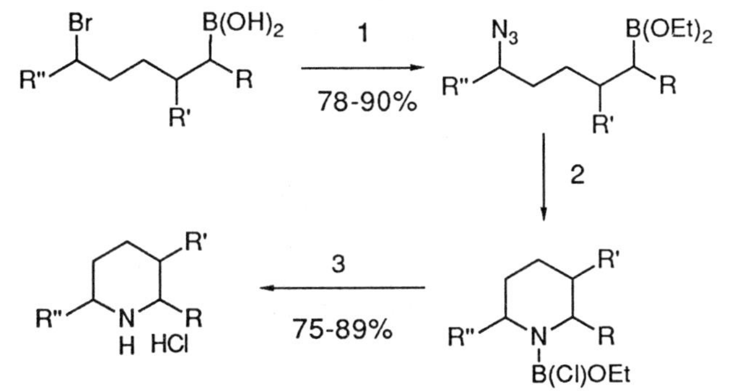

1. NaN_3, EtOH, heat; 2. BCl_3, CH_2Cl_2, -78-25°C; 3. EtOH, r.t.

SCHEME 7

Piperidones have been synthesized by the radical cyclization of allylic haloacetamides (**Scheme 8**). When tin hydride was used as the hydrogen atom donor, 65% **6** and 24% **7** were obtained, however, with germanium hydride, a less reactive hydrogen atom donor, only cyclization to **6** (85%) was found [89H(28)723].

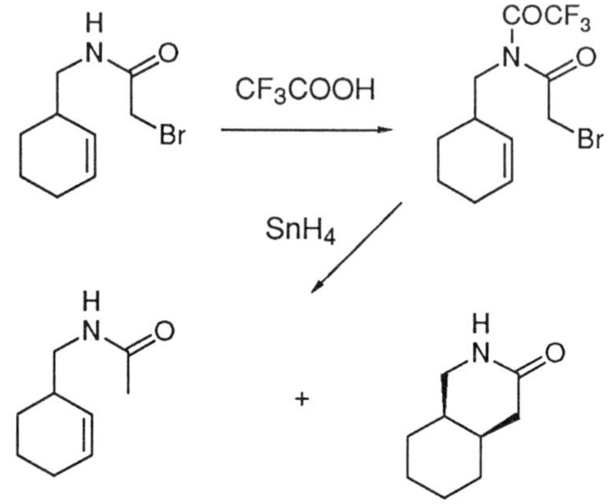

7 **6**

SCHEME 8

5.1.1.2 Quinolines

Reverse electron demand Diels-Alder reactions of 1,2,4-benzotriazines are a route into the quinoline ring system. Upon reaction with norbornadiene, 1,2,4-benzotriazine derivative **8** afforded quinoline derivative **9**. However, treatment with propargyl bromide afforded diquinoline lactone **10**, as shown in Scheme 9. When the substituents were chlorine atoms, the yields of **9** or **10** were 55% or 37%, respectively. Replacing the chlorine atoms with hydrogen atoms gave very poor yields of the corresponding products [89H(29)2253].

N-Substituted trifluoroacetimidoyl chlorides, upon reaction with active methylene compounds such as dimethylmalonate, afforded enamines in good yields. Thermal cyclization of these enamines led to fluorinated quinolines in respectable yields (**Scheme 10**, 66-70%) [89TL4821].

A photochemical cyclization plays an integral part in the formation of diphenylquinoline derivatives from substituted 1-azabuta-1,3-dienes. A series of aryl-*N*-benzoyloxy-2,3-diphenyl-1-azabuta-1,3-dienes underwent a photochemical transformation affording quinoline derivatives (**Scheme 11**). An advantage of this route is the entry to 7-substituted quinoline derivatives, which can be difficult to obtain through other synthetic schemes[89JCS(P1)1623].

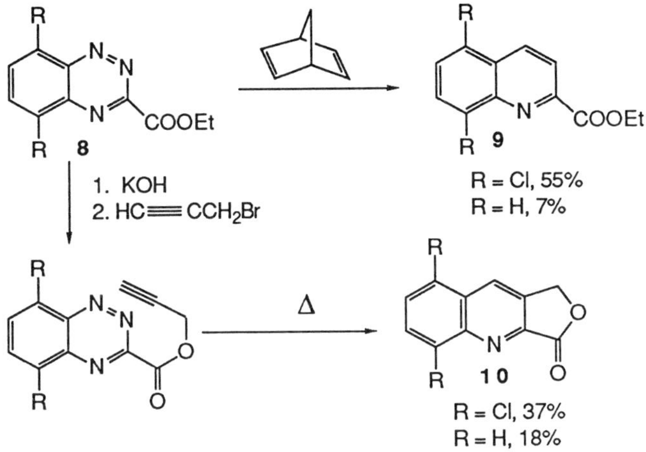

SCHEME 9

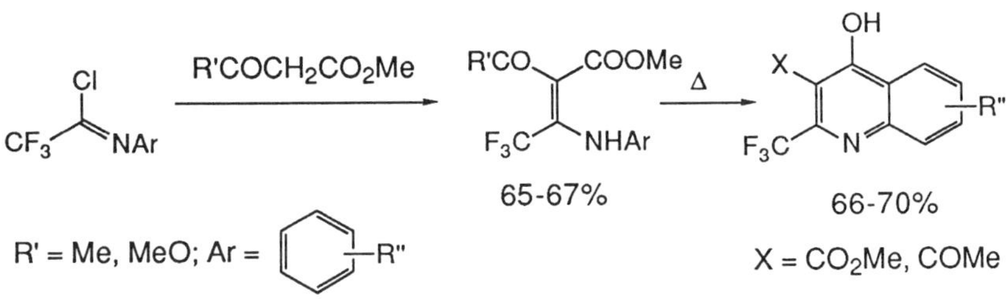

SCHEME 10

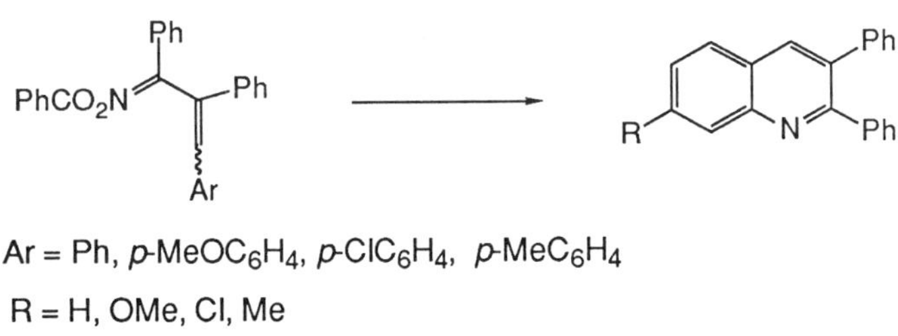

SCHEME 11

5.1.1.3 Isoquinolines

Isoquinolines have been obtained in respectable yields from a thermal cyclization reaction of 1-azahexa-1,3,5-trienes as seen in Scheme 12. When the phenyl ring was unsubstituted, low yields (16-39%) were found [89H(28)275].

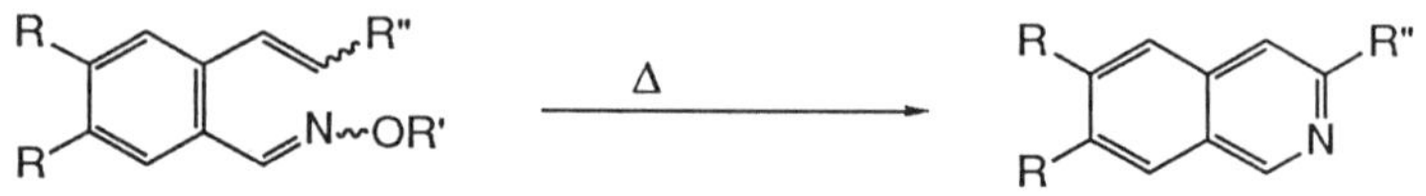

R = OMe, H; R' = H, Me; R" = H, Me, CH_2Ph

SCHEME 12

Benzocyclobutenyl ketoximes underwent a thermolysis reaction to give substituted isoquinolines in yields of 26-79% (**Scheme 13**). The 3,4-substitution pattern obtained in this sequence is not readily obtained by the more conventional methods [89H(28)39].

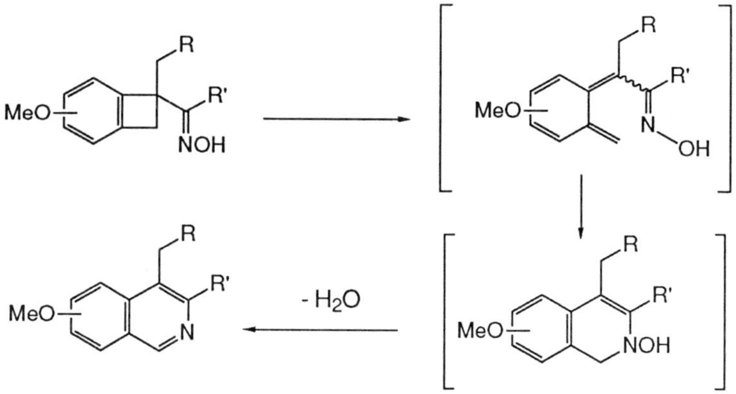

R = H, Me, CH_2Ph; R' = Me, *n*-Bu, Aryl; R,R' = $-(CH_2)_3-$

SCHEME 13

A convenient one-pot preparation of dihydroisoquinolines and β-carbolines, from arylalkylamines and indoline derived alkylamines, has been described. A moderate yield (40-78%) of the products was obtained from the Bischler-Napieralski cyclization of *N*-acyl-2-arylethylamines with triphenylphosphine and carbon tetrachloride (**Scheme 14**). Lower yields of products (20-46%) were obtained from the cyclization of carboxylic acids and amines [89JCR(S)228].

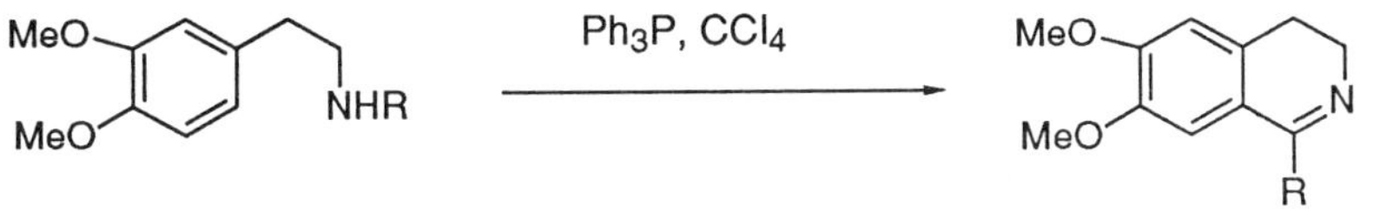

R = MeCO, PhCO, $PhCH_2CO$

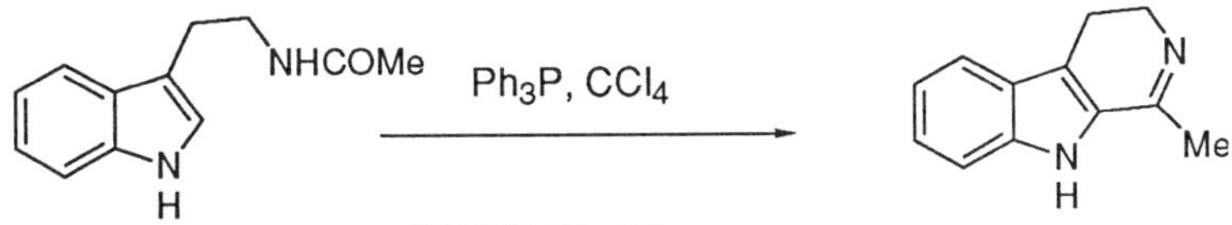

SCHEME 14

5.1.1.4 Acridines and other annelated pyridines

Polycyclic pyridines and isoquinolines can be readily obtained from dienamides in a cyclization/dehydration reaction sequence. Scheme 15 depicts a tricyclic ring system, however, additional ring annelation to other polycyclic ring systems proceeds in good yield [89TL6865].

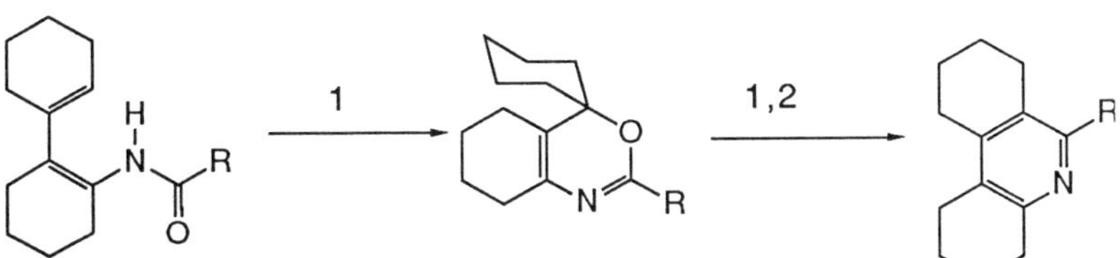

1. $POCl_3$; 2. $NaHCO_3$, H_2O 69-92%

SCHEME 15

Cycloaddition reactions of benzyne adducts and vinyl isocyanates afforded annelated ring systems as shown in Scheme 16 . The benzyne is generated in situ from the lead acetate oxidative decomposition of 1-aminobenzotriazole. Reaction with vinyl isocyanates provides a route to phenanthridinones and substituted derivatives [89JOC4019].

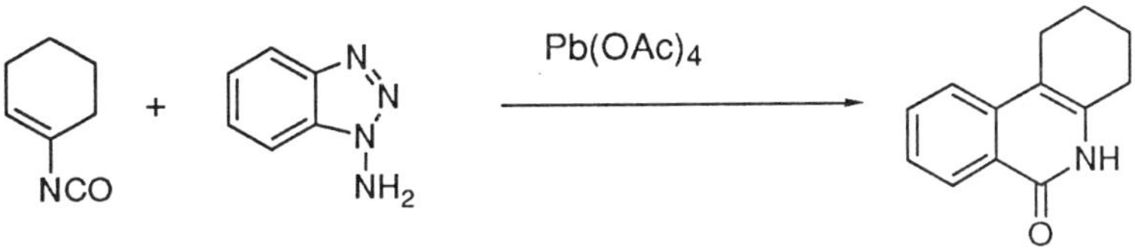

SCHEME 16

5.1.2 REACTIONS

5.1.2.1 Pyridines

Regiospecific (and stereospecific) elaboration of pyridines (and piperidines) continues to be an active area of research. A simple but highly regiospecific procedure for the synthesis of *N*-silyl-1,4-dihydropyridines (**12**) has been developed from pyridines bearing electron withdrawing groups in the 3- (or 3,5-) positions on reaction with silyl ketene acetals (**11**) in the presence of a montmorillonite supported Fe or Al catalyst. Other Lewis acid promoters proved to be less effective. 2- And 4-cyanopyridine, as well as quinoline, did not react whereas 3-cyanopyridine formed the dihydropyridines (**12**) in high yield. Products arising from attack at the 2- or 6-position were not observed. Isoquinoline gave the 1,2-adduct [89TL747].

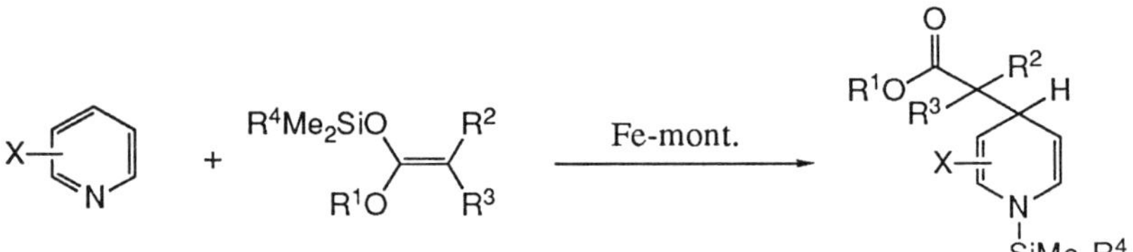

11 **12**

X = 3-CN, COOEt

R^1= Me, Et; R^2,R^3= H, Me, Ph; R^4= Me, *t*-Bu 0-95%

SCHEME 17

Fluorine may be introduced into the 2-position of the pyridine ring by base induced decomposition of certain *N*-fluoropyridinium salts (**13, Scheme 18**). The tetrafluoroborates, hexafluorophosphonates and hexafluoroantimonates worked well, but the pyridine-F_2 complexes gave lower yields (9-47% yield). Interestingly, 3-cyanopyridine gave only the 2-fluoroisomer. Whereas, ethyl nicotinate gave a mixture of isomers (7:1; 2,3-: 2,5-). The strength and amount of base employed were deemed important and the intermediacy of a carbene was proposed [89JOC1726].

Radicals generated from organomercurials [89H(28)143], carboxylic acids [89JHC45, 89TL4569], alkyl iodides [89JOC5224] and perfluoroalkyl peroxides [89JCS(P1)909] can lead to selective substitution of pyridines in good yield (**Scheme 19**). Use of *N*-methoxypyridinium salts (**14**) has enabled 2-hydroxymethylation (and occasionally 4-) to be achieved in moderate to good yields [89SC317]. α-Hydroxylation of the pyridine ring via Rozen's use of acetyl hypofluorite has recently been reviewed and the reaction has been extended to include substitution of quinoline [89H(28)249]. Direct α-hydroxylation has been observed when certain Cu-pyridine complexes were treated with iodosylbenzene and an oxidant [89CC447].

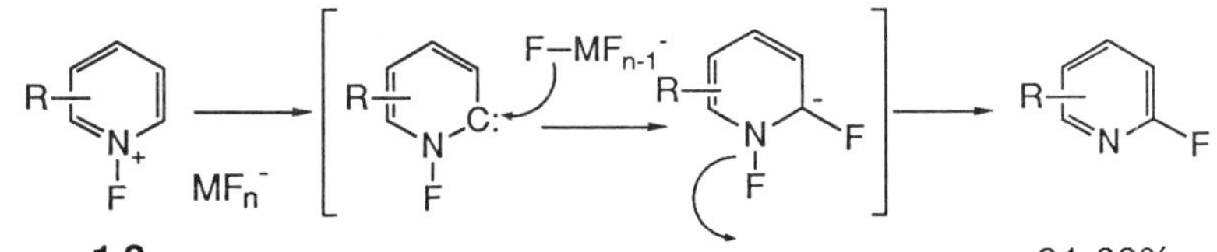

13 21-99%

MF_n^- = BF_4, SbF_6, PF_6

R= alkyl, alkoxy, phenyl, COOEt, Cl, CF_3, CN, NO_2

SCHEME 18

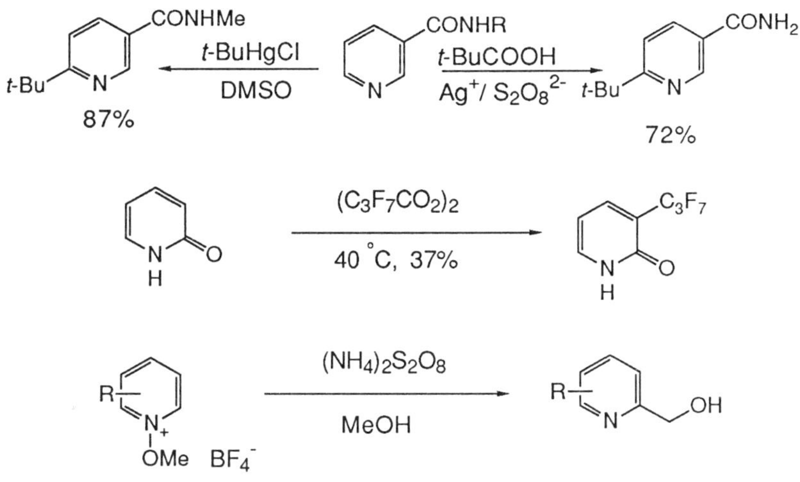

14 R = Cl, CN, OMe 16-71%

SCHEME 19

Nitro- [89JOC5094] and trichloromethylpyridines (**15**) [89JCS(P1)283] undergo vicarious nucleophilic substitution (VNS) with the trihalomethyl and methoxide ion respectively. The pattern of substitution and the product ratio are controlled by the nature and position of the substituent.

The metallation of 2-, 3- and 4-pivaloylaminopyridine derivatives (**16-18**) offers access to 2,3-, 3,4- and 2,3,4-substituted pyridines [89JHC105]. Yields in excess of 80% were reported in some cases. Extension of this work to pyridine carboxylic acids allows for the synthesis of additional 2,3,4-trisubstituted pyridines [89T7469].

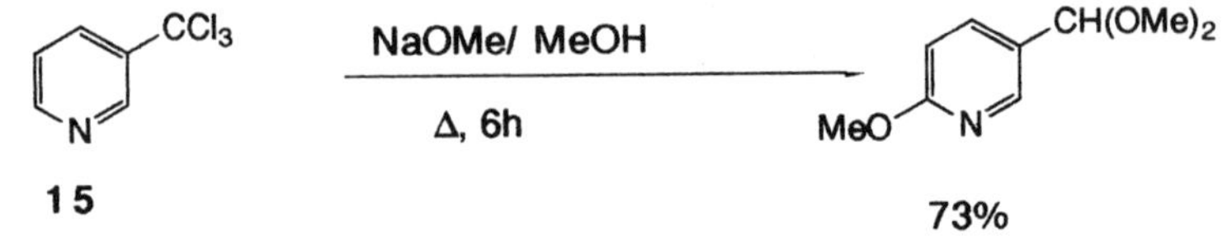

SCHEME 20

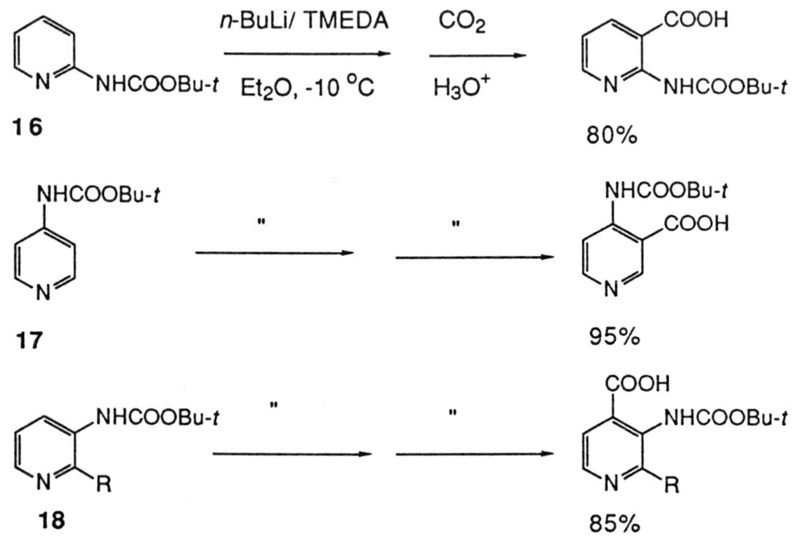

SCHEME 21

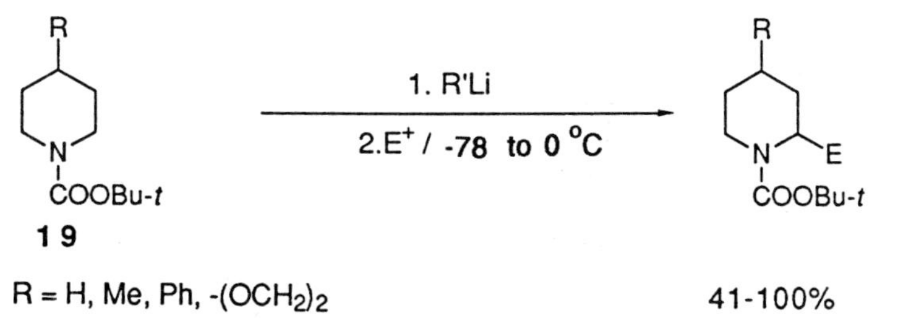

SCHEME 22

The α-functionalization of piperidines (and amines in general) via lithiation of a protected, activated amine is a useful synthetic procedure. The degree of stereocontrol and the procedural ease with which each of the steps is accomplished continues to be studied. Deprotonation of the *t*-BOC protected piperidine (**19**, *s*-BuLi, -78 °C) followed by addition of a variety of electrophiles (carbon, silicon, tin and sulfur) allow for α-functionalization in moderate to quantitative yield. Dipole stabilization, easy addition and removal are the major advantages to the use of the *t*-BOC group [89TL1197].

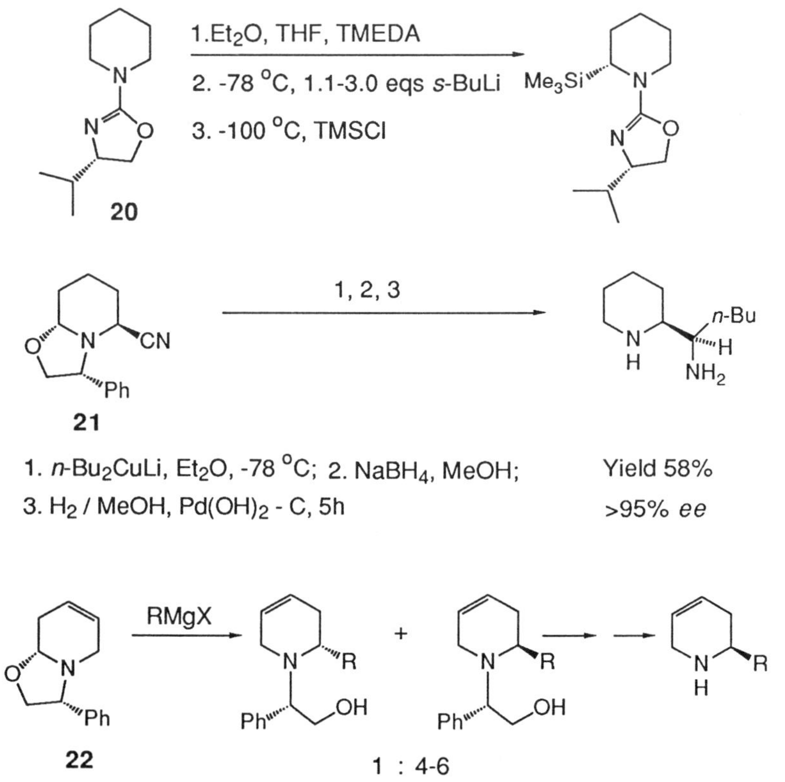

SCHEME 23

The deprotonation of valine derived chiral piperidinooxazolines (**20, Scheme 23**) generates a single organolithium diastereomer which gives 100% stereoselective alkylation with trimethylsilyl chloride [89JOC175]. However, <2% *de* was observed in most other cases and transmetallation with cuprous cyanide caused a loss of stereoselectivity although oxidation products (tetrahydropyridines) were suppressed. Allylic and benzylic systems were also investigated [89JA2211] (see Isoquinolines). With the (3,4-dehydropiperidino)oxazoline, γ-alkylation predominated. Synthetically useful transformations of 2-cyano-6-phenyloxazolopiperidine (**21**) have been demonstrated. Reaction of the cyano group with organocuprate and lithium reagents (as opposed to nucleophilic displacement or α-cyano carbon functionalization) leads to highly stereospecific formation of 1,2-diamines. Selectivity is explained in terms of a preferred facial attack by borohydride [89TL5137]. The oxazolidine (**22**) provides ready access to enantiomeric 2-alkyl-1,2,3,6-tetrahydropyridines (both *R*- and *S*-). Addition of the Grignard reagent occurs selectively and is controlled by the nature of the chiral auxiliary (phenyl glycinol) employed, *ee*'s of up to 75% were obtained [89CC1185].

Use of 1*S*, 2*R*-(-) *N*,*N*-dibutyrylnorephedrine (DBNE) as the controlling chiral alcohol allowed for the enantioselective addition of dialkylzincs to 3-pyridinecarboxaldehyde (**23**, **Scheme 24**). The view was advanced that the use of 1*R*, 2*S*-(+)-DBNE may well give predominantly the other isomer. The molar amount of DBNE used was important [89H(29)2065].

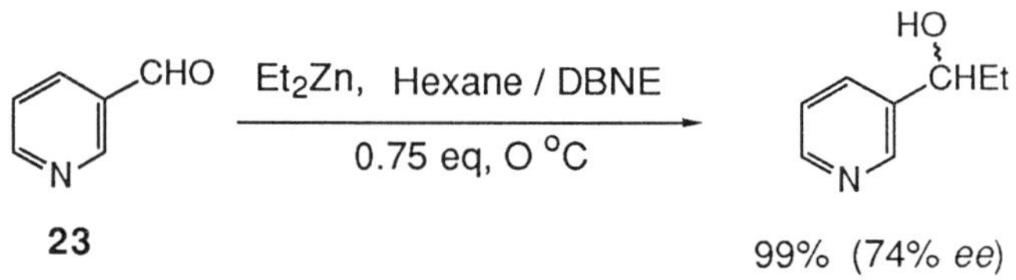

SCHEME 24

The deoxygenation of pyridine-*N*-oxides by $TiCl_4/SnCl_2$ has recently been expanded and shown to be effective in the presence of sensitive functionalities [89SC897]. Catalytic hydrogen transfer, using ammonium formate and Pd, similarly provides for a rapid reduction of *N*-oxides to the parent heterocycle in high yield (81-98%, [89S645]. 2,6-Disubstituted pyridine-*N*-oxides (**24**, **Scheme 25**) provide a stable oxygen source for the synthesis of epoxides from olefins catalyzed by ruthenium porphyrins. The advantage is in the lower activity of the oxidant leading to reduced byproduct formation [89TL6545].

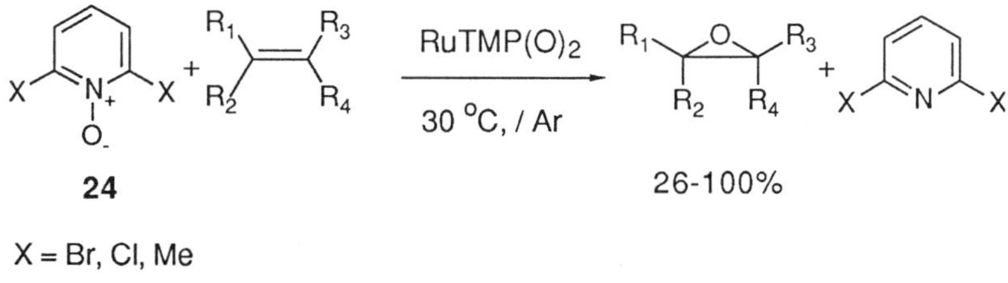

SCHEME 25

Polyvinylpyridines (PVP) are useful supports for oxidation and halogenation reagents. Osmium tetroxide on PVP catalyzes the hydroxylation of olefins [89S47] and PVP permanganate selectively oxidizes alcohols to carbonyl compounds [89MI1]. PVP Dichloroiodate provides a good source of electrophilic iodine useful for functionalization of a variety of aromatic substrates (68-90%) [89JCS(P1)2279].

Mercury photosensitized dehydrodimerization has been shown to functionalize piperidine at the 2-position (**Scheme 26**). The dimerization of piperidine occurs in high yield, piperidine with cyclic hydrocarbons and ethers give analogous products [89H(28)121].

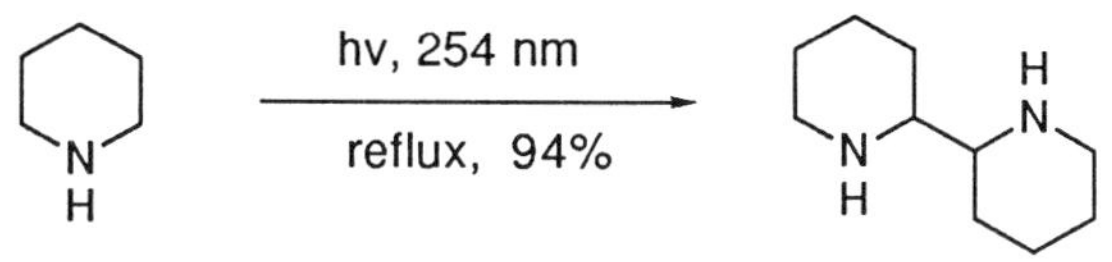

SCHEME 26

Immobilized *N*-alkylaminopyridinium salts have proved useful reagents for the conversion of activated chlorobenzenes into the fluoro derivatives (sulfolane, 189-220°C, 68-98%). The catalyst was recycled seven times without an apparent decrease in activity, higher temperatures, however, did lead to some deactivation of the catalyst [89TL7199].

5.1.2.2 Quinolines

During attempted oximation of some alkyl quinoline-*N*-oxides (**25**) with isopropyl nitrite the major product proved to be the 2-aminoquinoline-*N*-oxides (**26**), along with smaller amounts of the anticipated oxime and 2,2'-biquinoline-1,1'-dioxide. This was extended to quinoline-*N*-oxide itself, giving a respectable yield of 2-aminoquinoline-*N*-oxide. Isopropyl nitrite was found to be superior to $KMnO_4$ and $NaNO_2$ as an oxidant in this application of the Chichibabin reaction [89H(29)1781]. No amination products were observed with the 4-alkyl quinolines under the same conditions.

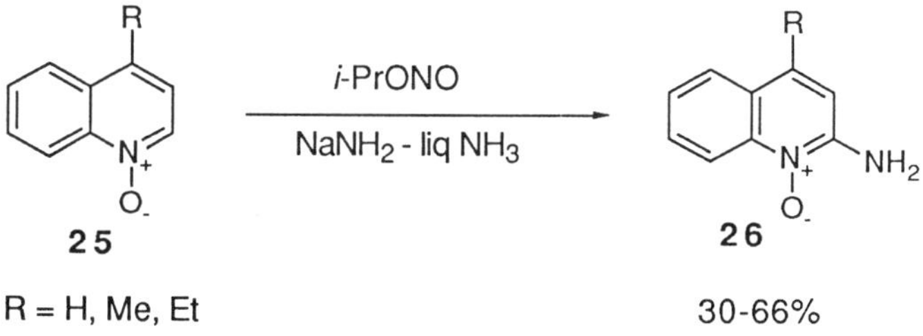

SCHEME 27

Extension of Katritzky's lithium carbamate protection method to the α-metallation of tetrahydroquinolines (**27**, **Scheme 28**) has been reported and a mixture of diastereoisomers is generated on addition of aldehydes and ketones.

Other common electrophiles gave complex mixtures which are most likely due to the *t*-BuLi/*t*-BuOK system required to metallate the intermediate carbamates. Use of this process offers a rapid approach to 2-alkyl-1,2,3,4-tetrahydroquinolines [89JCS(P1)17].

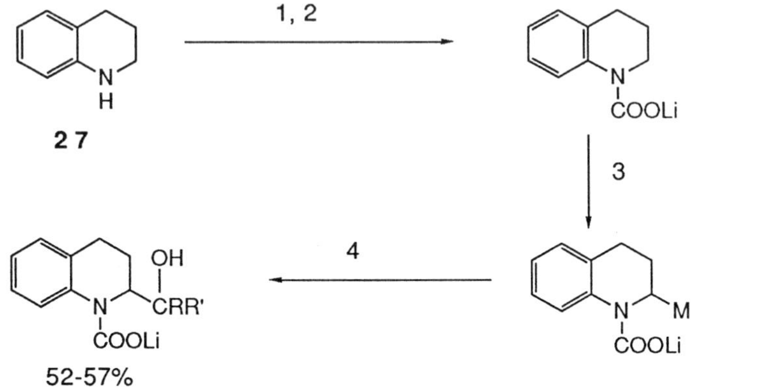

1. *n*-BuLi; 2. CO_2; 3. *t*-BuLi / *t*-BuOK, THF, -50 °C; 4.RCOR'; 5. H_3O^+

SCHEME 28

A large array of 2,3-disubstituted quinolines can be synthesized by virtue of the directing metallation capabilities of the chloro function. Thus 2-, 3- and 4-chloroquinolines give rise to 2,3-, 3,4- and 2,3-disubstituted quinolines in good yields with a variety of carbon, silicon, halogen and boron electrophiles (**Scheme 29**) [89JHC1589].

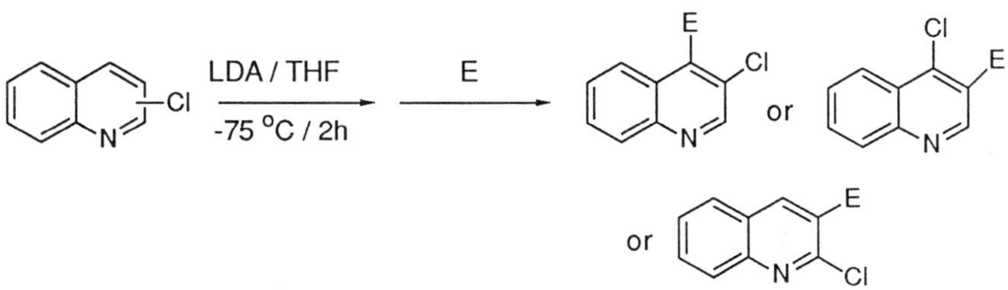

SCHEME 29

5.1.2.3 Isoquinolines

Tetrahydroisoquinolines bearing chiral oxazolines (**28**) are intermediates in a high yielding stereoselective route to 1-alkyl-1,2,3,4-tetrahydroisoquinolines (**29**). Diastereomer ratios in the range of 15-40/1 are recorded with yields typically in excess of 90%. This work has recently been extended to include unactivated systems (see Pyridines) and a mechanism proposed [89JA2211]. The chiral auxiliaries are available from a variety of chiral alcohols (via amino acid reduction) and are recoverable.

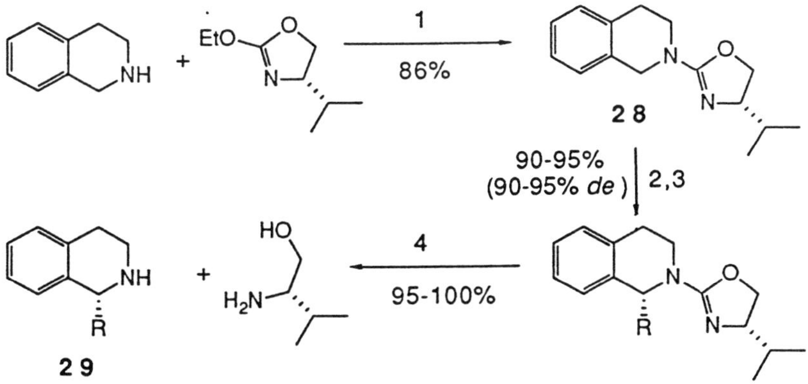

1. H^+, PhH / Δ; 2. BuLi, -78 °C , THF; 3. RX, -100 °C;
4. N_2H_4, H_3O^+, EtOH.

SCHEME 30

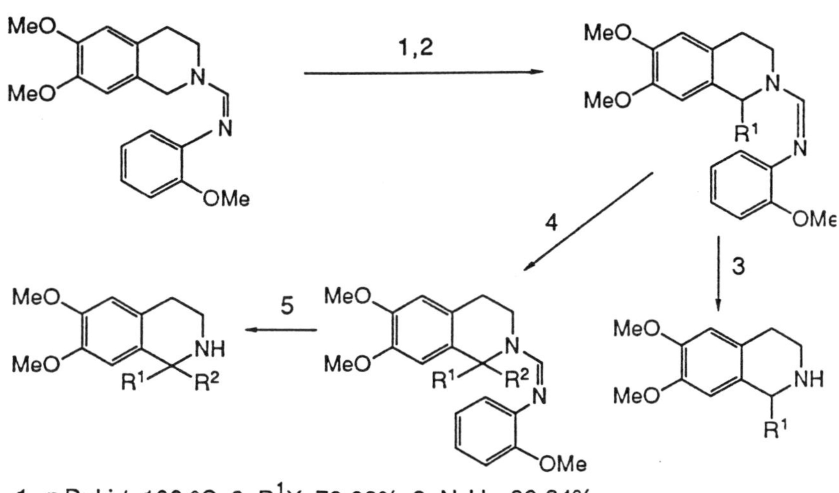

1. *n*-BuLi / -100 °C; 2. R^1X, 72-98%; 3. N_2H_4, 82-84%;
4. *n*-BuLi, R^2X; 5. N_2H_4, 53-91%

SCHEME 31

The development of the *o*-methoxymethylaniline (MMA) derived formamidine as an activating and chelating group for the benzylic metallation of tetrahydroisoquinolines has enabled α-quaternary substitution to be achieved [89TL47]. Previous attempts with stronger bases were unsuccessful (**Scheme 31**).

ACKNOWLEDGEMENTS

The authors would like to thank their colleagues at Reilly Industries, Inc. for their helpful comments. They are most grateful to Dr. Yantao Zhang for his assistance with the preparation of this manuscript.

REFERENCES

89CC142	J. M. Jego, B. Carboni, M. Vaultier, and R. Carrie; *J. Chem. Soc., Chem. Commun.*, **1989**, 142.
89CC447	M. Reglier, E. Amadei, R. Tadayoni, and B. Waegell; *J. Chem. Soc., Chem. Commun.*, **1989**, 447.
89CC1185	M. Mehmandoust, C. Marazano, and B. C. Das; *J. Chem. Soc., Chem. Comm.*, **1989**, 1185.
89CHE428	M.-G. A. Shvekhgeimer, K. I. Kobrakov, S. S. Sychev, and V. K. Promonenkov; *Chem. Heterocycl. Compd. (Engl. Transl.)*, **1989**, 428.
89H(28)39	K. Shishido, K. Hiroya, K. Fukumoto, and T. Kametani; *Heterocycles*, **1989**, *28*, 39.
89H(28)121	R. R. Ferguson, C. G. Boojamra, S. H. Brown, and R. H. Crabtree; *Heterocycles*, **1989**, *28*, 121.
89H(28)143	G. A. Russell, D. Guo, W. Baik, and S. J. Herron; *Heterocycles*, **1989**, *28*, 143.
89H(28)249	S. Rozen and D. Hebel; *Heterocycles*, **1989**, *28*, 249.
89H(28)275	S. Hibino, E. Sugino, Y. Adachi, K. Nomi, K. Sato, and K. Fukumoto; *Heterocycles*, **1989**, *28*, 275.
89H(28)723	G. Stork and R. Mah; *Heterocycles*, **1989**, *28*, 723.
89H(29)1781	Y. Tagawa, T. Yoshida, N. Honjo, and Y. Goto; *Heterocycles*, **1989**, *29*, 1781.
89H(29)2065	K. Soai, H. Hori, and S. Niwa; *Heterocycles*, **1989**, *29*, 2065.
89H(29)2253	M. Sagi, O. Sato, S. Kanno, and H. Yamanaka; *Heterocycles*, **1989**, *29*, 2253.
89JA1517	D. L. Boger and A. M. Kasper; *J. Am. Chem. Soc.*, **1989**, *111*, 1517.

89JA2211 R. E. Gawley, K. Rein, M. Goicoechea-Pappas, T. V. Anklekar, G. C. Hart, and G. A. Smith; *J. Am. Chem. Soc.*, **1989**, *111*, 2211.

89JCR(S)228 A. Bhattacharjya, P. Chattopadhyay, M. Bhaumik, and S. C. Pakrashi; *J. Chem. Res. (S)*, **1989**, 228.

89JCS(P1)17 A. R. Katritzky and S. Sengupta; *J. Chem. Soc., Perkin 1*, **1989**, 17.

89JCS(P1)283 R. S. Dainter, T. Jackson, A. H. H. Omar, H. Suschitzky, B. J. Wakefield, N. Hughes, A. J. Nelson, and G. Varvounis; *J. Chem. Soc., Perkin 1*, **1989**, 283.

89JCS(P1)909 M. Yoshida, T. Yoshida, M. Kobayashi, and N. Kamigata; *J. Chem. Soc., Perkin 1*, **1989**, 909.

89JCS(P1)1623 D. Armesto, M. G. Gallego, and W. M. Horspool; *J. Chem. Soc., Perkin Trans. 1*, **1989**, 1623.

89JCS(P1)2279 B. Sket, P. Zupet, and M. Zupa; *J. Chem. Soc., Perkin 1*, **1989**, 2279.

89JHC45 M. Tada and Y. Yokoi; *J. Heterocycl. Chem.*, **1989**, *26*, 45.

89JHC105 L. Estel, F. Linard, F. Marsaus, A. Godard, and G. Quéguiner; *J. Heterocycl. Chem.*, **1989**, *26*, 105.

89JHC1589 F. Marsais, A. Godard, and G. Quéguiner; *J. Heterocycl. Chem.*, **1989**, *26*, 1589.

89JOC175 R. E. Gawley, G. C. Hart, and L. J. Bartolotti; *J. Org. Chem.*, **1989**, *54*, 175.

89JOC1249 E. C. Taylor and J. E. Macor; *J. Org. Chem.*, **1989**, *54*, 1249.

89JOC1726 T. Umemoto and G. Tomizawa; *J. Org. Chem.*, **1989**, *54*, 1726.

89JOC4019 J. H. Rigby, D. D. Holsworth, and K. James; *J. Org. Chem.*, **1989**, *54*, 4019.

89JOC4275 S. Kagabu, S. Naruse, Y. Tagami, and Y. Watanabe; *J. Org. Chem.*, **1989**, *54*, 4275.

89JOC5094 M. Makosza and Z. Owczarczyk; *J. Org. Chem.*, **1989**, *54*, 5094.

89JOC5224 F. Minisci, E. Vismara, and F. Fontana; *J. Org. Chem.*, **1989**, *54*, 5224.

89MI1 K. J. John and V. N. Rajasekharan Pillai; *J. Polymer. Sci.*; *Pt. A: Polymer Chem.*, **1989**, 2279.

89S47 G. Cainelli, M. Contento, F. Manescalchi, and L. Plessi; *Synthesis*, **1989**, 47.

89S645 R. Balicki; *Synthesis*, **1989**, 645.

89SC317 R. B. Katz, J. Mistry, and M. B. Mitchell; *Synth. Commun.*, **1989**, *19*, 317.

89SC897 R. Balicki, L. Kaczmarek, and M. Malinowski; *Synth. Commun.*, **1989**, *19*, 897.

89T6511 W. A. W. Stolle, A. T. M. Marcelis, A. Koetsier, and H. C. van der Plas; *Tetrahedron*, **1989**, *45*, 6511.

89T6891 A. E. Frissen, A. T. M. Marcelis, W. C. Melger, and H. C. van der Plas; *Tetrahedron*, **1989**, *45*, 6891.

89T7469	J. Epsztajn, A. Bieniek, M. W. Plotka, and K. Suwald; *Tetrahedron*, **1989**, *45*, 7469
89TL47	M. A. Gonzalez and A. I. Meyers; *Tetrahedron Lett.*, **1989**, *30*, 47.
89TL747	M. Onaka, R. Ohno, and Y. Izumi; *Tetrahedron Lett.*, **1989**, *30*, 747.
89TL1197	P. Beak and W.-K. Lee; *Tetrahedron Lett.*, **1989**, *30*, 1197.
89TL1229	W. Schulz, H. Pracejus, and G. Oehme; *Tetrahedron Lett.*, **1989**, *30*, 1229.
89TL2481	M. Teng and F. W. Fowler; *Tetrahedron Lett.*, **1989**, *30*, 2481.
89TL3037	J. Sisko and S. M. Weinreb; *Tetrahedron Lett.*, **1989**, *30*, 3037.
89TL3259	A. Padwa, B. Harrison, and B. H. Norman; *Tetrahedron Lett.*, **1989**, *30*, 3259.
89TL4569	F. Minisci, E. Vismara, F. Fontana, and M. C. N. Barbosa; *Tetrahedron Lett.*, **1989**, *30*, 4569.
89TL4821	K. Uneyama, O. Morimoto, and F. Yamashita; *Tetrahedron Lett.*, **1989**, *30*, 4821.
89TL4973	T.-K. Yang, S.-M. Hung, D.-S. Lee, A.-W. Hong, and C.-C. Cheng; *Tetrahedron Lett.*, **1989**, *30*, 4973.
89TL5137	J. Zhu, J.-C. Quirion, and H.-P. Husson; *Tetrahedron Lett.*, **1989**, *30*, 5137.
89TL6283	I. Ojima and A. Korda; *Tetrahedron Lett.*, **1989**, *30*, 6283.
89TL6545	T. Higuchi, H. Ohtake, and M. Hirobe; *Tetrahedron Lett.*, **1989**, *30*, 6545.
89TL6865	A. Couture, C. Bochu, and P. Grandclaudon; *Tetrahedron Lett.*, **1989**, *30*, 6865.
89TL7199	Y. Yoshida, Y. Kimura and M. Tomoi; *Tetrahedron Lett.*, **1989**, *30*, 7199.

CHAPTER 5.2

Six-Membered Ring Systems: Diazines & Benzo Derivatives

T J KRESS & D L VARIE
Eli Lilly & Co, Indianapolis, IN, USA

5.2.1 INTRODUCTION

This chapter will focus on the chemistry of pyridazine, pyrimidine, pyrazine, and their benzo derivatives. Fused-ring derivatives are not covered. Each ring system will be considered with respect to new methods of ring construction, introduction and transformation of functional groups, and compounds possessing useful or unique biological activity. Articles that appeared late in 1988 are highlighted here and any similar paper from 1989 will appear in the next volume.

Several reviews containing relevant information on diazines should be noted. Participation of the diazines in the Chichibabin reaction has been surveyed <88AHC(44)1>. A chapter describing the nucleophilic, electrophilic, and free radical substitution reactions on the ring carbon atoms of the diazines has been updated <88AHC(44)199>. The use of ketene dithioacetals in the preparation of pyrimidines has been reviewed <89H1409>. The aromatic energies of the diazines and a number of other heterocycles have been estimated <89H1135>. The differences in energy between the diazines were too small to allow any ranking.

5.2.2 STRUCTURE

In the majority of publications on dihydropyrimidines the location of the tautomeric double bonds is not known and these compounds are arbitrarily represented as either 1,4- or 3,4-dihydropyrimidines. Recently, several groups have prepared dihydropyrimidines wherein each tautomer was observable by NMR spectroscopy. The position of the equilibrium in benzene solution for **1** (R=CF_3) was approximately 2:1 (**1a:1b**). When the substituent was an electron donating group (**1**, R=$N(Me)_2$) , the 3,4-dihydro form **1b** prevailed exclusively <88JA4832>. The dihydropyrimidine **2** was determined to be a mixture of tautomers. However, on reaction with ethyl chloroformate, the 1,4-dihydro derivative **3** was obtained as the only product <89JOC5898>.

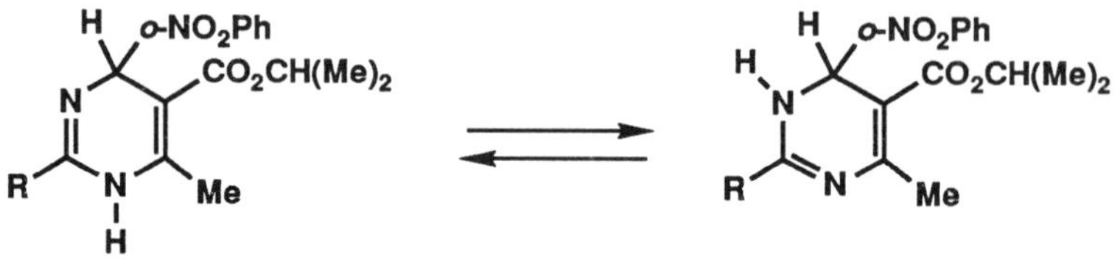

1a, 1,4- **1b, 3,4-**

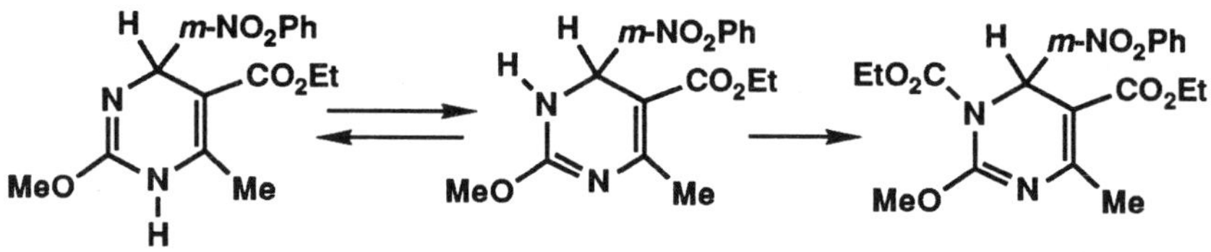

2a, 1,4- **2b, 3,4-** **3**

According to NMR data, compound **4** exists exclusively as the benzoylmethylene tautomer <89CB919>. The X-ray crystal structure of **5** reveals the C-3 and C-6 bridging bond angles are distorted out of the plane of the pyridazine ring by 17 degrees <89TL4649>.

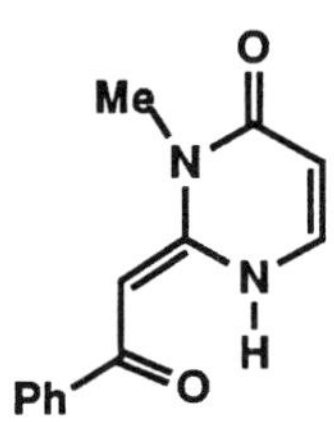

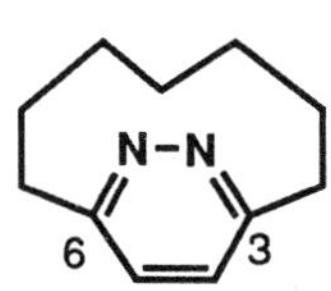

4 **5**

5.2.3 PYRIDAZINE, CINNOLINE

5.2.3.1 Synthesis

Cycloaddition reactions using 1,2,4,5-tetrazines as dienes have been employed to synthesize azocines **6** <89CZ254> and indolyl substituted pyridazines such as **7** <89HCA65>.

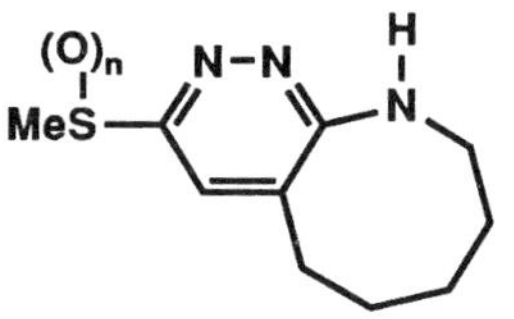

6 n = 0,1,2

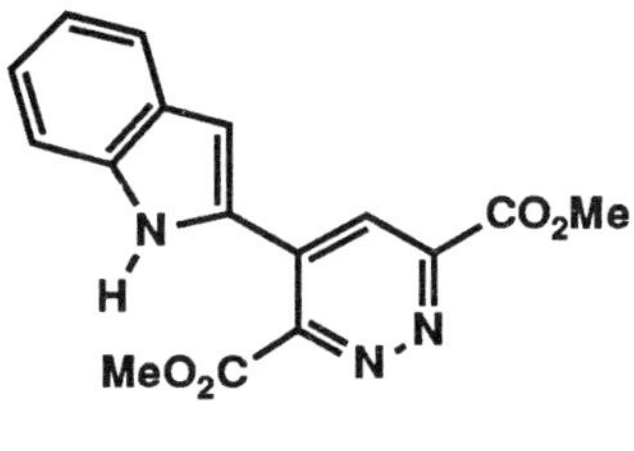

7

The cycloaddition of **8** and **9** led to the first synthesis of 4,5-di-t-butylpyridazine **10** <89H1241>. Pyridazines with symmetrical

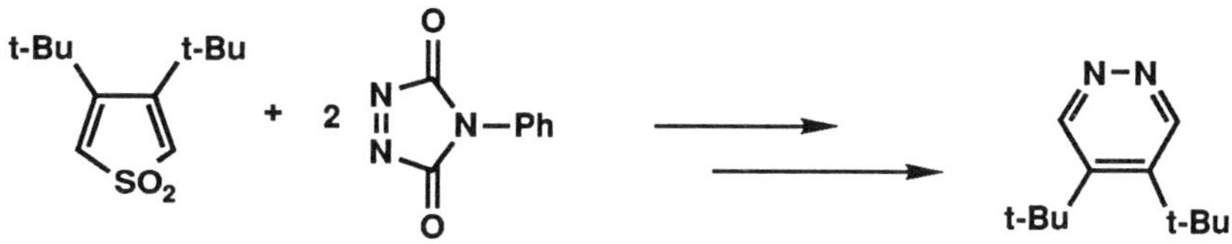

8 9 10

3,6-disubstitution (**13**) have been prepared from sulfides **12** derived from alpha-halo ketones <89BCJ2608>. The 2-aryl-3-cinnolinones **15** were

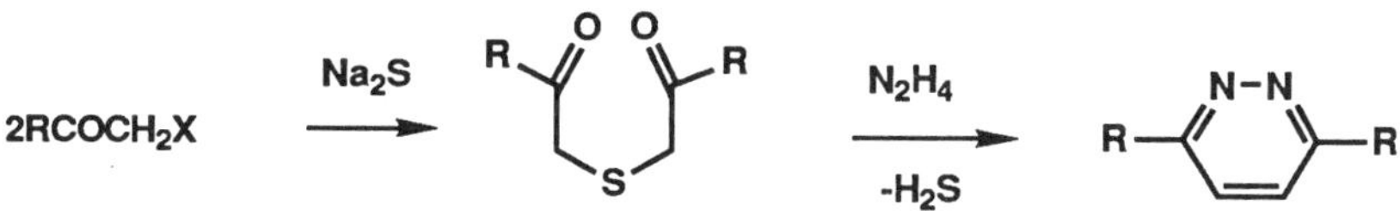

11 R = Ar, tBu, adamantyl 12 13

obtained **via** cyclization of azo phenylacetic acids **14** <89TL371>. The reaction was successful when one of the aryl groups of **14** contained an electron donating substituent. The naturally occuring amino acid **18** was

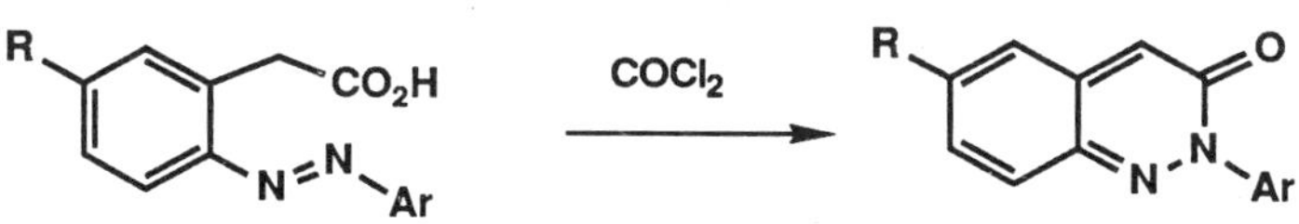

14 R = H, OH, OMe 15

prepared in optically active form from epoxy alcohol **16** <89JOC3260>.

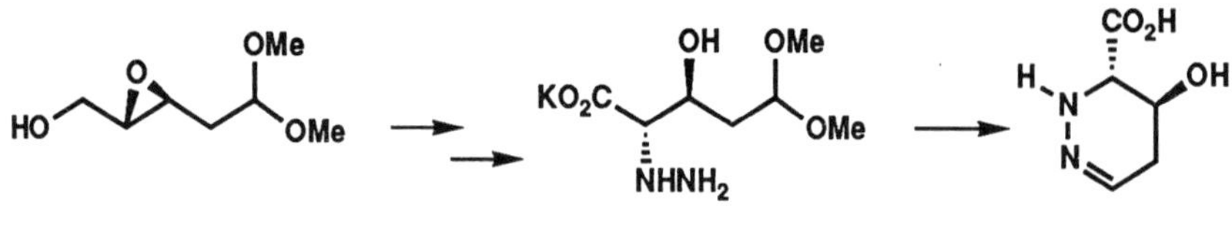

16 17 18

5.2.3.2 Reactions

A remarkable solvent effect was observed in the addition of alkoxides and thiolates to pyridazinones **20** <88JHC1757>. For example, sodium methoxide reacted with **20** in polar solvents (DMSO, DMF, MeOH) to give exclusively the 5-methoxy product **19.** The same reaction in less polar solvents (CCl_4, THF, t-BuOH) gave only the 4-methoxy product **21.**

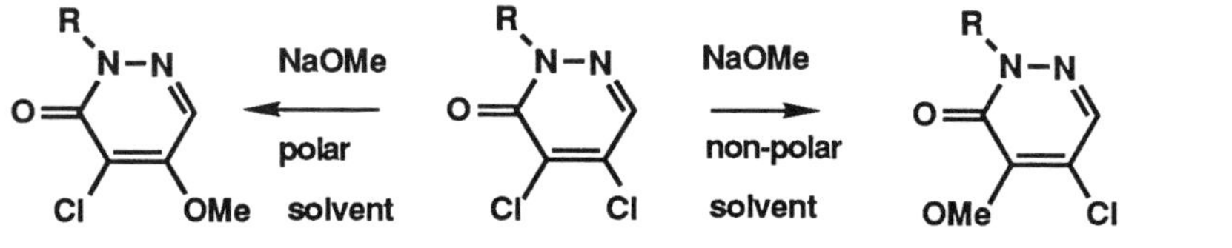

19 **20** R = Ph, Me **21**

The reaction of pyridazinones **22** with hydrazine hydrate was reported to give 4-aminopyridazinones **23** in good yields <89H1077>. A wide variety of

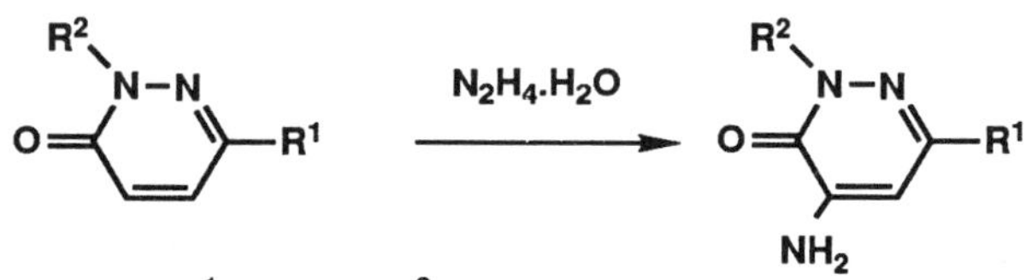

22 R^1 = Ar, Me R^2 = H, Me **23**

aminopyridazines containing easily reducible groups, such as **26,** were prepared from azides **via** treatment with triphenylphosphine followed by hydrolysis of the phosphazine <89S666>. Preparation of the ^{13}C labelled

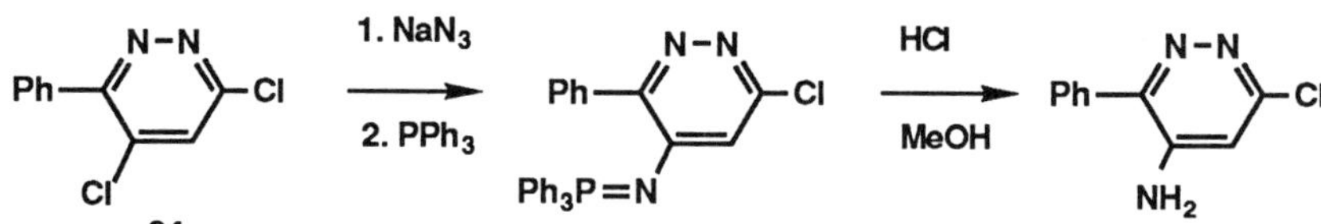

24 **25** **26**

pyridazinyl methanol **27** has led to the proposal of a mechanism for the rearrangement of **27** to diazole **28** <89JHC1009>.

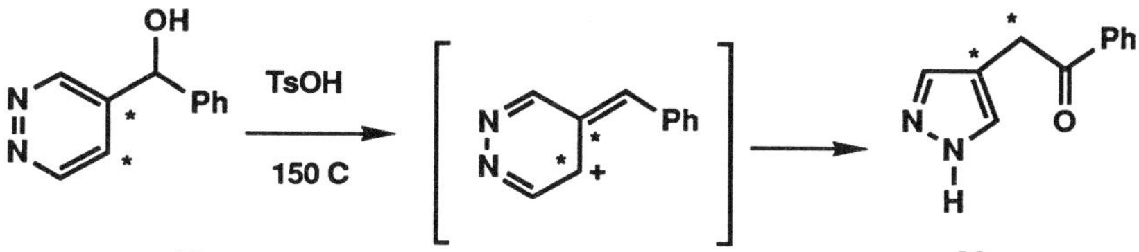

27 * = ^{13}C 28

The conversion of pyridazinones such as **29** to pyrrole diones **31** was shown to occur following dichlorination of the phenyl group of **29**. The pyrrole dione **31** could be converted back to a pyridazinone **32** with aqueous HCl <89JHC1649>. The synthetic equivalent of pyridazinone anion **35** was

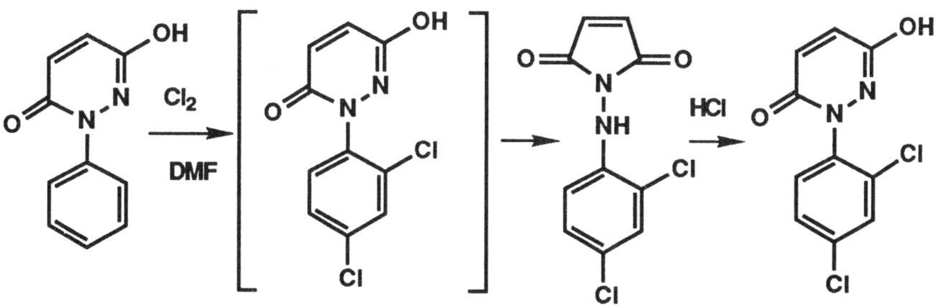

29 30 31 32

prepared by metalation of **33** <89JOMC14>. Anion **34** was successfully added to an iron-arene complex. The novel samarium complex **36** was the

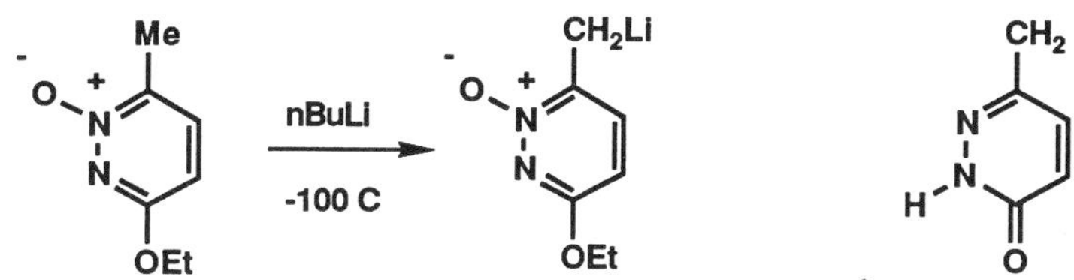

33 34 35

result of the reductive coupling of pyridazine with $(C_5Me_5)_2Sm(THF)_2$ <89JA3329>.

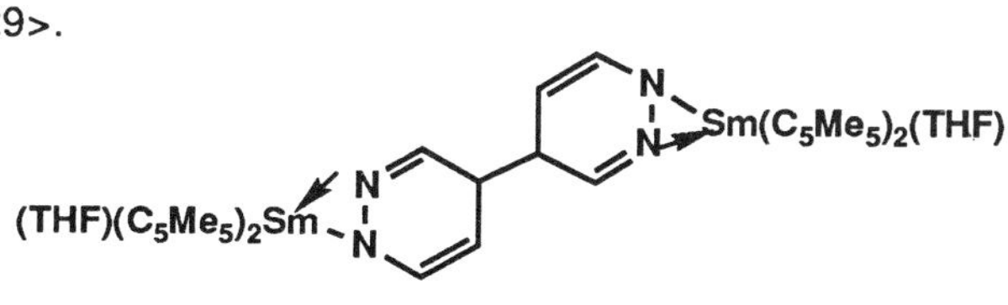

36

5.2.3.3 Applications

Pyridazinone **37** was shown in guinea pigs to inhibit bronchoconstriction induced by the slow releasing substance of anaphylaxis <89MI411>. Antihypertensive activity was reported for hydrazinopyridazines **38** <89AP331> and pyridazinones **39** <89JMC2277>.

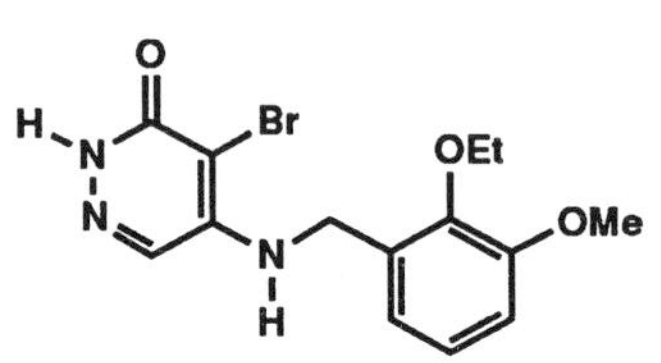

37

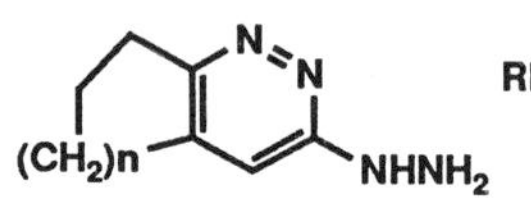

38

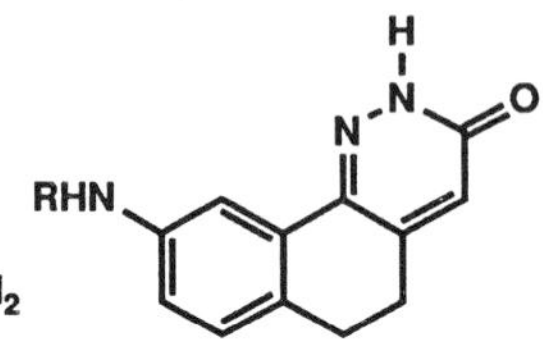

39

5.2.4 PYRIMIDINE, QUINAZOLINE

5.2.4.1 Synthesis

Several new methods are now available for the synthesis of reduced pyrimidines. The 1,3-diazadiene **40** undergoes [4+2] cycloaddition with the electron rich enamine **41** affording **42** in 80% yield as a single diastereomer. Acid hydrolysis of **42** followed by dehydration gave the 3,4-dihydropyrimidine **43** in 81% overall yield <89TL4573>.

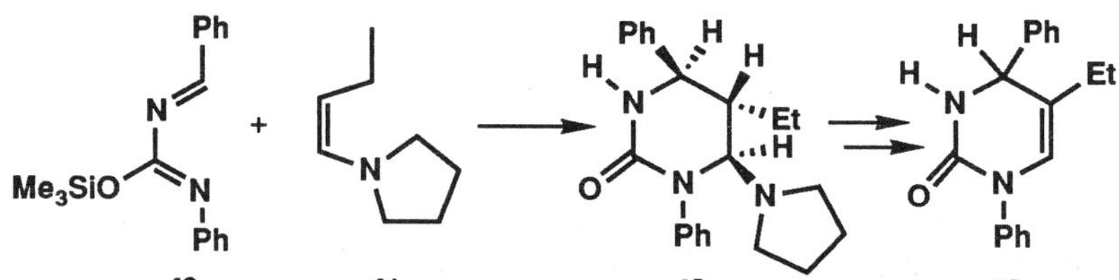

40 **41** **42** **43**

Intermediate **44**, obtained by condensation of methyl 3-aminopropionate with diphenylcyanocarbonimidate, reacts with benzylamine affording **45**. Surprisingly, treatment of **45** with bromine in acetic acid resulted not in bromination but in decyanation to the dihydropyrimidine **46** which was isolated as the hydrobromide salt <89JOC1062>.

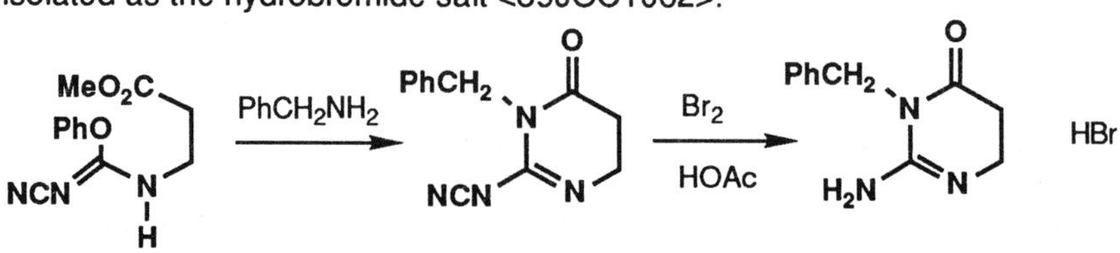

44 **45** **46**

Hexahydropyrimidines **47a** (R-1,2=H) and **47b** (R-1=Me, R-2=H) were obtained in good yield through condensation of 1,3-diaminopropane or 1,3-diamino-2,2-dimethylpropane with 40% aqueous glyoxal. The structures were confirmed by single crystal X-ray analysis of the tetranitro derivative **47c** (R^1=H, R^2=NO$_2$) <89JOC4771>. The "Biginelli" compounds **48** (R=H, Me),

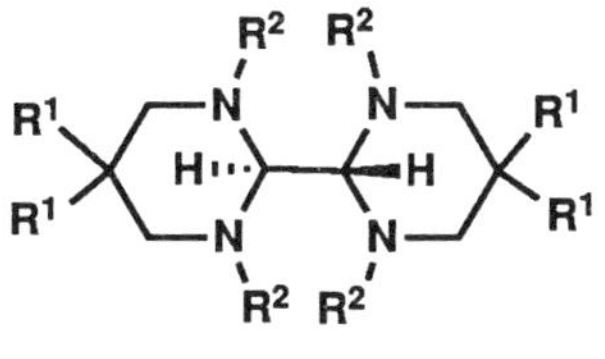

47a,b,c

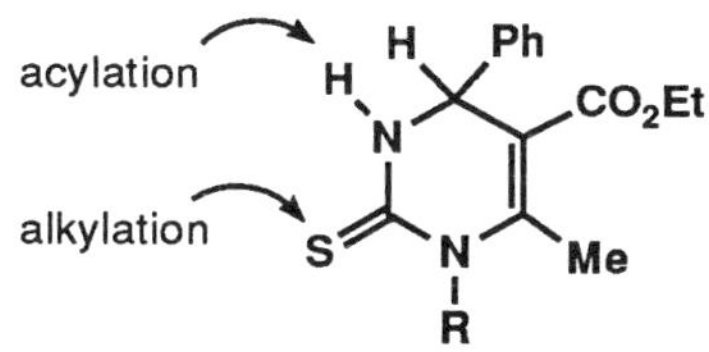

48

prepared in one step by condensation of benzaldehyde, thiourea or N-methylthiourea, and ethyl aceatate were examined with regard to their positioıı of acylation and alkylation. Reaction of **48** with methyl iodide resulted in exclusive S-methylation while treatment with acetic anhydride afforded only N-3 acylation <89JHC55>. Diethyl malonate **49** condenses with two moles of trifluoroacetonitrile **50** in a one-pot procedure to give an 88% yield of **51** <89JHC7>. Chalcone epoxides have been used as synthetic

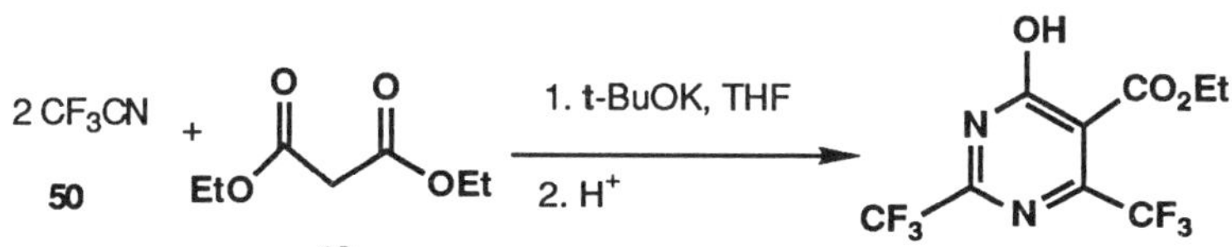

50 **49** **51**

equivalents to 1,3-diketones. Treatment of **52** (R = *p*-MeOPh) with guanidine carbonate gave the 2-aminopyrimidine **53** which was nitrosated at C-5 with concomitant loss of nitrogen at C-2. Reduction of the nitroso compound afforded the 5-aminopyrimidine **54** <89JHC947>.

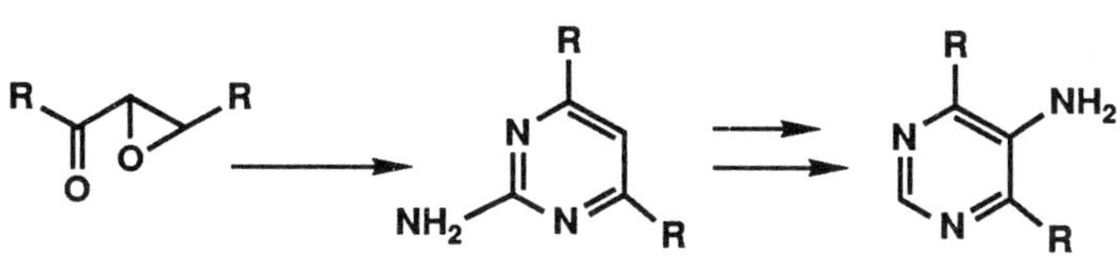

52 **53** **54**

Interest in the biological activity of an analog of the antibacterial Trimethoprin has led to a novel synthesis of 6-benzylpyrimidines. Condensation of O-methyl isourea sulfate **55** with the beta - ketoester **56** afforded **57** (Ar = 3,4,5-(OMe)$_3$Ph) in 78% yield. Reaction of **57** with phosphorous oxychloride followed by ammonia gave iso-Trimethoprin **58** <89JHC883>.

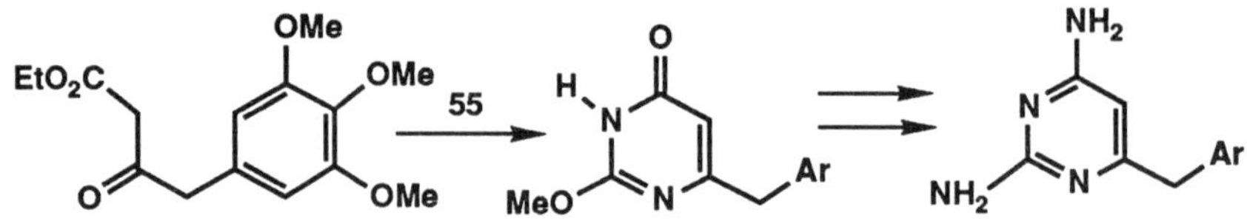

56 57 58

A new synthetic route to quinazolines based on [4+2] cycloaddition reactions allows access to a variety of structures. For example, reaction of the electron rich diene **59** with methyl vinyl ketone or dimethyl fumarate proceeds by formation of a 1:1 adduct, 1,4-elimination of dimethyl amine, and oxidative aromatization to give exclusively **60a** (R-1= H, R-2= COMe) and **60b** (R-1,2 = CO_2Me) <89CB1673>.

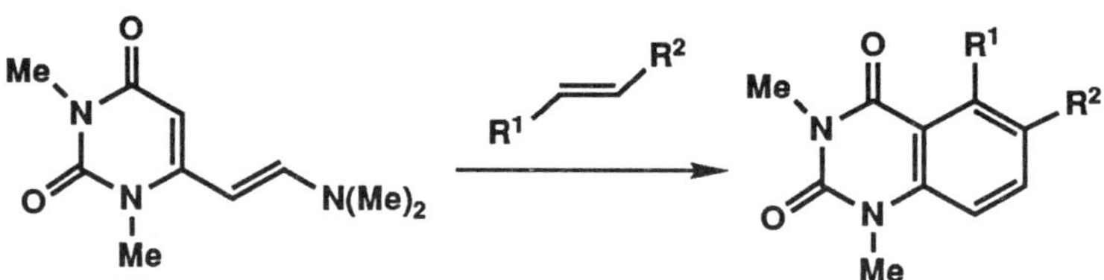

59 60a,b

Two groups have independently reported a new and general method for the preparation of quinazolinones based on an intramolecular aza-Wittig cyclization reaction. The iminophosphorane **62**, readily made by reaction of *o*-azidobenzamide **61** with triphenylphosphine (TPP), when treated with carbon dioxide or carbon disulfide gave **63** (**X** = S,O) in high yield <89T4263>. Similarly, reaction of **64** with TPP resulted in greater than 95% yields of **65** (R = alkyl, cyclopropyl) <89TL3313>.

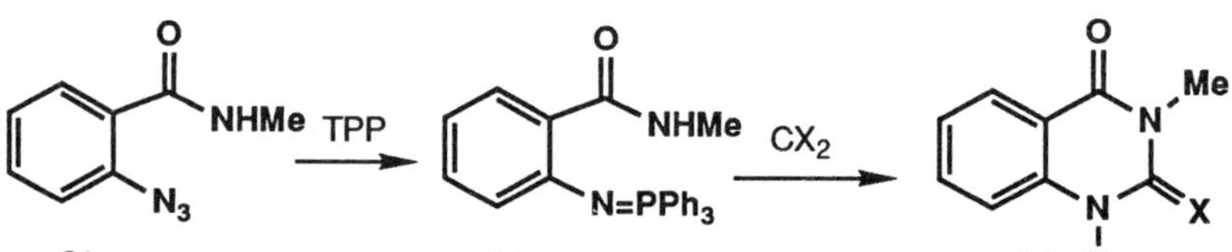

61 62 63

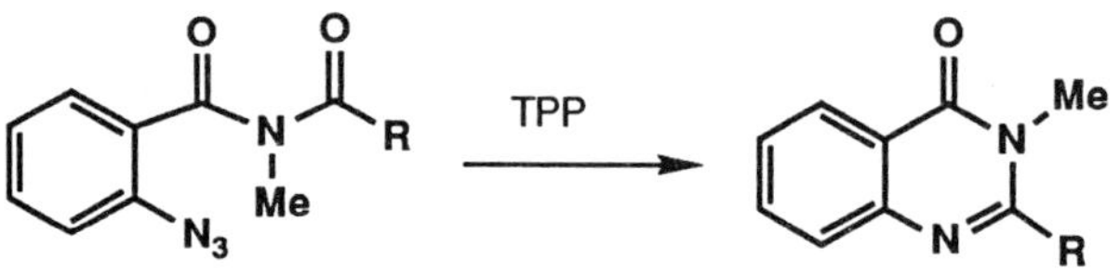

64 65

A simple route to a new class of dihydroquinazolines **67** (R-1,2,3 = alkyl or aryl) has been described and involves reaction of the Schiffs base **66** with an organolithium reagent resulting in direct intramolecular cyclization of the resulting anion <89JHC923>.

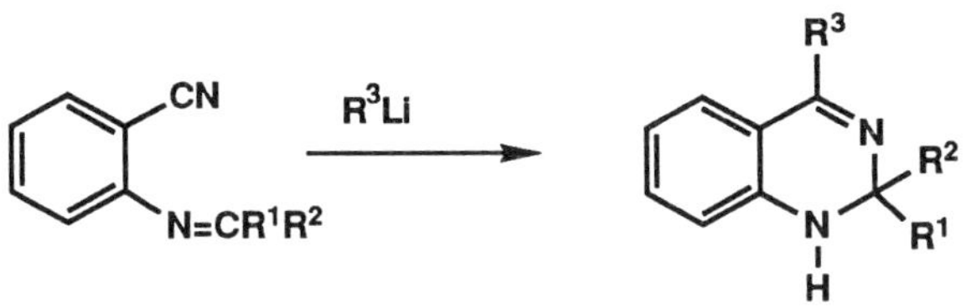

66 67

5.2.4.2 Reactions

Inverse electron demand Diels-Alder reactions of azadienes have continued to receive considerable attention. 5-Nitropyrimidine **68** undergoes regioselective cycloaddition with cyclic and non-cyclic enamines across the N-1 and C-4 positions, followed by loss of both HCN and the secondary amine leading to the formation of 3-nitropyridines. For example, treatment of enamine **69** with **68** provides a convenient route to the annelated pyridine **70** <89T2693>. An assortment of 2- and 5- substituted pyrimidines bearing

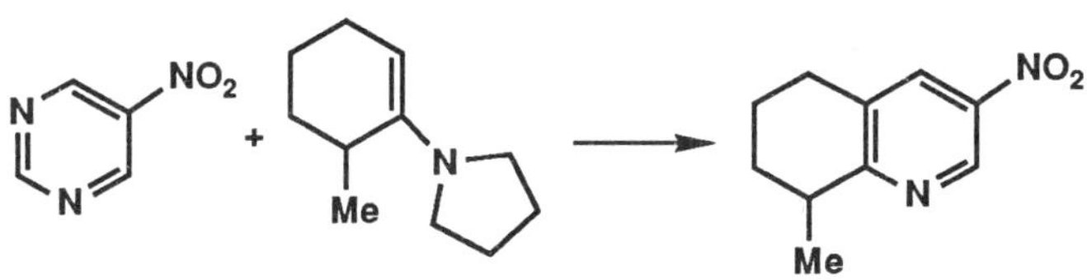

68 69 70

electron rich dieneophiles undergo intramolecular cyclization giving a novel route to a variety of annelated heterocycles <89T6511>. The 2-substituted pyrimidines **71** (**R**=H, Me, Ph, NO_2, **X**= O, S, NH) on heating in nitrobenzene afford fused pyridine products **72** in good to excellent yield <89T803>. Additional studies on these reactions have shown that electronic

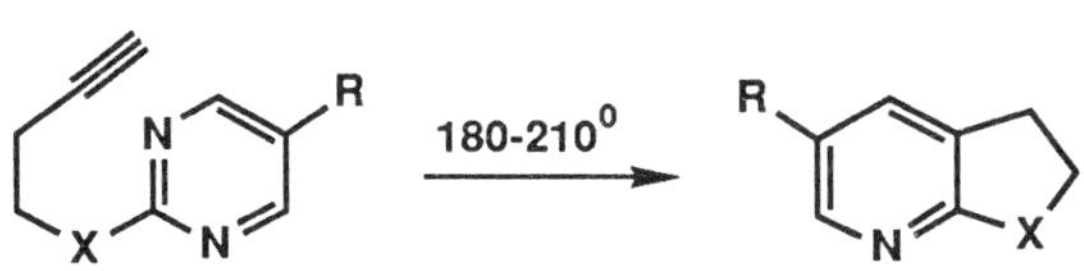

71 72

effects of ring substituents have little influence on the rate of reaction and that steric effects are most important in governing the reactivity towards cycloaddition <89T5151,6891>.

Palladium catalyzed reactions of halopyrimidines with carbon reagents have developed to be an extremely versatile and efficient method for formation of new carbon-carbon bonds. Ready access to 2-, 4-, and 5-aryl and vinylpyrimidines **73**,**74**,**75** has been achieved by treatment of the respective halopyrimidines (**73**, and **74** R= Cl, for **75** R = I was required) with aryl and vinylbutyl stannanes in the presence of $PdCl_2(PPh_3)_2$ <89CPB2814>.

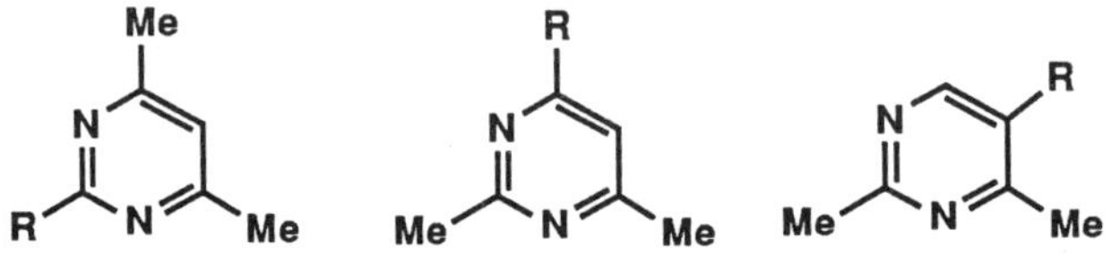

73 **74** **75**

In the reverse fashion, stannylated pyrimidines **76** undergo coupling with heteroaryl bromides (e.g. Ar = 2-bromofuran) affording biheteroaryls (**77**) in high yield <89ACS(B)684, 89T993>.

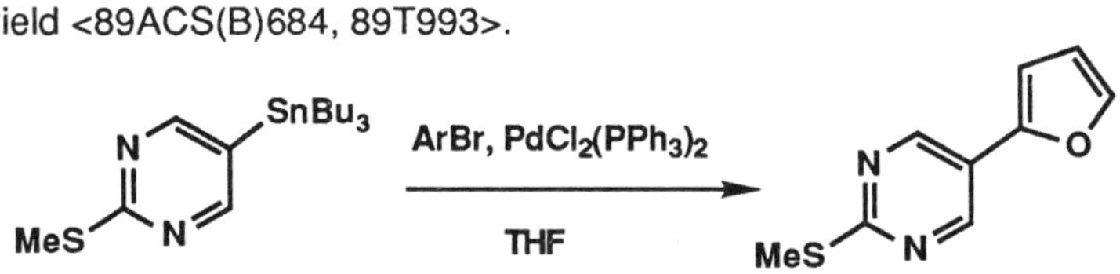

76 **77**

Because of its lower basicity, 5-pyrimidylcerium(III)chloride **79** is reported to give superior yields over the corresponding lithium reagent on reaction with enolizable aldehydes and ketones. Lithiation of **78** at -90°, addition of $CeCl_3$, followed by acetophenone gave an 83% yield of **80** while only a 50% yield was obtained using the pyrimidyl lithium directly <89ACS(B)816>.

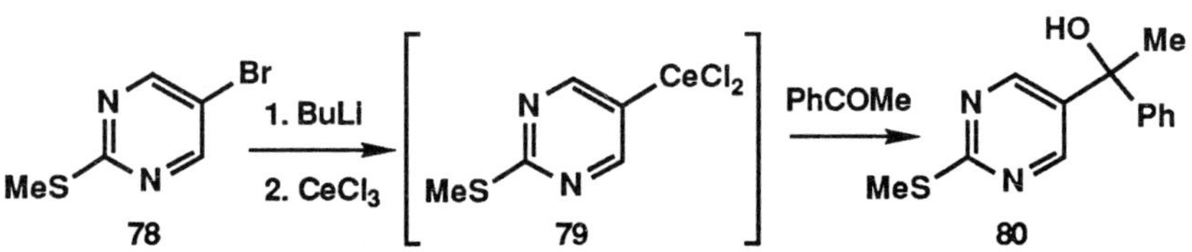

78 **79** **80**

Numerous 2-substituted 1,2-dihydro and 1,6-dihydropyrimidines (**81**, **82**, **X** = Me, Ph, $N(Me)_2$) have been prepared and a comparison made between their IR, UV, and NMR spectra <89JHC251>. Compound **82** (**X** = SMe) readily reacts with methoxide affording **83** <88JHC1713>.

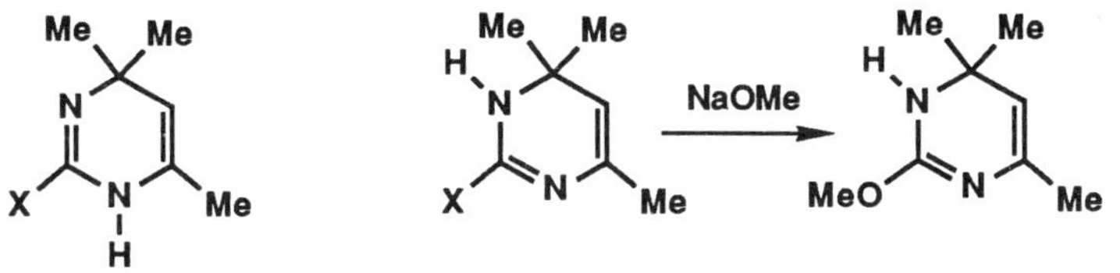

81 82 83

The mechanism for conversion of 5-acylpyrimidines into 4-acylpyrazoles is different from that of previously reported ring contractions which proceed through attack of hydrazine at C-4 or C-6 of the pyrimidine ring. In this case, the initial step is hydrazone formation giving **84** and subsequent intramolecular attack at C-6 by the exocyclic N-H group with C-6 and N-1 bond fission affording pyrazole **85** which is hydrolysed to **86** under the reaction conditions <89H1583>. Pyrimidine-5-carboxaldehydes containing strongly

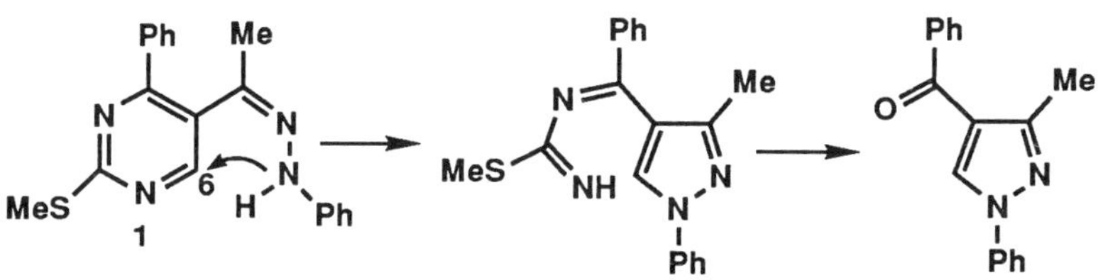

84 85 86

electron donating substituents (e.g. **87**) undergo facile loss of the 5-formyl moiety in the presence of strong acid <88JHC1697>. Quinazoline forms the

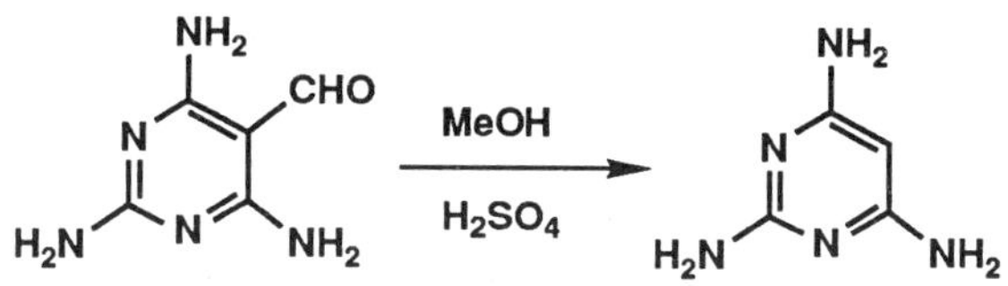

87 88

di-Reissert compound **89** on treatment with trimethylsilyl cyanide and acetyl chloride. Reaction of **89** with base followed by methyl iodide leads to alkylation at C-4 only. The position of alkylation was proven by regeneration of the parent ring system affording the known 4-methylquinazoline **91** <89JHC1357>.

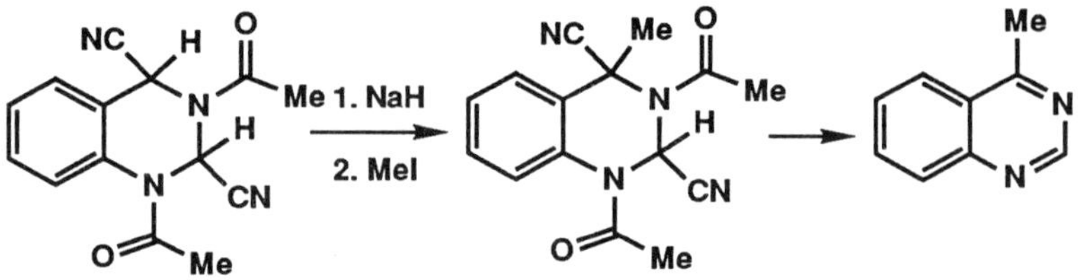

89 90 91

5.2.4.3 Applications

Pyrimidinylhydrazones **92,** a new class of fungicides with curative activity against a variety of fungi pathogenic to plants, continue to be of interest <89MI189, 351>. Several analogs of Sparsomycin (**93**), prepared by a novel route starting with L-alanine, were demonstrated to be strong inhibitors of protein synthesis and tumor growth <89JMC2002>.

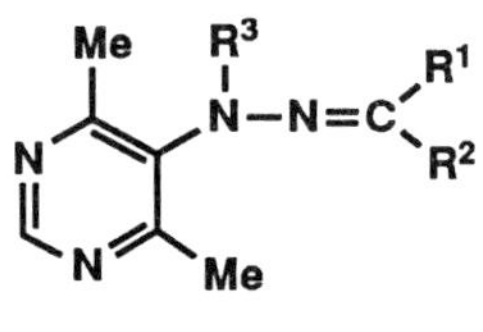

92

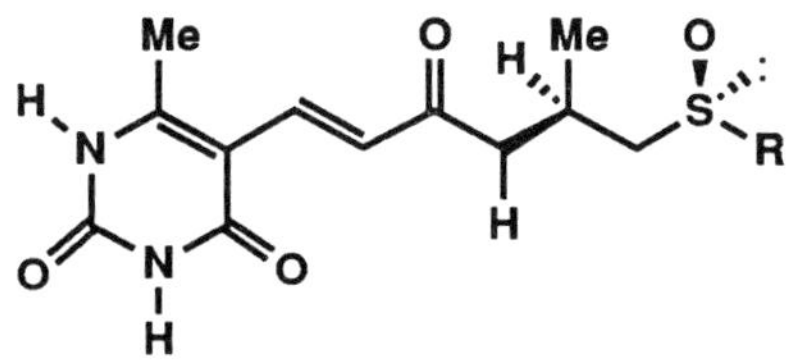

93

5.2.5 PYRAZINE, QUINOXALINE, PHENAZINE

5.2.5.1 Synthesis

The regiochemistry of the condensation of *o*-diamines **94** with phenylglyoxal **95** was found to be dependent on both the nature of the functional group in **94** and the conditions of the reaction. As an example, when **94** (R = NO_2) was treated with **95** under neutral or acidic conditions, a 5:1 ratio (79% yield) of **96** to **97** was obtained. However, under neutral conditions **94** (R = OMe) gave a 1:8 ratio of **96** to **97**. These changes in regioselectivity were ascribed to differences in the basicity of the amino groups in **94** <89MI202>.

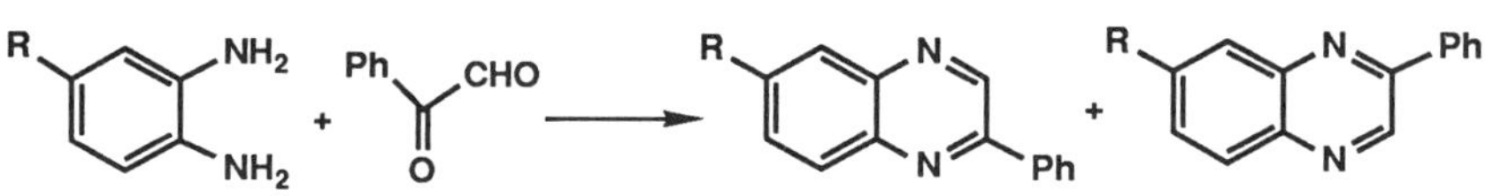

94 95 96 97

5.2.5.2 Reactions

Irradiation of pyrazine **98** in methanol under oxygen gave an 85% chromatographed yield of endoperoxide **99** as a stable solid. Thermolysis of **99** afforded the alpha-oxo-imide **100** in 95% yield <88JCS(P1)2921>. In a

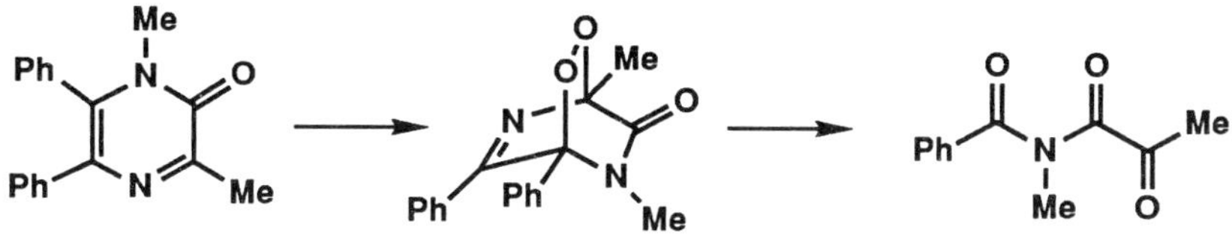

98 99 100

parallel fashion, **101** undergoes facile addition of singlet oxygen to form the remarkably stable endoperoxide **102.** Deoxygenation of **102** with triphenylphosphine caused skeletal rearrangement to the imidazole **103** <89JCS(P1)453>.

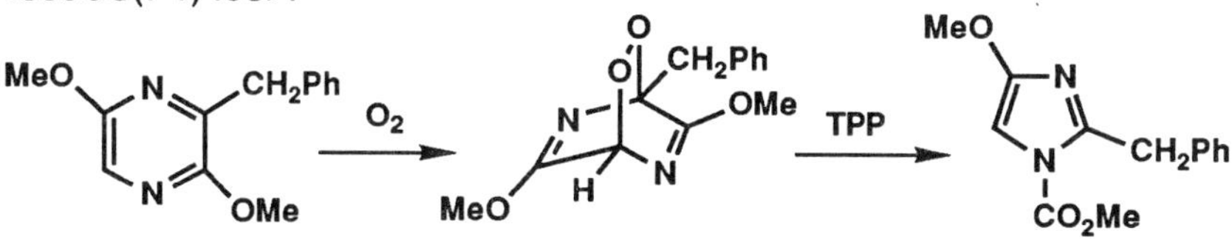

101 102 103

Similar to the reaction of alkynyl pyrimidines (see section **6.2.4.2**), alkynyl pyrazines also undergo the Diels-Alder reaction to give annelated pyridines <89T6211,6519>. Conditions for cyclization are considerably milder for quaternary salts than for the corresponding free base. Two representative examples of this series are conversion of **104** (**X** = O,S) to **105** in 65 and 71% yields, respectively. A key building block for the synthesis of 2,4-diamino-

104 105

pteridines, 2-amino-3-cyanopyrazine (**107**) is now available in 80% yield by treatment of the N-oxide **106** with trimethylsilyl cyanide which was generated **in situ** from sodium cyanide and trimethylsilyl chloride<89JHC817>. Homolytic acylation of pyrazine **108** (R-2 = OMe, Cl) gave predominantly ortho substitution affording **109** (R-3 = Et, i-Pr), whereas

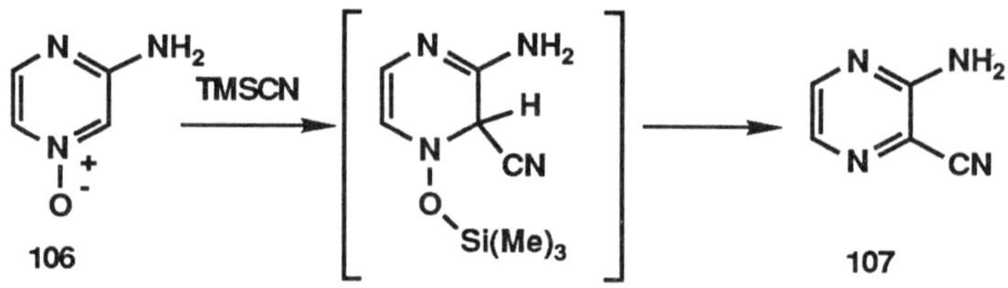

106 107

when R-2 was acetyl, carboxamide or carbethoxy, primarily *para* acylation occurred giving **110** <89JOC640>.

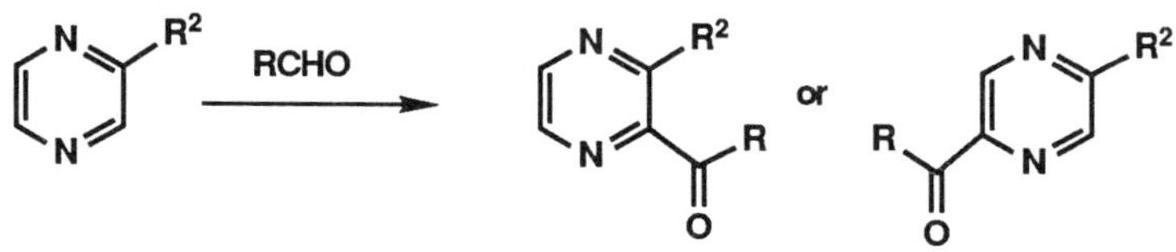

108 109 110

The alkynyl pyrazine **113**, prepared by palladium mediated coupling of alkyne **111** (R = alkyl) with **112**, undergoes hydration to give a novel route to alkyl ketone **114** <89JOC1218>. Alkyl and aryl pyrazines may be prepared in

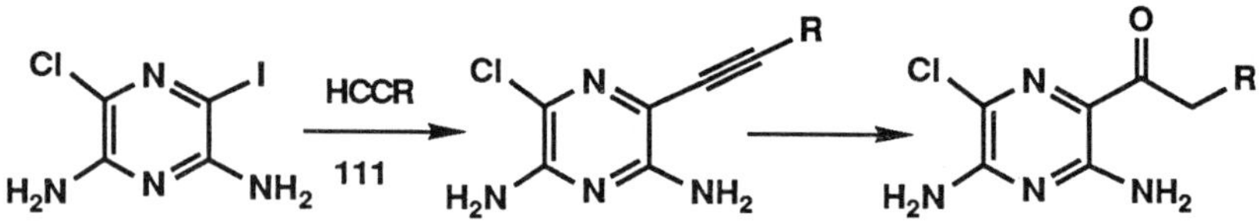

112 113 114

high yield by palladium catalyzed cross coupling reactions between chloropyrazines with organotin and organoboron reagents. As an example, **117** was prepared in 80% yield by treatment of **116**, made *in situ* from pentylmagnesium bromide and tin (IV) chloride, with chloropyrazine **115** <89H123>. In a similar manner, the ant pheromones **118** were prepared by reaction of the corresponding organoboron reagents with 2-chloro-3,6-dimethylpyrazine in 44 and 41% yields, respectively <89H939>.

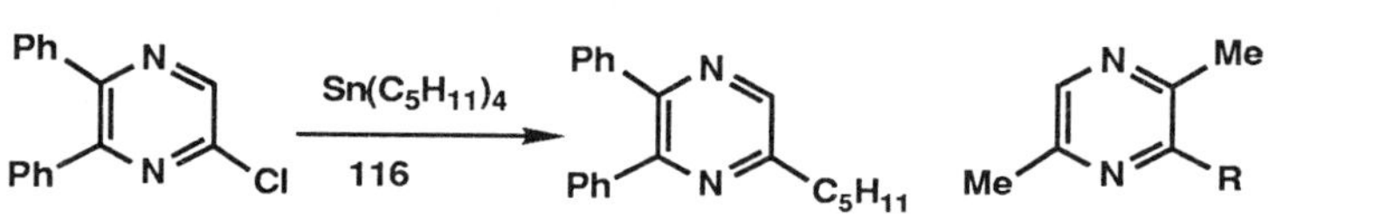

115 117 118 R = $CH_2CH_2CH(CH_3)_2$
R = $CH(CH_3)CH_2CH(CH_3)_2$

5.2.5.3 Applications

The structure and absolute configuration of a novel antitumor antibiotic phenazinomycin **119** , isolated from a strain of streptomyces from a soil sample, has been determined <89TL3151>. Quinoxaline **120** has shown significant antihypertensive activity <89JHC1623>.

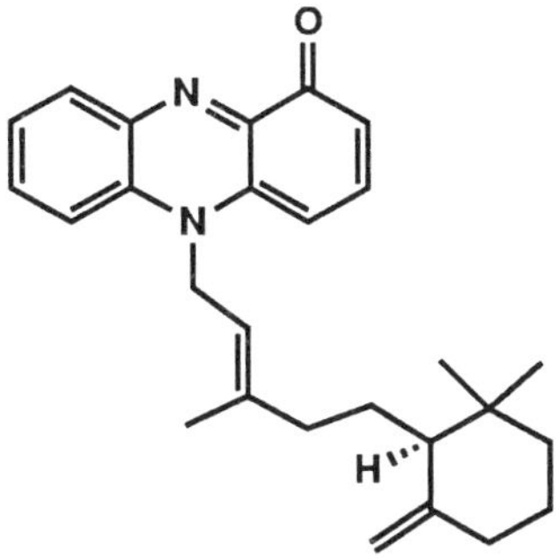

119

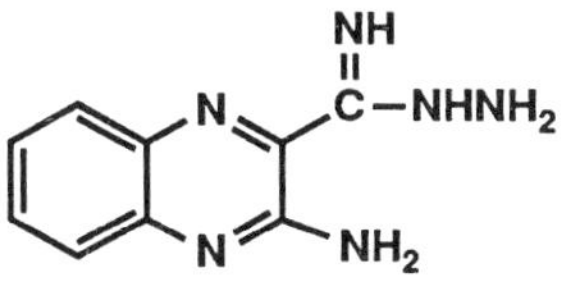

120

Two (1,2-dihydroxyethyl)pyrazines **121a** (R -1 = H, R-2 = Me) and **121b** (R-1 = Me, R-2 = H) were synthesized and identified as one of the chemical constituents of caramels <89MI59>.

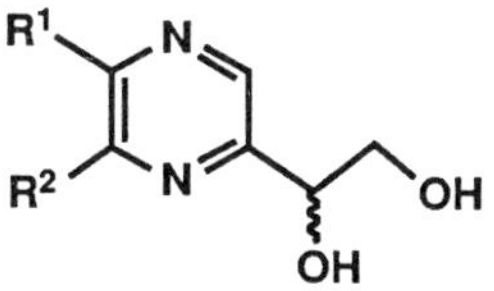

121a,b

REFERENCES

88AHC(44)1	C. K. McGill and A. Rappa, *Adv. Heterocycl. Chem.*,1988, **44**, 1.
88AHC(44)199	D. L. Comins and S. O'Connor, *Adv. Heterocycl. Chem.*, 1989, **44**, 199.
88JA4832	H. Cho, T. Iwashita, M. Ueda, A. Mizuno, K. Mizukawa and M. Hamaguchi, *J. Am. Chem. Soc.*, 1988, **110**, 4832.
88JCS(P1)2921	T. Nishio, N. Tokunaga, M. Kondo and Y. Omote, *J. Chem. Soc., Perkin Trans. 1*, 1988, 2921.
88JHC1697	T. J. Delia, B. Polenz and M. C. Martin, *J. Heterocycl. Chem.*, 1988, **25**, 1697.
88JHC1713	C. Kashima, M. Shimizu, K. Harada and Y. Omote, *J. Heterocycl. Chem.*, 1988, **25**, 1713.
88JHC1757	J. W. Lyga, *J. Heterocycl. Chem.*, 1988, **25**, 1757.
89ACS(B)684	J. Sandosham and K. Undheim, *Acta Chem. Scand., Ser. B,* 1989, **43**, 684.
89ACS(B)816	J, Zheng and K. Undheim, *Acta Chem. Scand., Ser. B*, 1989, **43**, 816.
89AP331	C. Teron, E. Ravina, L. Santana, N. Garcia-Dominguez, G. Garcia-Mera, J. A. Fontenla, F. Orallo and J. M. Calleja, *Arch. Pharm.*, 1989, **322**, 331.
89BCJ2608	J. Nakayama, T. Konishi, A. Ishii and M. Hoshino, *Bull. Chem. Soc. Jpn.*, 1989, **62**,2608.
89CB919	J. Elguero, P. Goya, A. Martinez and I. Rozas, *Chem. Ber.*, 1989, **122**, 919.
89CB1673	E. B. Walsh and H. Wamhoff, *Chem. Ber.*, 1989, **122**, 1673.
89CPB2814	Y. Kondo, R. Watanabe, T. Sakamoto and H. Yamanka, *Chem. Pharm. Bull.*, 1989, **37**, 2814.
89CZ254	V. G. Szitz and J. Richter, *Chem. -Ztg.*, 1989, **113**, 254.
89H123	T. Watanabe, K. Hayashi, J. Sakurada, M. Ohki, N. Takamatsu, H. Hirohata, K. Takeuchi, K. Yuasa and A. Ohta, *Heterocycles*, 1989, **29**, 123.
89H939	A. Ohta, R. Itoh, Y. Kaneko, H. Koike and K. Yuasa, *Heterocycles*, 1989, **29**, 939.
89H1077	W. J. Coates and A. McKillop, *Heterocycles*, 1989, **29**, 1077.
89H1135	M. J. S. Dewar and A. J. Holder, *Heterocycles*, 1989, **29**, 1135.
89H1241	J. Nakayama and A. Hirashima, *Heterocycles*, 1989, **29**, 1241.
89H1409	Y. Tominaga, S. Kohra, H. Honkawa and A. Hosomi, *Heterocycles*, 1989, **29**, 1409.
89H1583	A. Bajnati, M. Hubert-Habart and K. Takagi, *Heterocycles*, 1989, **29**, 1583.
89HCA65	U. Pindar, L. Pfeuffer and M. Kim, *Helv. Chem. Acta*, 1989, **72**, 65.
89JA3329	W. J. Evans and D. K. Drummond, *J. Am Chem. Soc.*, 1989, **111**, 3329.
89JCS(P1)453	I. M. Dawson, A. J. Pappin, C. J. Peck and P. G. Sammes, *J. Chem. Soc. Perkin Trans. 1*, 1989, 453.
89JHC7	Y. L. Sing and L. F. Lee, *J. Heterocycl. Chem.*, 1989, **26**, 7.

89JHC55	C. O. Kappe and P. Roschger, *J. Heterocycl. Chem.*, 1989, **26**, 55.
89JHC251	C. Kashima, M. Shimizu and Y. Omote, *J. Heterocycl. Chem.*, 1989, **26**, 251.
89JHC817	N. Sato, *J. Heterocycl. Chem.*, 1989, **26**, 817.
89JHC883	M. Botta, M. Artico, S. Massa and A. Gambacorta, *J. Heterocycl. Chem.*, 1989, **26**, 883.
89JHC923	L. Strekowski, M. T. Cegla, S. Kong and D. B. Harden, *J. Heterocycl. Chem.*, 1989, **26**, 923.
89JHC947	R. A. Osisanya and J. O. Oluwadiya, *J. Heterocycl. Chem.*, 1989, **26**, 947.
89JHC1009	G. Heinisch, T. Huber and C. Langes, *J. Heterocycl. Chem.*, 1989, **26**, 1009.
89JHC1357	J. Hahn and F. D. Popp, *J. Heterocycl. Chem.*, 1989, **26**, 1357.
89JHC1623	A. Monge, J. A. Palop, I. Urbasos and E. Fernandez-Alvarez, *J. Heterocycl. Chem.*, 1989, **26**, 1623.
89JHC1649	C. Yaroslavsky, P. Bracha and R. Glaser, *J. Heterocycl. Chem.*, 1989, **26**, 1649.
89JMC2002	L. A. G. M. van denBroek, E. Lazaro, Z. Zylicz, P. J. Fennis, F. A. N. Missler, P. Lelieveld, M. Garzotto, D. J. T. Wagener, J. P. G. Ballesta and H. C. J. Ottenjeijm, *J. Med. Chem.*, 1989,**32**,2002.
89JMC2277	G. Cignarella, D. Barlocco, G. A. Pinna, M. Loriga, M. Curzu, O Tofanaetti, M Germini, P. Cazzulani, E. Cavalleti, *J. Med. Chem.*, 1989, **32**. 2277.
89JOC640	Y. Houminer, E. W. Southwick and D. L. Williams, *J. Org. Chem.*, 1989, **54**, 640.
89JOC1062	P. J. Garratt, C. J. Hobbs and R. Wrigglesworth, *J. Org. Chem.*, 1989, **54**, 1062.
89JOC1218	M. J. Chapdelaine, P. J. Warwick and A. Shaw, *J. Org. Chem.*, 1989, **54**, 1218.
89JOC3260	P. Hughes and J. Clardy, *J. Org. Chem.*, 1989, **54**, 3260.
89JOC4771	D. S. Black, D. C. Craig, O. Giitsidis, R. W. Read, A. Salek and M. A. Sefton, *J. Org. Chem.*, 1989, **54**, 4771.
89JOC5898	K. S. Atwal, G. C. Rovnyak, B. C. O'Reilly and J Schwartz, *J. Org. Chem.*, 1989, **54**, 5898.
89JOMC14	R. C. Cambie, S. J. Janssen, P. S. Rutledge and P. D. Woodgate, *J. Organomet, Chem.*, 1989, **359**, C14.
89MI59	A. J. Boulton, A. McKillop and P. M. Rowbottom, *J. Chem. Research (S)*, 1989, 59.
89MI189	K. Konishi, T. Kuragano and T Tsujikawa, *J. Pesticide Sci.*, 1989, **14**, 189.
89MI202	M. Loriga, A. Nuvole and G. Paglietti, *J. Chem. Research (S)*, 1989, 202.
89MI351	K. Konishi and T. Kuragano, *J. Pesticide Sci.*, 1989, **14**, 351.
89MI411	M. Hibi, K, Shikada, T. Iwama, A. Yamamoto, M. Sakashita and S. Tanaka, *Jap. J. Pharmacol.*, 1989, **51**, 411.
89S666	T. Kappe, A. Pfaffenschlager and W. Stadlbauer, *Synthesis.* 1989, 666.

89T803	A. E. Frissen, A. T. M. Marcelis, and H. C. van der Plas, *Tetrahedron*, 1989, **45**, 803.
89T993	A. J. Majeed, O. Antonsen, T. Benneche and K Undheim, *Tetrahedron*, 1989, **45**, 993.
89T2693	A. T. M. Marcelis and H. C. van der Plas, *Tetrahedron*, 1989, **45**, 2693.
89T4263	P. Molina, M. Alajarin and A. Vidal, *Tetrahedron*, 1989, **45**, 4263.
89T5151	A. E. Frissen, A. T. M. Marcelis, G. Guertsen, D. A. de Bie and H. C. van der Plas, *Tetrahedron*, 1989, **45**, 5151.
89T6211	M. Biedrzycki, D. A. de Bie and H. C. van der Plas, *Tetrahedron*, 1989, **45**, 6211.
89T6511	W. A. W. Stolle, A.T. M. Marcelis, A. Koetsier and H. C. van der Plas, *Tetrahedron*, 1989, **45**, 6511.
89T6519	B. Geurtsen, D. A. de Bie and H. C. van der Plas, *Tetrahedron*, 1989, **45**, 6519.
89T6891	A. E. Frissen, A. T. M. Mardelis, W. C. Melger and H. C. van der Plas, *Tetraheadron*, 1989, **45**, 6891.
89TL371	M. G. Hutchings and D. P. Devonald, *Tetrahedron Lett.*, 1989, **30**, 371.
89TL3151	S. Funayama, S. Eda, K Komiyama and S. Omura, *Tetrahedron Lett.*, 1989, **30**, 3151.
89TL3313	H. Takeuchi and S. Eguchi, *Tetrahedron Lett.*, 1989, **30**, 3313.
89TL4573	J. Barluenga, M. Tomas, A. Ballesteros and L. A. Lopez, *Tetrahedron Lett.*, 1989, **30**, 4573.
89TL4649	P. G. Gassman and G. S. Boardman, *Tetrahedron Lett*, 1989, **30**, 4649.

CHAPTER 5.3
Six-Membered Ring Systems: Triazines, Tetrazines and Fused Ring Polyaza Systems

DEREK T HURST
Kingston Polytechnic, Kingston-upon-Thames, UK

5.3.1 Synthesis

5.3.1.1 *Triazines and Benzoderivatives*

A synthesis of 1,2,4-triazines uses nitrile groups as heterodienophiles in Diels-Alder reactions with inverse electron demand. Thus the triazines (1) are obtained by reacting dimethyl tetrazine-3,6-dicarboxylate with RR^1CNCN (88CZ281)

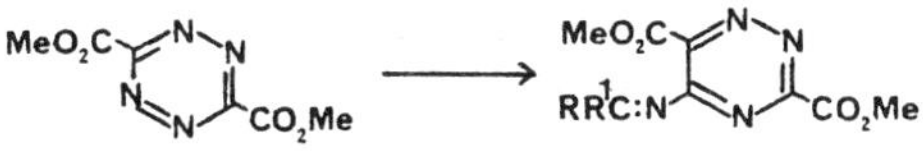

(1) $R=R^1$ = OPh,SMe, and others

A simple, regioselective, synthesis of 5,6-dihydro-1,3,5-triazine-2,4(1H,3H)-diones involves the cycloaddition of 2-trimethylsilyloxy- and 2-trimethylsilylthio-1,3-diaza-1,3-butadienes with isocyanates (Scheme 1 (89S228),

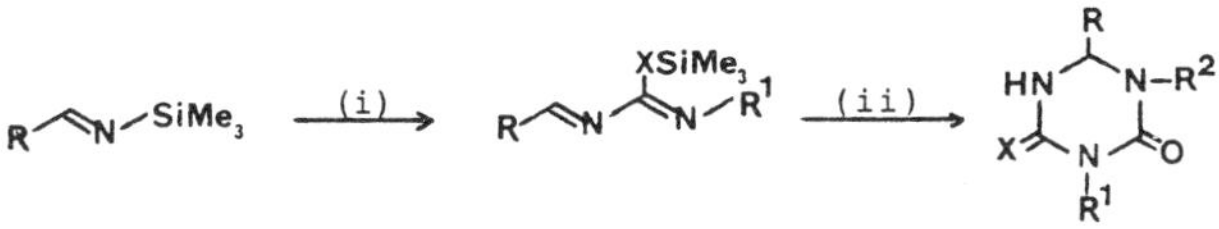

Reagents (i) R^1NCX,CH_2Cl_2,r.t.(X=O),reflux (X=S),8h (ii)R^2NCO,$CH_2Cl_2$25°, overnight then MeOH,1h ($R=R^1=R^2$, various, X=O,S)

Scheme 1

A novel pyrimidine to 1,2,4-triazine rearrangement involves the reaction of aryl diazonium ions with pyrimidine-4,6-diols which results in the formation of 1,2,4-triazine-5(2H)-ones (Scheme 2) (89TL3719).

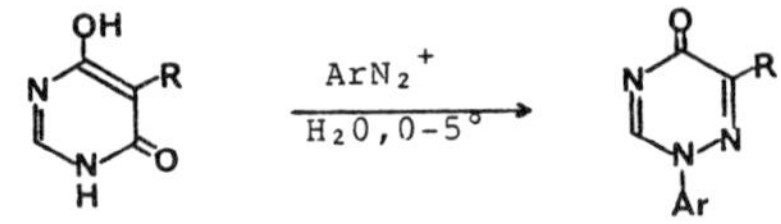

R=Me,Et;Ar=4-BrC$_6$H$_4$,4-ClC$_6$H$_4$,4-NO$_2$C$_6$H$_4$

Scheme 2

The reaction of N-(trimethylsilyl)-benzaldimine with 3,6-disubstituted-1,2,4,5-tetrazines results in the formation of dihydrotriazines which can be subsequently oxidised to give 3,5,6-trisubstituted triazines (Scheme 3) (89H1379). Both of the last reactions can be seen as cycloadditions to the N-heterocycle which acts as the diene, or as a two step process with subsequent elimination of HNCO or N_2 and TMS respectively.

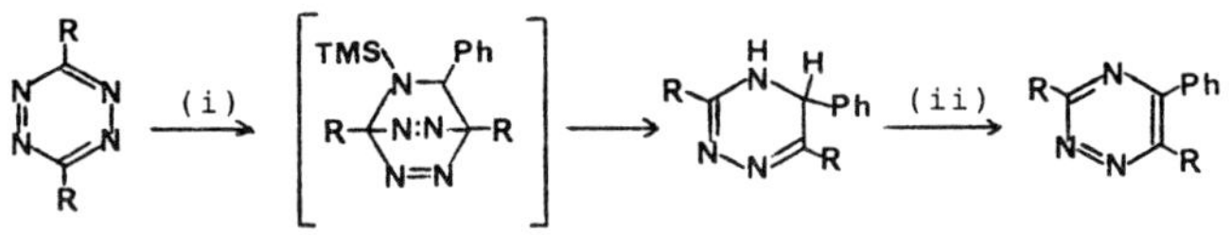

R=2-,3-,4-pyridyl, phenyl

Reagents (i) TMS-N=CHPh, toluene,N_2 reflux, 12-50h (ii) KMnO$_4$, Me$_2$CO, r.t. 6-12h

Scheme 3

The novel, simple-substituted, 1,2,3-benzotriazinone 1-oxides (2) can be obtained by the oxidation (mcpba) of the 3-alkyl- and 3-methoxy-1,2,3-benzotriazin-4(3H)-ones (89JCS(P1)5431).

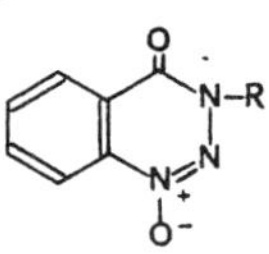

R=H,Me,CH$_2$Ph,CH$_2$OMe, CHMeOEt, THP, Ph

(2)

5.3.1.2 *Tetrazines and Benzoderivatives*

Benzene annelated tetrazines are obtained when N-aminopyridinium salts having ester, aroyl, and nitrile functions at the 2-position undergo dimerisation on treatment with hydroxide or alkoxide ion (Scheme 4). However the unannelated pyridinium salt (3) does not give such tetrazine products. Several compounds of type (4) were thus obtained (89JOC3062).

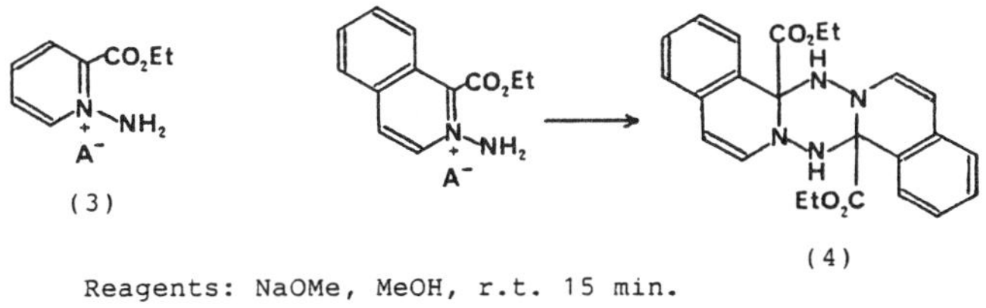

Scheme 4

1,2,3,4-Tetrahydro-s-tetrazines can be obtained from nitrile imines and alkanal methylhydrazones. These products exist in solution in tautomeric equilibrium with acyclic compounds, the position of the equilibrium being controlled by the nature of the substituents (Scheme 5) (89H1163;88H917).

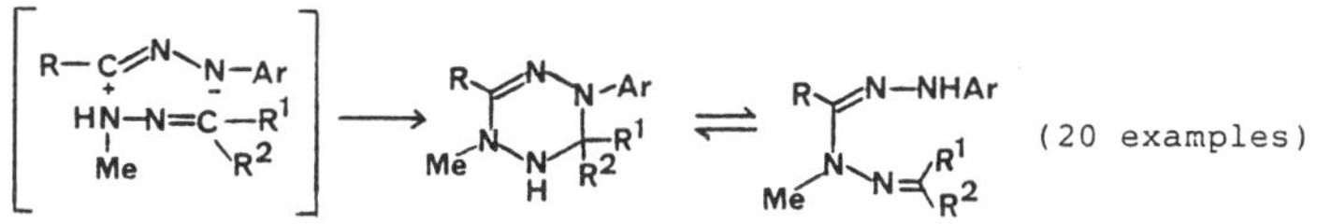

Scheme 5

5.3.1.3 *Purines and Pyrazolopyrimidines*

Modified pyrimidine and purine nucleosides continue to be of great interest as potential antiviral and antitumour agents.

2', 3'-Dideoxyadenosine (5), already shown to have promising in vitro antiviral activity can now be synthesised in multigram quantities in a 3 step synthesis from adenosine (Scheme 6) (89JOC4780).

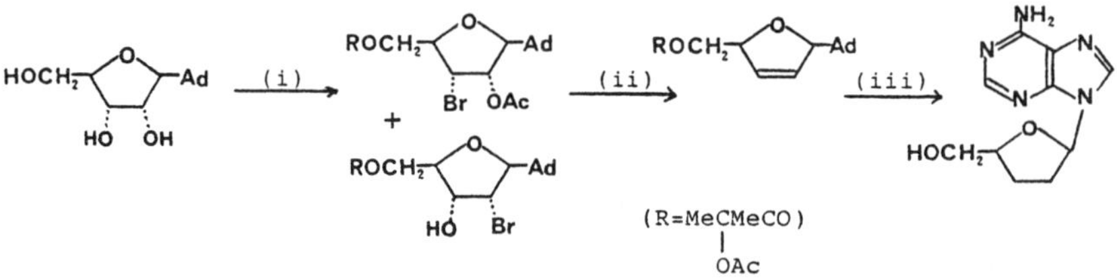

Reagents (i) 2-acetoxyisobutyryl bromide, MeCN,H_2O
(ii) Zn/Cu (iii) Pd/C, aq.EtOH

Scheme 6

A novel cyclobutane nucleoside, SQ-33054(6), has been synthesised from 2-amino-6-benzyloxypurine (89TL6453). This product has potent antiviral activity and is related to the recently discovered antibiotic oxetanocin (7) (87MI1077).

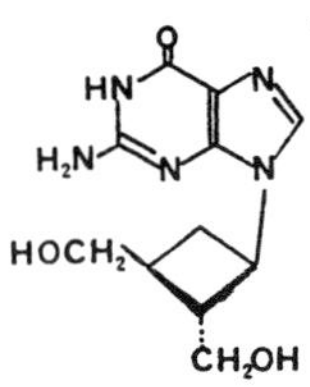

(6)

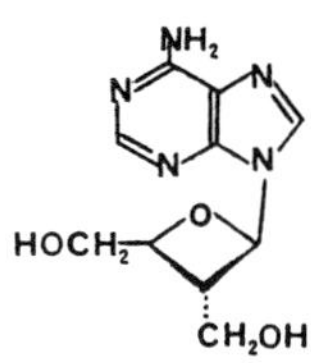

(7)

5.3.1.4 *Pteridines, other diazinodiazines, and Benzoderivatives.*

The readily available 3-chloro-4-pyridazinecarbonitrile (8) can be used as the starting material for the synthesis of pyrimido[4,5-c]pyridazine-5,7-(6H,8H)-diones (9), pyrimido[4,5-c]pyridazin-5(8H)-ones (10), and dihydropyrimido[4,5-c]pyridazin-5-(6H)-ones (11) (Scheme 7). Also synthesised during this work was the tetrazolo[1,5-b]-pyridazine (12) (88JHC119).

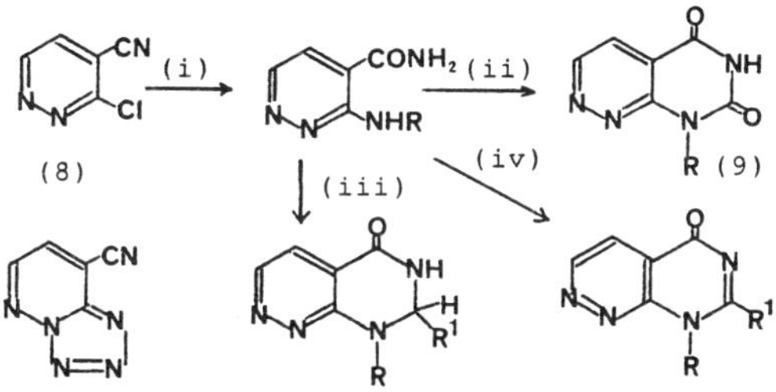

(8) (9) R=Ph,PhCH$_2$

(12) (11) R=Ph,PhCH$_2$, Me$_2$CH; R^1=Ph,3-Pyridyl (10) R=Ph,PhCH$_2$; R^1=H,Me,Et.Ph

Reagents (i) RNH$_2$,EtOH,reflux, then NH$_3$,H$_2$O,reflux
(ii) urea, 200° (iii) RCHO,180°
(iv) R^{1}C(OEt)$_3$, reflux

Scheme 7

Convenient and efficient syntheses of 2,4-diaminopteridine-6-carboxylic acid (13a) and pterin-6-carboxylic acid (14a) have been developed using mild permanganate oxidation of the corresponding 2-furyl derivatives (13b, 14b) (88JHC1737).

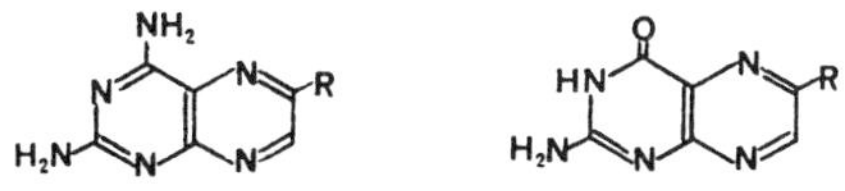

(13) (a) R=CO$_2$H (14)
(b) R=2-furyl

A new approach has been developed for the synthesis of pyrimido[4,5-d]pyrimidines which involves the synthesis of ethyl 1,3-dimethyl-6-(triphenylphosphoranylideneamino)-uracil-5-carboxylate (15). This reacts with isocyanates via an uracil-carbodiimide derivative to give 7-ethoxypyrimido[4,5-d]-pyrimidines (16) which can be converted into the corresponding 7-amino or 7-oxo-derivatives (Scheme 8) (88S919).

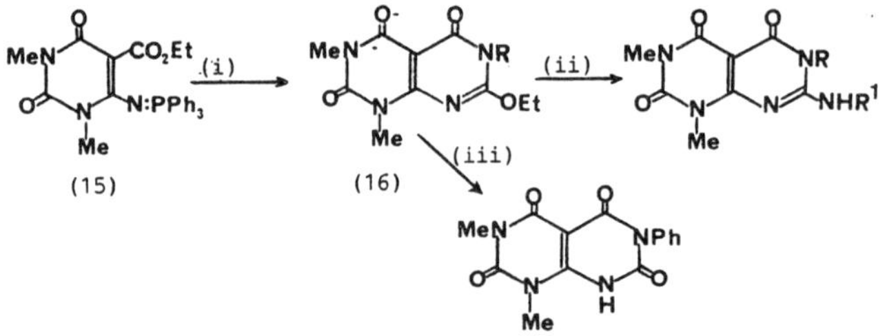

Reagents (i) RNCO,MeCN,reflux,4-8h (R=Ph,$PhCH_2$, and 4 other examples) (ii) R^1NH_2, reflux, various times (R^1=$PhCH_2$,*i*-Pr) (iii) 2M HCl,AcOH, reflux, 4h (for 12,R=Ph)

Scheme 8

The structure of the naturally occurring pteridine leucettidine has been revised to (17) and its structure proved by unambiguous synthesis (88T3373).

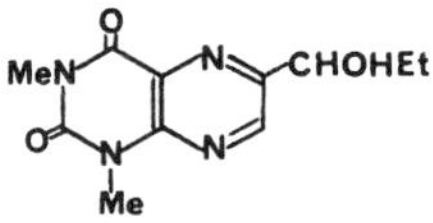

(17)

Although not strictly simple diazino-diazines, it is convenient at this point to record the continuing work on analogues of folic acid and related compounds. Compounds related to the folate cofactor 5,10-methylene-tetrahydrofolate (18) are important as potential antitumour agents. The new heterocyclic ring system 8-benzyl-1,3-diamino-5,7,8,9-tetrahydro-1*H*-pyrrolo[3[1],4[1]:4,5]pyrido[2,3-d]pyrimidine-6-one (19) can be obtained by the condensation of 2,4,6-triaminopyrimidine with ethyl *N*-benzyl-4-oxo-3-pyrrolidinecarboxylate in diphenylether at 190 - 200° (87% yield)(88JHC1597).

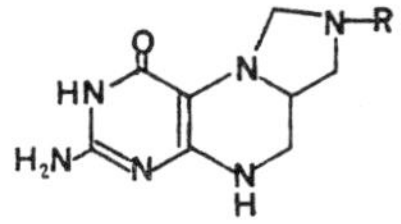

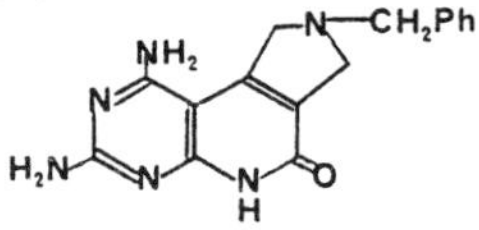

(18) (19)
R=benzoylglutamate

The synthesis of this ring system has also been achieved by using a pyridopyrimidine or pyrrolopyridine starting material to give compounds (20) and (21) respectively (89JOC220).

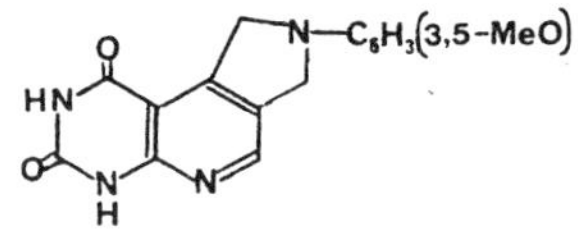

(20)

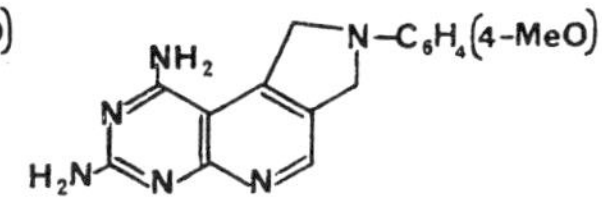

(21)

A folate analogue of current interest is 5,10-dideaza-5,6,7,8-tetrahydrofolic acid (22), and a new convergent and efficient synthesis of this compound has been designed (88JOC3618).

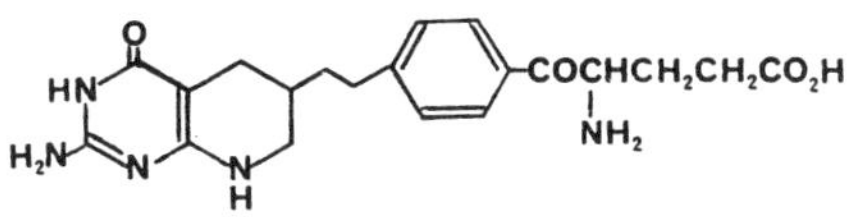

(22)

A series of novel flavinophanes (24) has been synthesised from the phenol-substituted flavin (23) as flavin coenzyme models, and some 5-deazaflavins and 5-deaza-10-oxaflavins (25) have been obtained by direct condensation of 6-chlorouracils with o-(substituted amino)benzyl alcohol and o-hydroxybenzyl alcohol (89CC44).

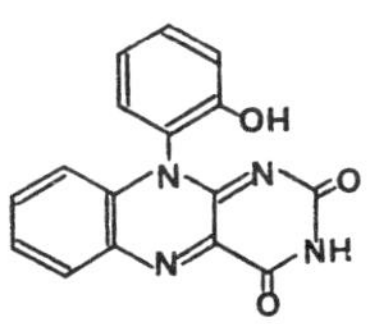

(23)

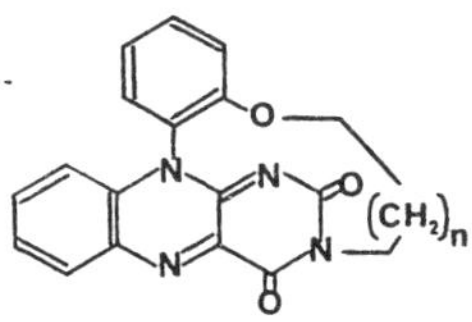

(24) n=6,7,8,10,12

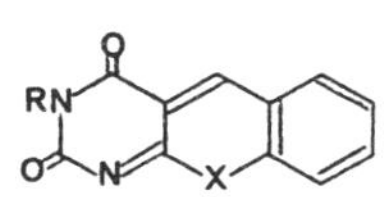

(25)
R=Me,H;X=O,NR^1
(R^1=Et,Ph,$4ClC_6H_4$)

Yoneda and coworkers have also synthesised "double 5-deazaflavins" having the potential for oxidoreduction, these compounds having either the linear (26) or angular (27) structures (88JCS(P1)1809,1813).

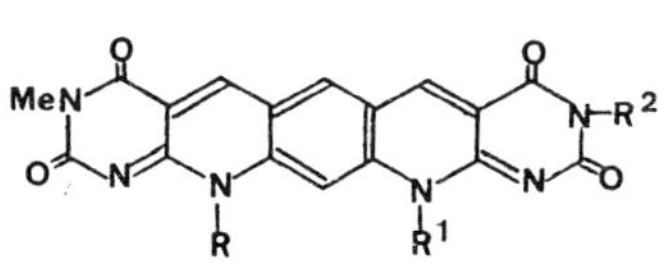

(26)

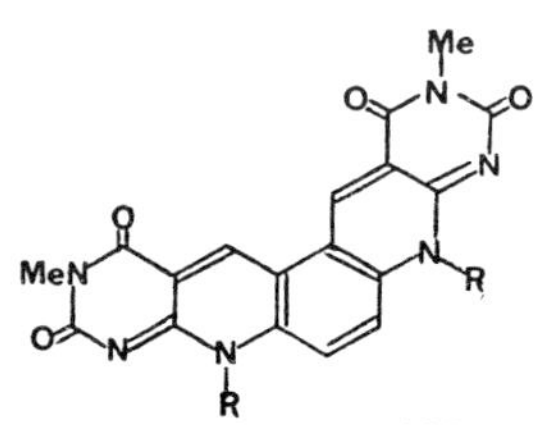

(27)

5.3.1.5 *Triazines with other fused rings*

The 3-alkylamino-5-phenylpyrazoles (28, R=Me,Et,$PhCH_2$) react with amines to give tetrahydropyrazolopyrimidines (29). However the 3-arylamino compounds (28, R=Ph,4-MePh) cyclise on the ring nitrogen to yield tetrahydropyrazolo[1,5-a]-triazines (30) in good yield (Scheme 9) (88JHC1387).

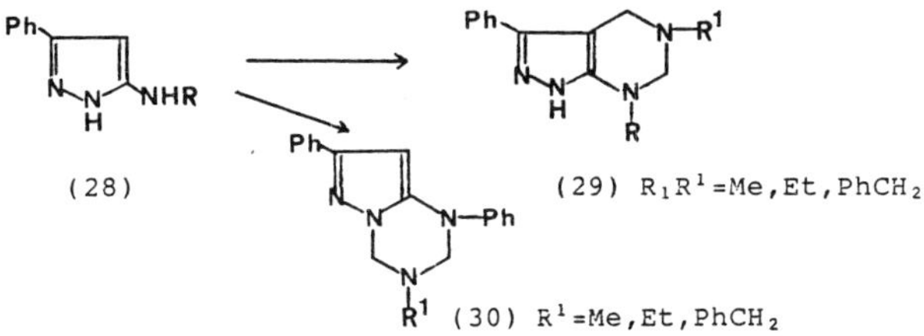

(28)

(29) R_1R^1=Me,Et,$PhCH_2$

(30) R^1=Me,Et,$PhCH_2$

Reagents: R^1NH_2,HCHO,MeOH, r.t. various times

Scheme 9

The new heterocyclic system [1,2,3]-triazolo[1,5-d][1,2,4]triazine (31) is obtained by the reaction of a variety of one carbon units (ortho esters, aldehydes, etc) with hydrazones formed by 4-benzoyl-1,2,3-triazole. The products are very easily hydrolytically cleaved by base back to substituted triazoles (88JHC743).

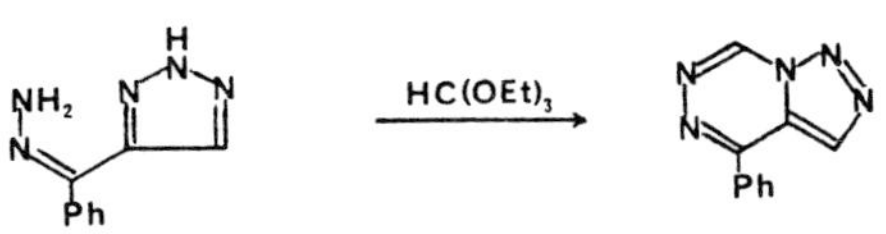

(31)

A new route to pyrazolo[3,4-e]-1,2,4-triazines (33) has been developed starting from the hydrazonopyrazolamines (32) (89LA501).

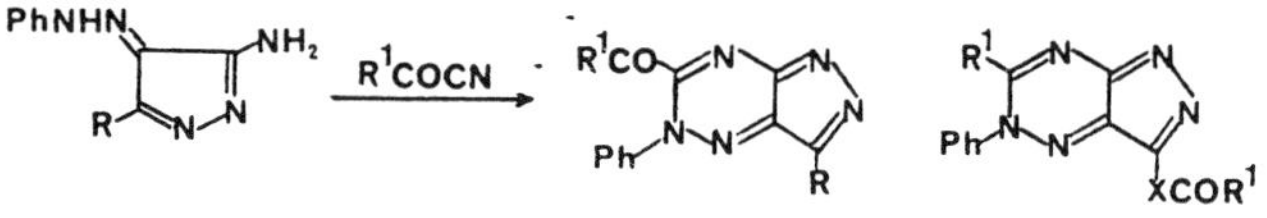

(32) (33)

The condensation of hydrazine hydrate with 1-acyl-2-aroyl-4,5-disubstituted-imidazoles (34) leads to the 5-phenyl(methyl)-2,3,8-trisubstituted-imidazo [1,2-d]-*as*-triazines (35) (88JIC784).

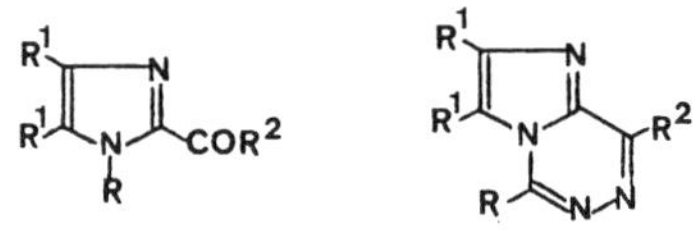

(34) R=Ac,PhCH$_2$
R^1=Ph,2-furyl;
R^2=Ph,4-ClC$_6$H$_4$

(35) R=Me,Ph

An unexpectedly easy synthesis of ethyl 1,2,3-triazino[5,4-b]indole-4-carboxylate 3-oxide (37) has resulted from the reaction of ethyl 1-methylindol-2-ylacetate (36) with excess of sodium nitrite in acetic acid. The reaction appears to proceed by nitrosation of both the methylene group and the 3-position of the indole, followed by cyclisation, the nitrous acid acting as a reducing agent (88JCR(S)344).

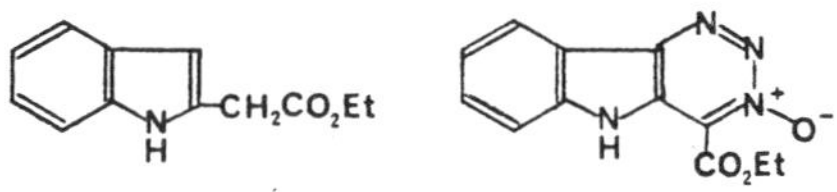

(36) (37)

Derivatives of the [1,2,4]triazine-[6,5-e]-1,2,4-triazine system (39) have been prepared via the triazine (38) which is obtained from the corresponding 5-sulfur-substituted triazines (88S877). Using related chemistry a number of 1,2,4-triazolo[4,3-d][1,2,4]triazines (40-42) have also been prepared (88S778).

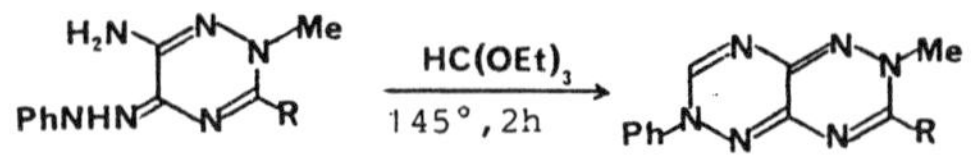

(38) (39)R=Me,Ph,MeS,4-ClC_6H_4,4-MeC_6H_4

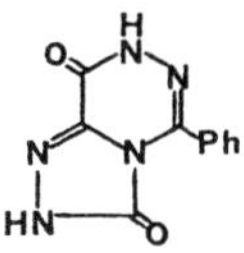

(40)

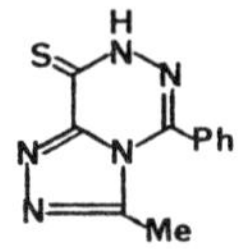

(41)

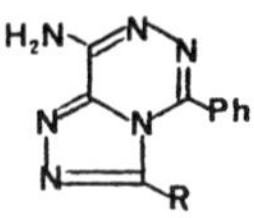

(42) R=H,Me,Ph

Some derivatives of the new heterocyclic system pyrimido[3,4-b]-1,2,4-triazine (44) have been synthesised as potential anticancer agents using the pyrimidinethione (43)(88MI805).

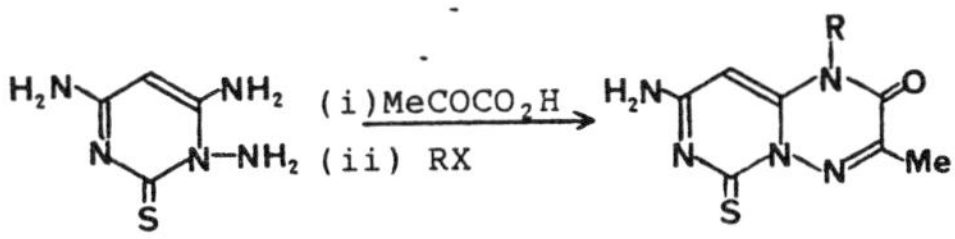

(43) (44) R=Me,Et,Pr,Bu,$PhCH_2$, 4-$NO_2C_6H_4CH_2$

Other new compounds which have also been developed as potential chemotherapeutic agents include the pyrazolo[1,5-a]pyrimidines (45) and pyrazolo[5,1-c]-1,2,4-triazines (46)(88AP851).

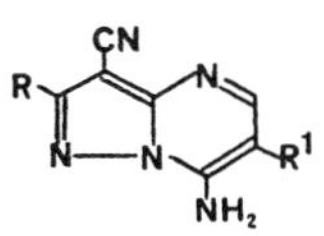

(45) R=Ph;R^1=CN,CO_2Et R=H;R^1=CN

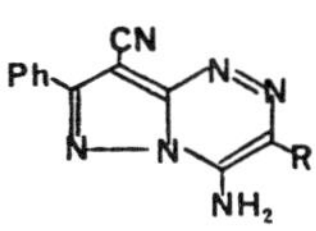

(46) R=CN,CO_2Et

The diazotisation of 5-amino-4-carbamoyl-1-(α-styryl)-1,2,3-triazole (47) initially forms 7-phenyl[1,2,3]triazolo[5,1-c][1,2,4]triazine-3-carboxamide (48). However this undergoes a Dimroth rearrangement so that the product which is isolated is 5-phenyl[1,2,3]triazolo[1,5-b]-[1,2,4]triazine-3-carboxamide (49)(88BSB179).

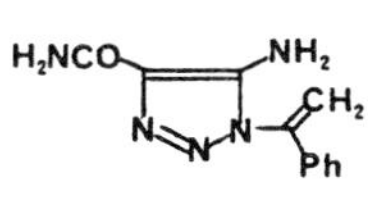

(47)

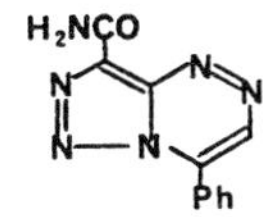

(48)

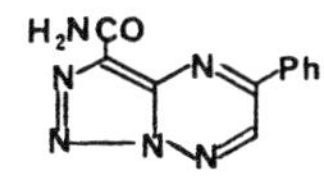

(49)

5.3.1.6 *Miscellaneous ring systems*

Some novel 2,3-dihydro-6-hydroxypyrimido[2,1-f]purine-4,8(1H,9H)-diones (51) have been obtained by the regiospecific reduction of the carbonyl group at the 2-position of 6-hydroxypyrimido[2,1-f]purine-2,4,8(1H,3H,9H)-triones (50) using $LiBH_4$ in refluxing dioxane (89JMC1118; 88JOC3265).

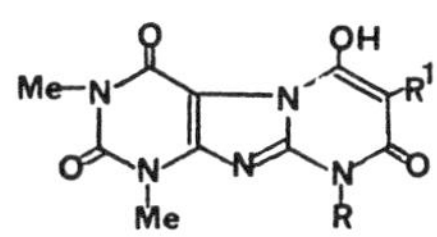

(50)

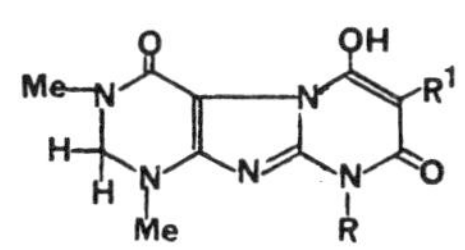

(51) (various R,R^1)

Pyrimido[4,5-d]pyridazines having good leaving groups at C-5 and C-8 can be reacted with aminoalkanols, followed by cyclisation, to yield imidazo [1,2-b]-pyrimido[4,5-d]pyridazines, imidazo [1,2-b]-pyrimido[5,4-d]pyridazines, dipyrimido-[1,2-b:5^1,4^1-d]pyridazines and dipyrimido-[1,2-b:4^1,5^1-d]pyridazines, and other systems (Scheme 10) (89T4485).

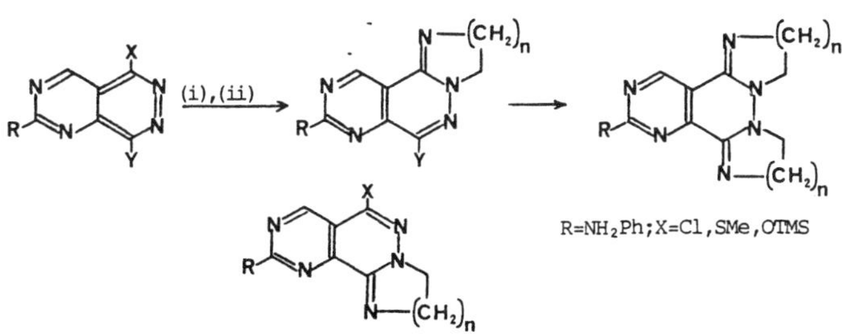

Reagents (i) EtOH, reflux, 3h (ii) $SOCl_2$ reflux, 2h.

Scheme 10

The reactions of 5,6-dichloropyrazine-2,3-carbonitrile with appropriate amines and with hydrazine has led to the synthesis of some new fused nitrogen-rich heterocycles. The systems (52-55) have been obtained (88LA1197).

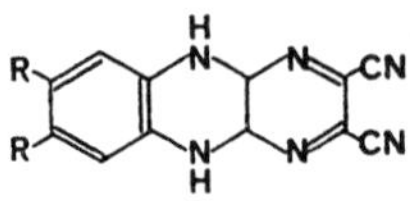
(52)

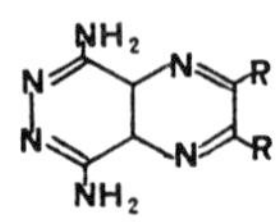
(53)

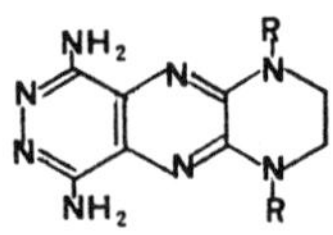
(54)

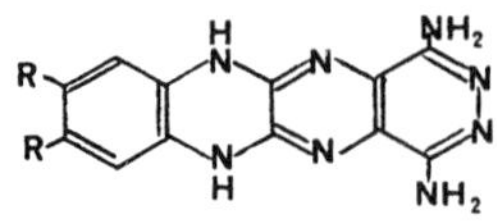
(55)

Another nitrogen-rich system has been obtained by the reaction of 3-tetrazolylpyridin-2-amines with carbonyl compounds which gives [3,2-e]tetrazolo[1,5-c]pyrimidines (56)(88CZ311).

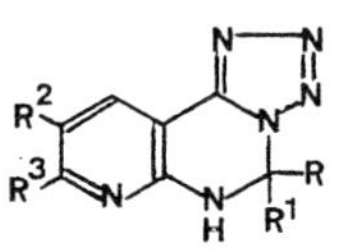
(56) (various Rs)

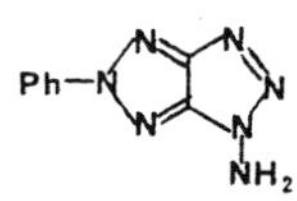
(57)

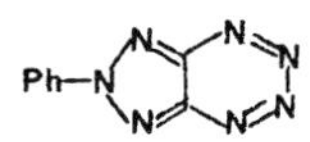
(58)

The first synthesis of a 1,2,3,4-tetrazine has been reported. The oxidation of the phenyltriazolotriazolamine (57) using lead tetraacetate in methylene chloride at 0° gives 6-phenyl[1,2,3]triazolo[4,5-e]-1,2,3,4-tetrazine (58) in 81% yield, the structure having been confirmed by X-ray crystallography (88CC1608).

Three new ring systems have been obtained by reactions of the substituted pyridines (59). These systems are 13*H*-4,6,7,13a,13c-pentaazabenzo[hi]chrysene (60a), 13*H*-4,7,13a,13c-tetraazabenzo[hi]chrysene (60b), and 7*H*-3,7,10,10b-tetraazacyclohepta[d,e]naphthalene (61) (88JHC1837).

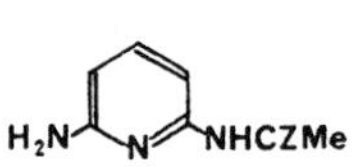
(59) Z=NCN, $C(CN)_2$

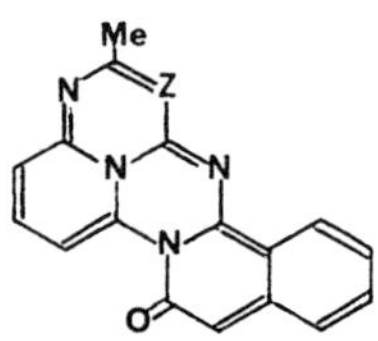
(60) (a) Z=N
(b) Z=CH

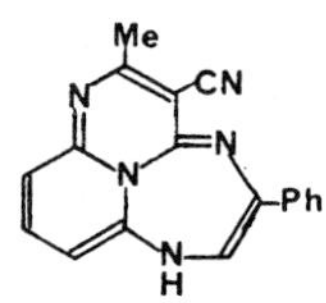
(61)

The new fused diazacyl[3.3.3]azine system (63) has also been obtained from the methylthioazacyl[3.3]azine (62) on reaction with methyl 3-oxobutanoate (89CPB1188). Some related compounds have also been obtained.

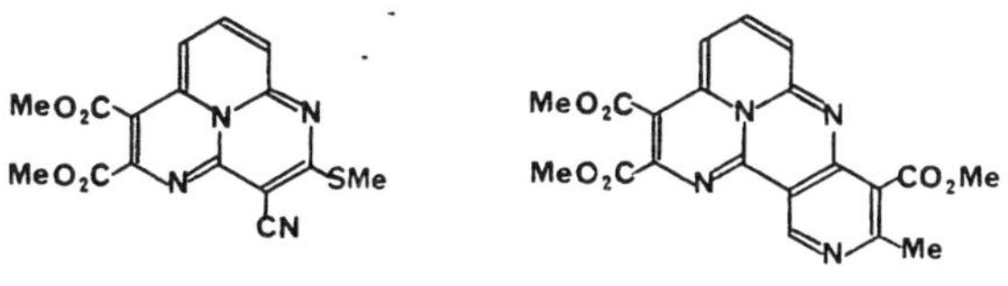

(62) (63)

5.3.2 Reactions

5.3.2.1 *Triazines and derivatives*

Attempts to desulfurise the 3,6-sulfur bridged hexahydro-1,2,4-triazin-5-ones (64) with Raney Nickel do not give triazinone products but give instead imidazolones (Scheme 11). The reaction may proceed via a diradical intermediate (89JCS(P1)33).

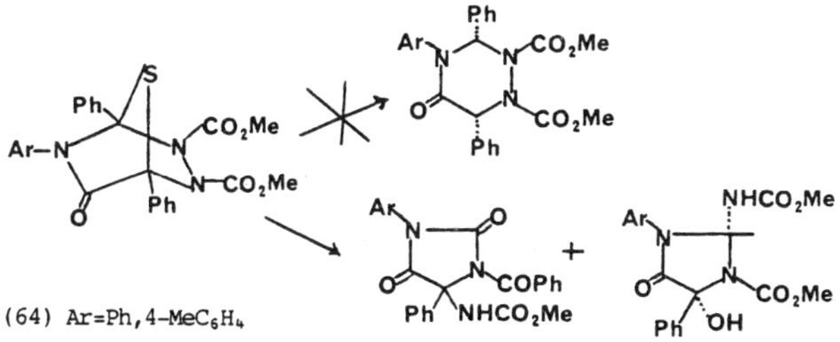

(64) Ar=Ph, 4-MeC_6H_4

Scheme 11

The cycloaddition of cyanamides to the 1,2,4-triazine (65) gives exclusively C-5, N-2 addition to give (66) as the non-isolable intermediate. The main products are 1,3,5-triazines (67) resulting from loss of trifluoroacetonitrile, but some 1,2,4-triazines (68) are also formed by loss of methylthiocyanate (89CB1381).

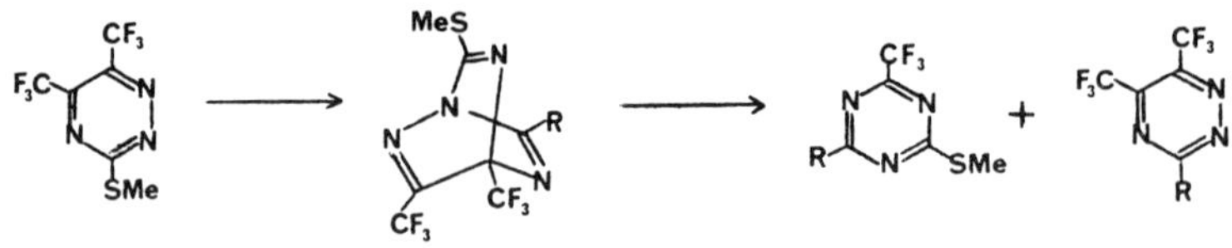

(65) (66) (67) (68)

(R=Me_2N,1-pyrrolidino, piperidino, morpholino

Triethyl 1,2,4-triazine-3,5,6-tricarboxylate (prepared in four steps from ethyl cyanoformate and diethyl oxosuccinate) undergoes cycloaddition-elimination reactions with vinyl pyrrolidines to give tricarbethoxypyridines, an example of an inverse-electron demand Diels-Alder reaction of an electron deficient azadiene (88OS142;88T3379).

The synthesis of trifluoro-1,2,3-triazine has been achieved by the reaction of the trichloro compound with KF at elevated temperatures. In the presence of hexafluoropropene and fluoride ion the trichlorotriazine gives hexafluoropropylfluorotriazines and also gives compound (69)(88T2583).

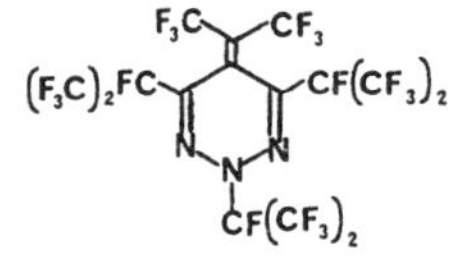

(69)

1,2,4-Triazines having electron-withdrawing substituents at the 3-position react with cycloocta-1,5-diene by an intermolecular Diels-Alder reaction followed by elimination of nitrogen to produce 2-azadienes which subsequently undergo intramolecular Diels-Alder reaction to give novel cage structures (70) (Scheme 12)(88CC1482).

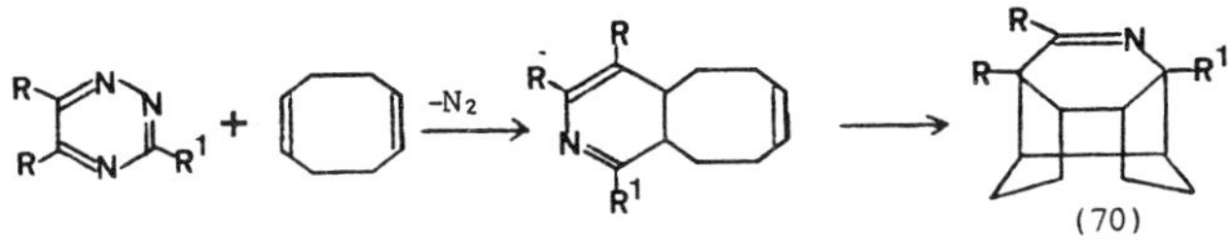

(70)

Reagents: xylene, 140°. R=R^1=Cl; R^1=CO$_2$Me,SO$_2$Me

Scheme 12

The thermal transformations of some 3-[{2^1-(2propynylthio)phenyl}amino]-1,2,4-triazines (71) have been investigated. The major reaction products are not the expected fused 1,5-benzothiazepines (72), which are obtained in less than 5% yield, but the pyrido-[3,2-b][1,4]benzothiazines (73) and the 3-(2H-1-benzothiopyran-8-ylamino)-1,2,4-triazines (74) (Scheme 13)(89JOC1456).

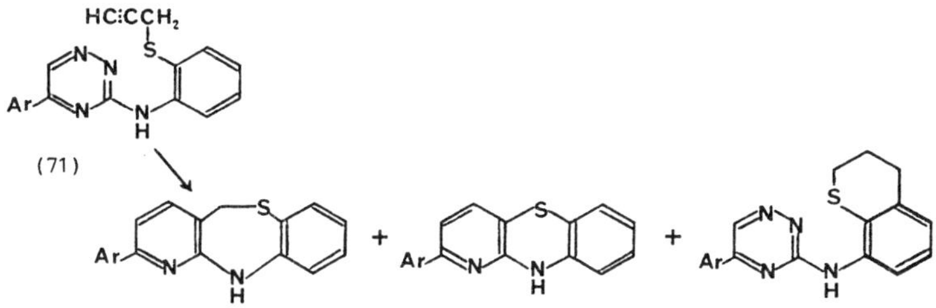

(71) (72) (73) (74)

The intramolecular Diels-Alder reaction of the triazine (75) has been used to make the thienopyridine (76) as the precursor to 2-amino-6,7-dihydrothieno-[3^1,2^1:5,6]pyrido-[2,3-d]pyrimidin-4-one (77)(88JHC1733).

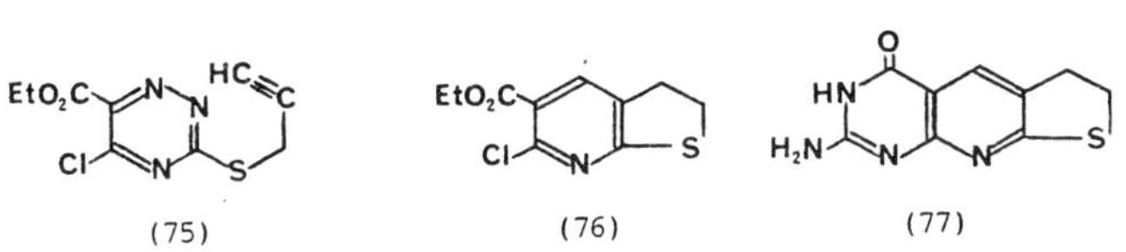

(75) (76) (77)

Further examples of intramolecular Diels-Alder reactions of 6-(alkynylthio)-1,2,4-

triazines have led to the synthesis of thieno-[2,3-c]pyridines (78) and 3,4-dihydro-2H-thiopyrano[2,3-c]pyridines (79)(89JOC4934) and the use of 3-alkynylthio-1-ethyl-1,2,4-triazinium salts leads to the formation of the thieno-[2,3-b]- and thiopyrano[2,3-b]pyridines under very mild conditions (89T6499).

(78) (79)

A new route to fused-ring 1,2,4-triazines involves the reaction of 1,2,4-triazinium salts with bifunctional nucleophiles (Scheme 13). The products (80)(81) can not be isolated as free bases. The nitrosation of products (81) gives the N-nitroso compounds (82). These reactions represent the first examples of orthocyclisation on the 1,2,4-triazine ring by the addition of dinucleophiles at C-5,6 (88TL1431).

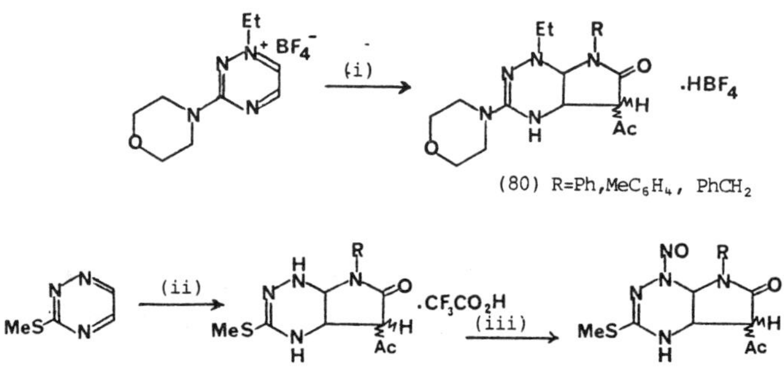

(80) R=Ph, MeC_6H_4, $PhCH_2$

(81) R=Ph, MeC_6H_4, HOC_6H_4 (82)

Scheme 13

Two general methods for the preparation of [1,2,4]triazolo[5,1-c]triazine and [1,2,4]-triazino[4,3-b][1,2,4,5]tetrazine ring systems have been described by the synthesis and subsequent aza-Wittig reaction of iminophosphoranes derived from 4-amino-1,2,4-triazines having a hydrazine substituent at the 3-position. Annelation occurs via a carbodiimide available from reaction of the iminophosphorane and an isocyanate (Scheme 14) (89JCS(P1)247). This work extends earlier studies reported in PHC,Vol.1(88T2249).

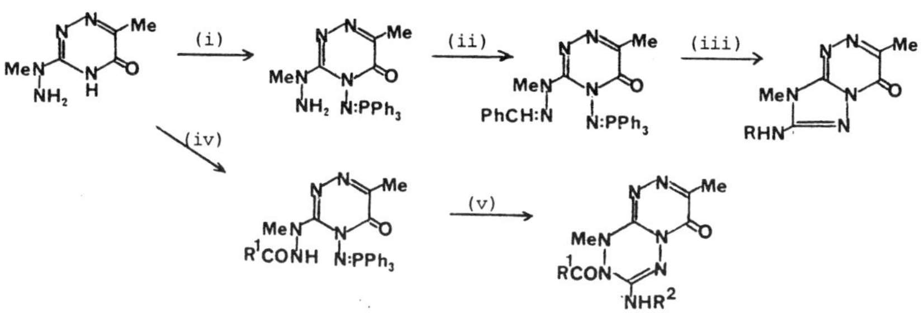

Reagents (i) Ph_3PBr_2, r.t. (ii) PhCHO, r.t. (iii) RNCO (R=Me,Et,and 5 other examples)
(iv) R^1COCl,Et_3N,r.t. (v) R^2NCO,CH_2Cl_2,r.t.(R^1=Me,Ph, and 6 other examples)

Scheme 14

The iminophosphorane (83) reacts with thioureas to give the 7-arylamino-3-methyl-8H-4-oxo[1,2,4]triazolo[5,1-c][1,2,4]-triazines (84) directly, these compounds also being obtained in good yields by the reaction of the triazine (85) with diarylcarbodiimides (Scheme 15) (89H1607).

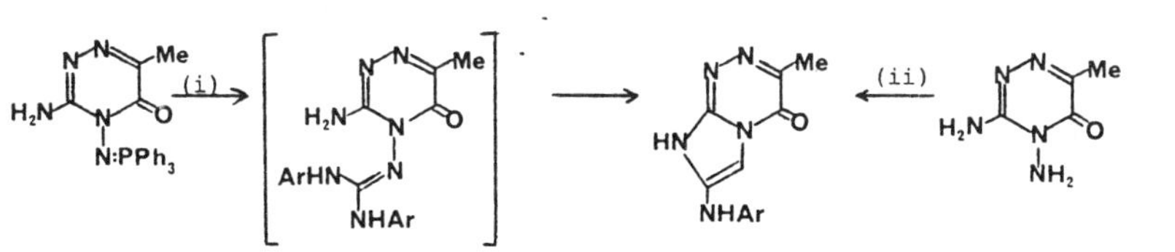

(83) (84) (85)

Reagents (i) ArNHCSNHAr, toluene, reflux 24h (ii) ArN=C=NAr,toluene, reflux 24h.

Scheme 15

Some 1,2,4-triazinyl-1,3-diazetidines (86) have been synthesised and have been converted to N^1,N^2,N^3,N^4,N^5-penta-substituted biguanides (87) having triazinyl substituents. Two examples of biguanide ($Ar=Ph, 4-ClC_6H_4$, $R=CH_2CH_2NH_2$) gave [1,2,4]-triazolo[5,1-c][1,2,4]triazines (88) on refluxing in dry toluene together with the 2-(triazinyl)imidazolidine (89) (Scheme 16) (89JOC1264). Some other transformations of (86) have also been described including the synthesis of some [1,3,4]thiadiazolo[2,3-c]-[1,2,4]triazines (89JCR(S)140).

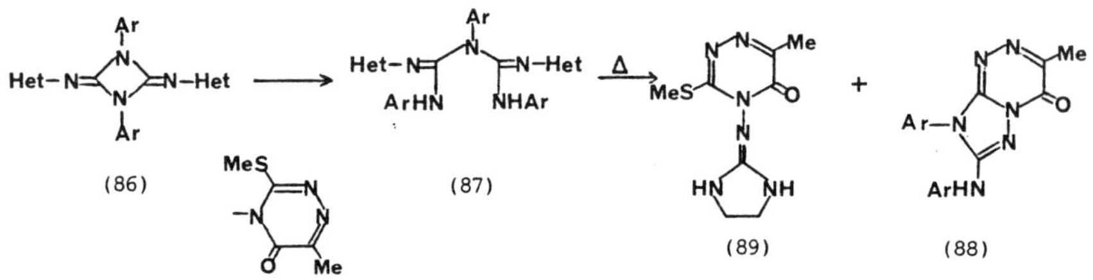

$Ar=Ph, 4-ClC_6H_4, 4-MeOC_6H_4$; R=Me, various

Scheme 16

5.3.2.2 *Pteridines and related heterocycles*

1,3-Dimethyllumazine (90) is converted to a 7-phthalamidomethyl derivative (91) by a radical reaction involving phthalamido-alkanoic acids in the presence of peroxodi-sulphate and silver ions. Although (91c) gives an expected 7-(3^1-aminopropyl)-1,3-dimethyllumazine on treatment with hydrazine, (91a) undergoes loss of the phthalamidomethyl residue to regenerate (90). Free-radical methylation of 91a) gives the 6-methyl derivative which also undergoes loss of the phthalamidomethyl residue to give 1,3,6-trimethyllumazine. These results suggest that the phthalamidomethyl group has a potential protecting role for the 7-position in pteridines (88JHC1295).

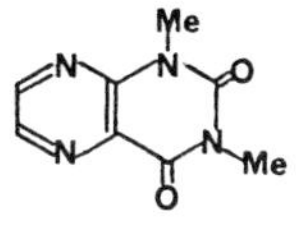

(90)

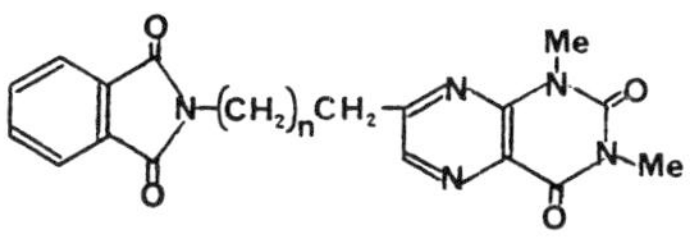

(91) (a) n=o (b) n=1 (c) n=2

5.3.2.3 *Ring-fused Triazines*

5<u>H</u>-3-Aroylmethylthio[1,2,4]triazino-[5,6-b]indoles (92) cyclise to yield the 3-arylthiazolo[3[1],2[1]:2,3][1,2,4]triazino-[5,6-b]indoles (93) and not the isomeric 1-arylthiazo[2[1],3[1]:3,4][1,2,4]triazino[5,6-b]-indoles (94). A similar reaction using dibromoalkanes $Br(CH_2)_nBr(n=2,3)$ gives similar linear and not angular products (88IJC(B)346,731).

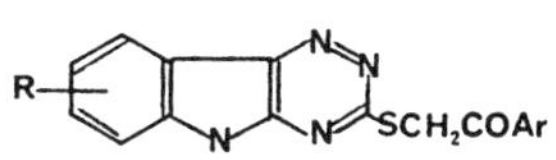

(92) R=7-Cl,8-Br; Ar=Ph,4-XC_6H_4 (X=Br,Cl,Me,Ph)

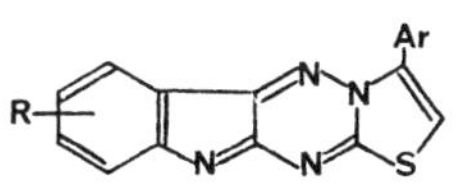

(93)

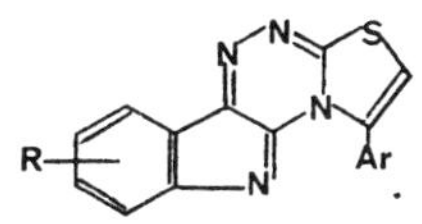

(94)

The first example of the direct amination of a 1,2,4-triazine ring is the amination using the Chichibabin procedure of isofervenulin (95) to give a dihydro product which can be subsequently oxidised to give the aromatised product (96). However the reaction using dimethylamine gives the pyrimidinylamidine (97). The reaction of fervenulin with KNH_2/NH_3, followed by permanganate, results in the formation of the product (98) (89CHE1404).

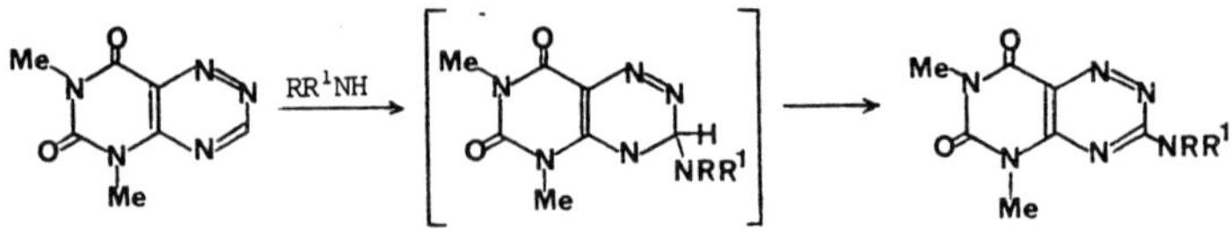

(95) (96) R=H,Me,Et

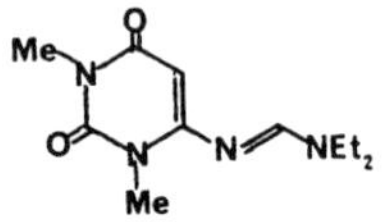

(97)

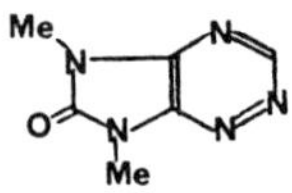

(98)

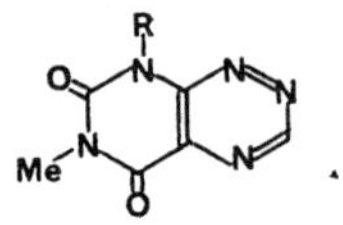

(99) (a) R=H (b) R=Me

Studies of the protonation of pyrimidotriazinediones have been carried out using ^{1}H and ^{13}C n.m.r. spectroscopy. These studies have established that the protonation of rheumycin (99a) and fervenulin (99b) takes place at N-2, whereas isofervenulin (95) takes place competitively at N-1, N-2, and O-6. The equilibrium constants for these processes have been measured (89CHE1271).

^{15}N and ^{14}N n.m.r. data have also been reported for a number of fused ring N-heterocycles leading to progress in the generalisation of the effects of N-N interactions on N-shieldings (89MI159).

The work of Neunenhoffer and coworkers on the synthesis of fused-ring triazines has continued with synthesis of [1,2,4]-triazino[6,5-e]-1,2,4-triazines (100), 1,2,4-triazolo[4,3-d][1,2,4]triazines (101), and triazines containing fused S-containing rings such as 1H-[1,2,4]triazino[5,6-e]-[1,3,4]thiadiazines (102) (88S778;88S877).

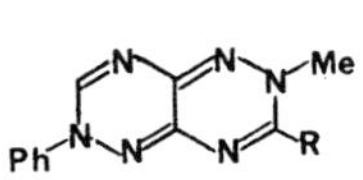

(100)

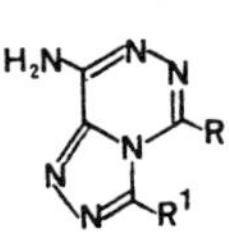

(101)

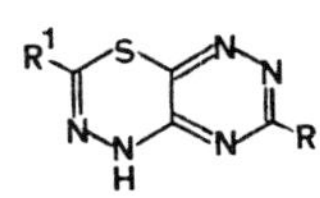

(102)

(R,R^{1}= various alkyl,aryl)

5.3.2.4 *Purines, and related systems*

8-Azidopurines carrying 7-alkynyl or alkenyl substituents undergo cyclisation reactions, with loss of nitrogen, on refluxing in toluene to yield pyrimido-[2,1-f]purines (103) in good yields (89S681). However the azidomethylene compound (104) gives the 1,2,3-triazolo[1^1,5^1:1,2]pyrazino-[5,4-f]purine structure (105) under similar conditions (88CCC319) (Scheme 17)

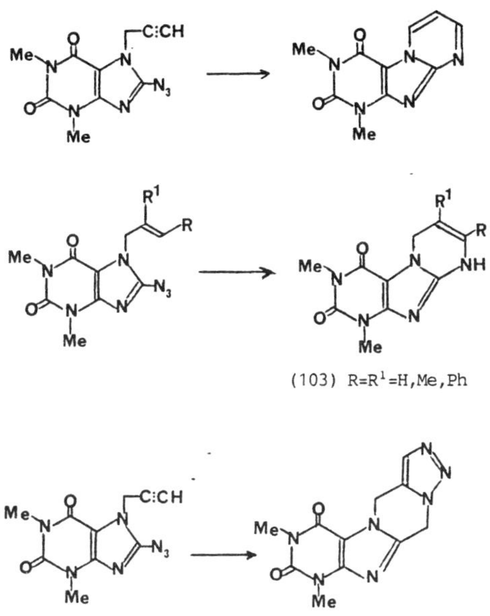

(103) R=R^1=H,Me,Ph

(104) (105)

Reagents: toluene, reflux, 1-2h

Scheme 17

REFERENCES

87MI1077	H.Hoshino, N.Shimizu, N.Shimada, T.Takita, and T.Takeuchi; J.Antibiotics, 1987,40, 1077.
88AP851	M.H.Elnagdi, K.U.Sadek, F.M.A.Galil, and S.M.E.Hassan; Arch.Pharm.(Weinheim), 1988, 321, 851; Chem.Abstr. 1989, 110, 114805.
88BSB179	G.L'Abbe, W.Meutermans, L.Van Meervelt, G.S.D.King, and A.T.H.Lenstra; Bull.Soc.Chim. Belg., 1988, 97, 179; Chem. Abstr., 1989, 111, 7350
88CC1482	I.Lantos, P.W.Sheldrake, and A.Wells; J.Chem.Soc.Chem. Commun., 1988, 1482.
88CC1608	T.Kaihoh, T.Itoh,K.Yamaguchi, and A.Ohsawa; J.Chem.Soc.Chem. Commun., 1988, 1608.
88CCC319	D.Hesek, F.Povazanec, A.Martvon, and J.Kovac; Collect.Czech.Chem.Commun., 1988, 53, 319.
88CZ281	G.Seitz and H.Wassmuth;Chem. -Ztg., 1988,112,281;Chem. Abstr;1989,110,135203.
88CZ311	W.Reid and M.Singh;Chem.-Ztg., 1988,112,311; Chem.Abstr., 1989,110, 212750.
88H917	M.M.El-Abadelah, M.R.Kamal, and K.H.Al-Ashami; Heterocycles,1988,27, 917.
88JCR(S)344	A.Monge, J.A.Palop, J.I.Recalde, and E.Fernandez-Alvarez; J.Chem.Res.Synop., 1988,344.
88JCS(P1)1809	F.Yoneda and M.Koga; J.Chem. Soc.Perkin Trans.1,1988,1809.
88JCS(P1)1813	F.Yoneda,M.Koga, and Y.Yano; J.Chem.Soc.Perkin Trans.1, 1988,1813.
88JHC119	N.Haider, G.Heinisch, and D.Lassnigg; J.Heterocycl. Chem.,1988,25,119.

88JHC743	M.Bianchi, A.Butti, and J.Perronet; J.Heterocycl. Chem.,1988,25, 743.
88JHC1295	M.Tada and S.Totoki; J.Heterocycl.Chem.,1988,25, 1295.
88JHC1387	J.N.Vishwarkarma, M.Mofizuddin, H.Ila, and H.Junjappa; J.Heterocycl. Chem., 1988,25,1387.
88JHC1597	A.Gangjee, A.Patel, and F.-T.Lin; J.Heterocycl. Chem., 1988, 25, 1597.
88JHC1733	E.C.Taylor, J.L.Pont, and J.C.Warner; J.Heterocycl. Chem., 1988, 25, 1733.
88JHC1737	N.Sata and N.Saito; J.Heterocycl.Chem.,1988,25, 1737.
88JHC1837	J.T.Shaw, W.L.Corbett, D.L.Layman, G.D.Cuny, and J.L.Kerschener; J.Hetero-cycl.Chem., 1988, 25, 1837.
88JIC784	E.K.Dora, B.Dash, and C.S.Panda; J.Indian Chem. Soc., 1988, 65, 784.
88JOC3265	D.J.Conn, J.J.Kaminski,and D.M.Solomon; J.Org.Chem., 1988, 53, 3265.
88LA1197	W.Reid and G.Tsiotis; Liebig's Ann.Chem., 1988, 1197.
88OS142	D.L.Boger, J.S.Panek, and M.Yasuda; Org.Synth., 1988, 66, 142.
88IJC(B)346	J.Mohan and G.S.R.Anjaneyulu; Indian J.Chem.,Sect.B, 1988, 27B,346.
88JIC(B)731	J.Mohan and G.S.R.Anjaneyulu; Indian J.Chem.,Sect.B, 1988, 27B, 731.
88MI805	T.G.Koksharova, L.N.Dianova, N.V.Volkova, O.S.Anisimova, G.M.Anoshina, and G.V.Schvechnikova; Khim.-Farm.Zh.,1988,22,805;Chem. Abstr.,1989,110,114803.

88S778	H.Neunhoffer, M.Dostert, and H.Hammann; Synthesis, 1988, 778.
88S877	H.Neunhoffer and D.Reichel; Synthesis, 1988, 877.
88S919	H.Wamhoff and J.Muhr; Synthesis,1988,919.
88T2249	P.Molina, M.Alajarin, A.Vidal, J.Elguero, and R.M.Claramunt; Tetrahedron, 1988,44,2249.
88T2583	R.D.Chambers, T.Shepherd, and M.Tamura; Tetrahedron, 1988,44,2583.
88T3373	W.Pfleiderer; Tetrahedron, 1988,44,3373.
88T3379	D.L.Boger and Q.Dang; Tetrahedron, 1988, 44, 3379.
88TL1431	S.G.Alexeev, V.N.Charushin, and O.N.Chupakhin; Tetrahedron Lett.,1988,29,1431.
89CB1381	G.Seitz and R.John; Chem. Ber.,1989,122,1381
89CC44	X.Chen, M.Nagata, K.Tanaka, and F.Yoneda; J.Chem.Soc. Chem.Commun., 1989, 44.
89CHE1271	S.V.Shorshnev, S.E.Esipov, and A.I.Chernyshev; Chem. Heterocycl.Comp.(Eng.Transl.) 1989,24,1271.
89CHE1404	A.F.Pozharskii,A.V.Gulevskaya and V.V.Kuz'menko;Chem. Heterocycl.Comp.(Eng.Transl.) 1989,24,1404.
89CPB1188	Y.Matsuda, H.Goton, K.Koton, and H.Matsumoto, Chem.Pharm. Bull., 1989,37,1188.
89H1163	A.Q.Hussein, M.M.El-Abadelah, and K.H.Al-Adhami; Heterocycles,1989,29,1163.
89H1379	M.Takahishi, Y.Hikita, and M.Fukui; Heterocycles, 1989, 29,1379.
89H1607	P.Molina, M.Alajarin, M.J. Perez de Vega, and A.Lopez; Heterocycles, 1989,29,1607.

89JAC4928 S.Shinkai, A.Kawase, T.Yamaguchi, D.Manabe, Y.Wada, F.Yoneda, Y.Ohta, and K.Nishimoto; J.Am.Chem.Soc., 1989,111,4928.

89JCR(S)140 P.Molina, M.Alajarin, and C.Lopez-Leonardo, J.Chem.Soc. Res.Synop.,1989,140.

89JCS(P1)33 T.Sheradsky and N.Itzhak; J.Chem.Soc.Perkin Trans.1, 1989,33.

89JCS(P1)247 P.Molina, M.Alajarin, and A.Vidal; J.Chem.Soc.Perkin Trans.1,1989,247.

89JCS(P1)543 A.J.Boulton, P.Devi, N.Henderson, A.A.Jarrar, and M.Kiss; J.Chem.Soc.Perkin Trans.1, 1989, 543.

89JMC1118 J.J.Kaminski, D.M.Solomon, D.J.Conn, S.-C.Wong, P.J.S.Chin, T.Massa, M.I. Siegel, and A.S.Watnick, J.Med.Chem,m 1989,32,1118.

89JOC220 T.L.Su and K.A.Watanabe; J.Org.Chem.,1989,54,220.

89JOC1264 P.Molina, M.Alajarin, C.Lopez-Leonardo, M.de la Concepcion Foces-Foces, F.H.Cano, R.M.Claramunt, and J.Elguero, J.Org.Chem., 1989, 54,1264,

89JOC1456 E.C.Taylor, J.L.Pont, and J.C.Warner; J.Org.Chem., 1989, 54,1456.

89JOC3062 S.Batori, G.Hajos, P.Sandor, and A.Messmer; J.Org.Chem., 1989,54,3062.

89JOC3618 E.C.Taylor and G.S.K.Wong; J.Org.Chem.,1989,54,3618.

89JOC4780 M.M.Mansuri, J.E.Starrett, J.A.Wos, D.R.Tortolani, P.R.Brodfuehrer, H.G.Howell, and J.C.Martin; J.Org.Chem., 1989,54,4780.

89JOC4984 E.C.Taylor and J.E.Macor; J.Org.Chem., 1989,54,4984.

89LA501 K.U.Sadek, K.Abouhadid, and A.H.H.Elghandour; Leibig's Ann.Chem., 1989,501.

89MI159	L.Stefaniak, M.Witanowski, V.Mahmoud, J.D.Roberts, and G.A.Webb; J.Crystallogr. Spectrosc.Res.,1989,19,159.
89S228	J.Barluenga, M.Tomas, A.Ballesteros, and L.A. Lopez; Synthesis,1989,228.
89S681	D.Hesek, M.Tegza, A.Rybar, and F.Povazanec; Synthesis, 1989, 681.
89T4485	K.J.Szabo, J.Csaszar, and A.Toro; Tetrahedron,1989,45, 4485.
89T6499	V.N.Charshin, B.van Veldhuizen, and H.C.van der Plas; Tetrahedron,1989,45,6499.
89TL3719	D.T.Hurst and N.S.Jennings; Tetrahedron Lett., 1989,30, 3719.
89TL6453	W.A.Slusarchyk, M.G.Young, G.S.Bisacchi, D.R.Hockstein, and R.Zahler; Tetrahedron Lett., 1989,30,6453.

CHAPTER 5.4
Six-Membered Ring Systems: With O and/or S Atoms

JOHN D HEPWORTH
Lancashire Polytechnic, Preston, UK

HETEROCYCLES CONTAINING ONE OXYGEN ATOM

5.4.1.1 Pyrans

Progress in the application of the hetero Diels Alder reaction to dihydropyran synthesis includes the use of 4,4,4-trichloro-3-oxobutanal as a formylacetic acid equivalent giving access to dihydropyran-5-carboxylates (89CB643). Increased diastereoselectivity is observed by applying high pressure to the reaction of enaminoketones with ethyl vinyl ether (89CB1179) and the use of (2R)-borane-10,2-sultam as a chiral auxilliary in the cycloaddition of 1,2-dicarbonyl compounds to methoxybutadiene leads to high asymmetric induction (89HCA482). All four enantiomers of 2,5-divinyltetrahydropyran have been obtained from the palladium catalysed enantioselective telomerisation of butadiene with formaldehyde (89CC1151).

Ring expansion of 2-(cycloprop-1-en-1-yl)ethanol derivatives (1) leads to either methylenetetrahydrofurans or dihydropyrans (2). The latter is the major product when a methyl group is present at C-2 of the cyclopropene (89TL253). 3-Methylenetetrahydropyrans result from the reaction of epoxides with a propenyl Grignard reagent (89TL4863).

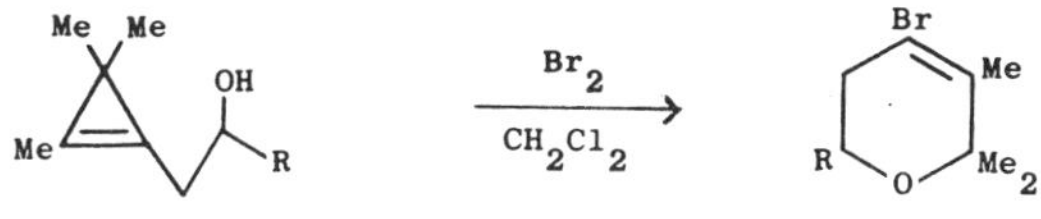

(1) (2)

Lewis acid catalysed reactions of silyl ethers with aldehydes affords di- or tetra-hydropyrans according to the degree of unsaturation in the ether chain (89TL4065, 89JOC5768). A silyl ether is also involved in the intramolecular nucleophilic addition of an alkoxide to a nitroalkene which leads to some novel derivatives of tetrahydropyran (89JOC4723).

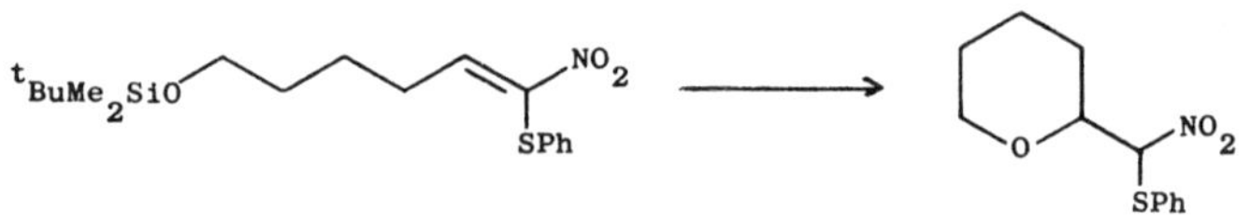

A study of the acid catalysed ring opening of hydroxy epoxides indicated that an electron rich double bond at the epoxide carbon remote from the hydroxyl function leads to tetrahydropyrans by an endo ring closure. The hydroxyl function can be incorporated into a cyclic structure allowing elaboration to a bicyclic system with the framework of the brevetoxins (89JA5330), a unit also accessible from tetrahydropyran-3-ones (89TL2437).

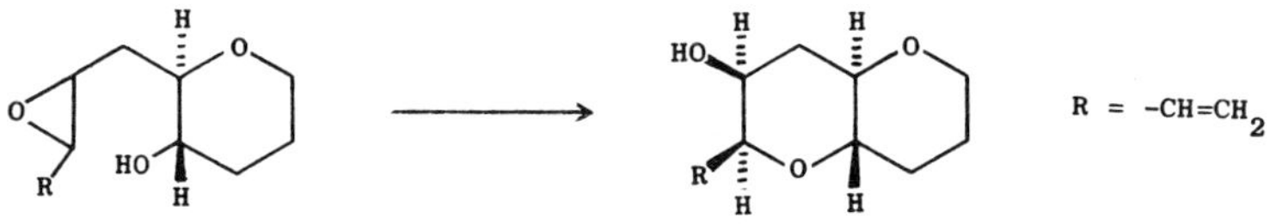

Utilising this methodology, the ABC, FG and IJK ring skeletons of brevetoxin have been synthesised (89JA6666, 6676, 6682). It remains to piece the three fragments together through the oxepane units to complete a total synthesis of this toxin associated with the red tide fish deaths in the Gulf of Mexico.

A range of heterocyclic derivatives of pseudomonic acid have been prepared either by cyclisation of monic acid derivatives or, better, by Peterson or Horner-Wittig methodology. Some of the compounds retain the antimicrobial activity of pseudomonic acid and show little evidence of in vivo breakdown (89JCS(P1)2047). Isotopic incorporation experiments indicate that five of the seven oxygen atoms of pseudomonic acid are derived from acetate and refute the proposed involvement of β-hydroxy-β-methyl glutarate in its biosynthesis (89JCS(P1)207).

Total syntheses of two potent poisons, (+)-pederine and palytoxin carboxylic acid have been reported (88T7063, 89JA7525). The latter involves the assembly of eight individually synthesised and protected fragments and the subsequent removal of 42 protecting groups (89JA7530).

There is an increasing interest in the syntheses and reactions of spiroketals because of their widespread occurrence and pharmacological significance and a review of their chemistry is timely and welcome (89CR1617).

Significant developments include total syntheses of avermectin A1a (89JA2967), (+)-milbemycin β1 (89T7161), milbemycin E (89CC1250) and (-)-talaromycin A (89JOC1157), whilst a new synthesis of the spiroketal fragment of avermectin B1b has been described (89TL5993). The acid catalysed rearrangement of 2-furyl ketones to pyranones is the key feature of the synthesis of 2-hydroxyspiroketals and trioxadispiroketals (89JOC2044), whilst aza analogues of spriocyclic ethers have been derived from ω- bromoamides via the 2-aminotetrahydropyrans (89TL6573).

5.4.1.2 Chromenes

Interest continues to be shown in the precocenes, alkoxy 2,2-dimethyl-2H-chromenes, because of their biological properties (89ACH149). The importance of chelation in aldol reactions of enolates derived from chroman-4-ones has been recognised, enabling 3-methylenechromanones and thence 3-substituted precocenes to be synthesised (89JCR(S)36).

The Iwai-Ide propargylation route to chromenes has been used to form 2-spiro-annulated derivatives (89TL2297) and to synthesise pyrano[2,3-c][1]benzopyran-5(3H)-one (3) (89JCS(P1)1901) and benzo[1,2-b:3,4-b']dipyran-4-ones (4) (89G481). The 4,8-diones corresponding to (4) are available by Knoevenagel methodology.

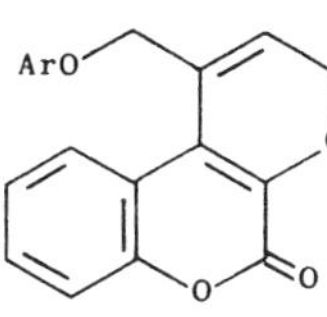

(3)

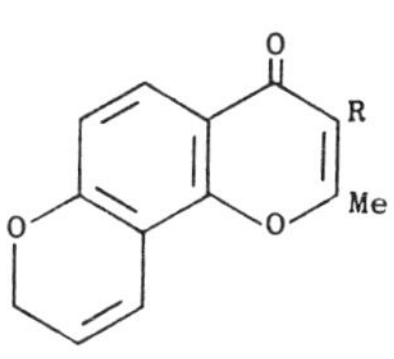

(4)

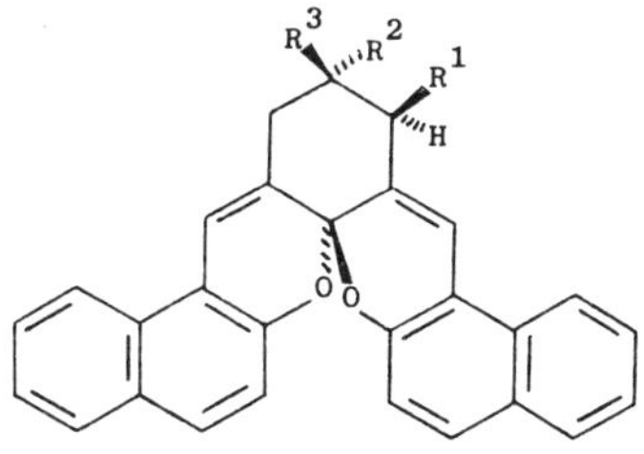

(5)

(R)-2,2'-Spirobibenz[d]chromene derivatives (5) are formed with good to excellent diastereoselectivity when optically active ketones react with 2-hydroxy-1-naphthaldehyde (89T6875).

The reaction of 2-hydroxybenzyl alcohols with β-functionalised enamines provides a new route to 4H-chromenes (89S69). Although 2H-chromenes are formylated in the aromatic ring under Vilsmeier conditions, 2,4-diaryl-4H-chromenes are attacked at C-3. In acetic anhydride, the 3-formyl derivatives yield benzo[b]indeno[2,1-e]pyrylium salts (6) involving an intramolecular electrophilic attack on the 2-phenyl substituent (89ZOK155).

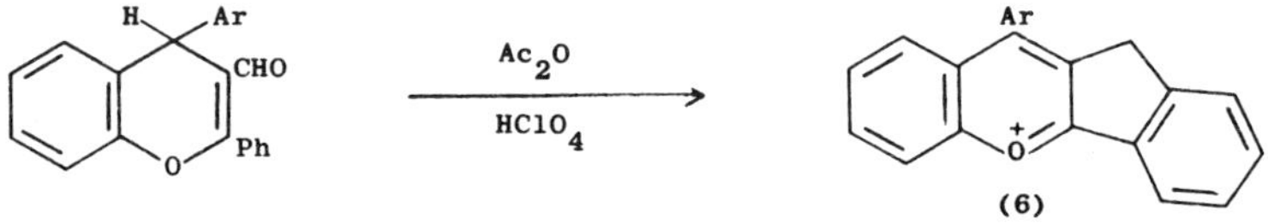

(6)

The reaction of phenanthrene-9,10-dione with acylmethylene phosphoranes leads to fused pyran derivatives (7) in good yield. The intermediate o-quinone methanide is trapped by a second mole of the phosphorane, but it can be trapped by other dienophiles. Isomerisation of the pyran has been noted (89JCS(P1)2329).

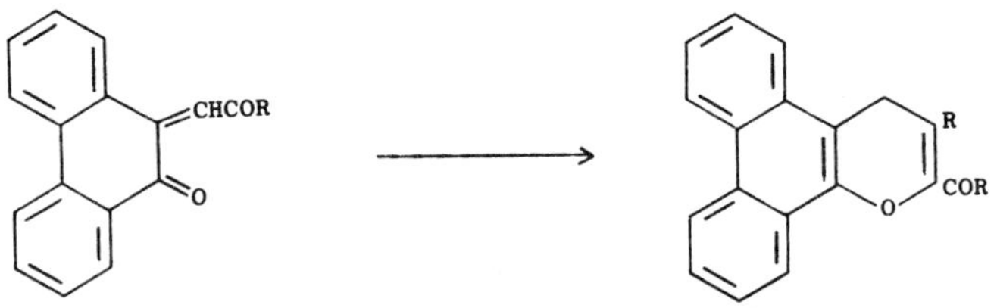

(7)

5.4.1.3 Chromans

Several novel radical initiated syntheses of the chroman ring system have been described. Phenylselenosalicylates (8; X=O) lose the phenylselenyl moiety on reaction with tributyltin radicals to yield chroman-4-one by a 6-exo-trig cyclisation (89H583). A similar cleavage occurs with selenoimidates (8;X=NR) yielding 4-iminochromans, which undergo a second cyclisation in the case of the N-aryl derivatives to give the chromanoquinoline (9) (89JA1886).

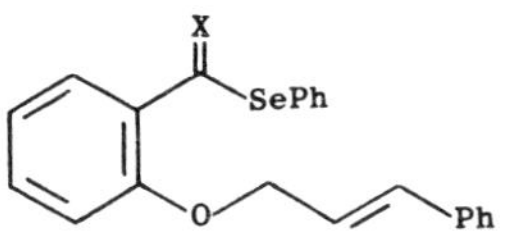

(8)

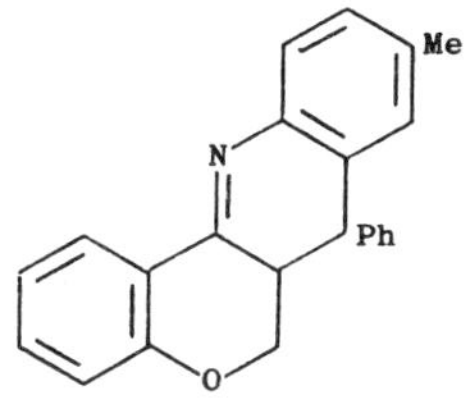

(9)

Aryloxymethylene radicals generated by photolysis of thiohydroxamate esters (10) undergo intramolecular addition to suitably placed double bonds to yield chromans (89TL4731).

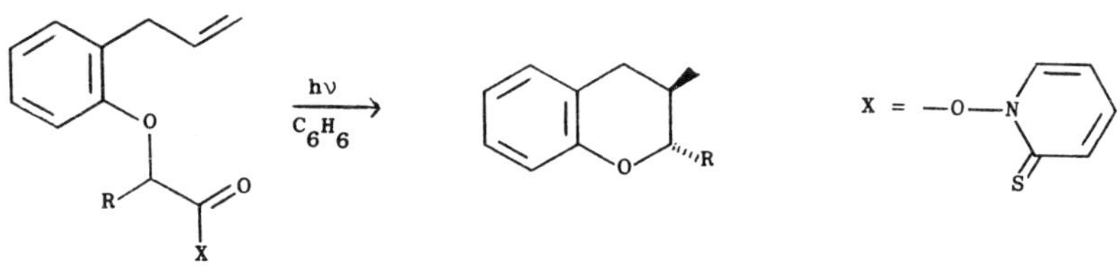

(10)

Interest in the robustadials continues. A stereoselective construction of the ring system starting from β-pinene relies on kinetic control of an intramolecular Michael reaction by careful choice of base (89TL6653). The total synthesis of robustadials A and B from (+)-nopinone utilises the Kabbe chromanone synthesis (78S886) and spectral analysis confirms their absolute stereostructures as (11, 4β-H and 4α-H), respectively (89JOC1562).

The ^{13}C nmr spectra of both natural (2R, 4'R, 8'R) and synthetic, all racemic, α-tocopherol have been completely assigned with the aid of some selectively deuterated analogues. Chirally induced chemical shifts are observed over five bonds from C-2 (89JOC560).

The structure of cryptosporin has been established by

the synthesis of its enantiomer (12) from a sulphonyllactone and nitroglycal or through the cylcoaddition of L-fucal to an isoquinolinium salt (89HCA1649).

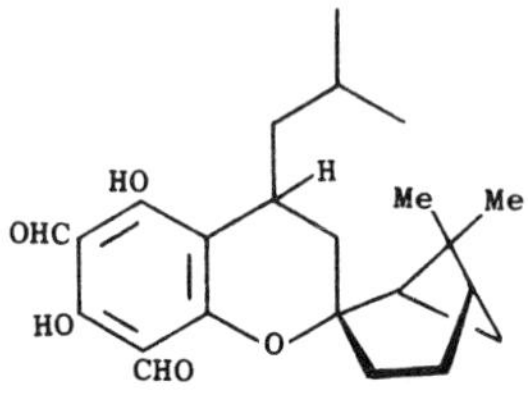

(11)

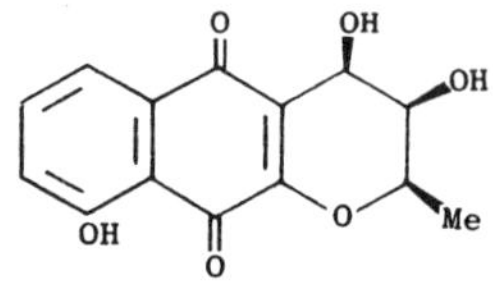

(12)

Although a 4-halogeno substituent in 2,2-dimethylchromans is known to be labile (83JCS(P1)827), reaction of a 3,3-disubstituted 4-bromochroman with a hindered nitrogen nucleophile leads to exclusive attack at C-5 (89CC1258). Chromans related to tocopherol are selectively methoxylated at C-4 on treatment with a basic methanolic solution of iodine via isomerisation of the 8a-methoxy compound (89JOC1987).

The reaction of various nucleophiles with 2-alkoxy and related chromans under Lewis acid catalysis leads to their incorporation at C-2, although ring opened products may also result depending on the nature of the leaving group. The choice of Lewis acid can also influence the course of the reaction, notably with the 2-spiro-annulated analogues (89JOC3282).

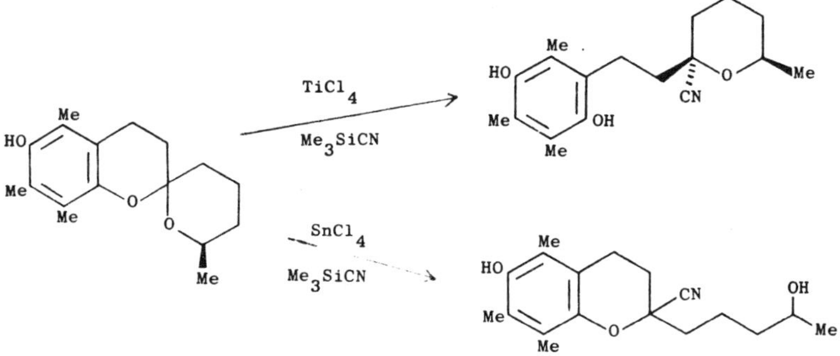

The synthesis and chemical and biological properties of isochromans have been reviewed up to 1985 (89RCR479) and a number of isochroman derivatives have been obtained by the palladium catalysed cyclisation of 2-iodobenzyl cycloalkenyl ethers (89H55).

A new strategy based on the aroylation of 4-phenylsulphonylchromans and which uses readily available starting materials has significant potential in the synthesis of rotenoids (89T5895). Rotenoids with the unnatural trans-fusion of the B/C rings result from the DIBAL reduction of 6a, 12a-dehydrorotenoids. The trans-isomers are readily epimerised to the more stable cis-series by acid (89JCS(P1)204).

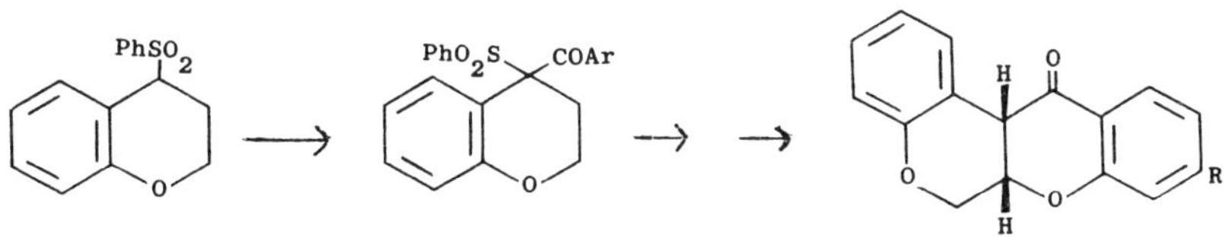

5.4.1.4 Pyranones

Alkynylalkoxycarbene complexes of Cr and W yield pyranyl metal complexes on reaction with ethyl diethoxyacrylate, presumably through a [2+2]cycloaddition, conrotatory ring opening of the cyclobutene and cyclisation at the metal. Removal of the metallic component is straightforward, affording a synthesis of 6-ethoxypyran-2-ones (89CC1560).

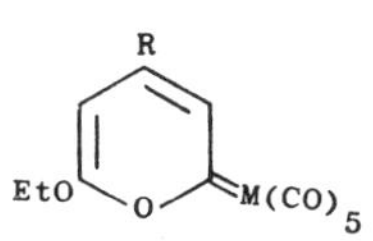

M = Cr or W

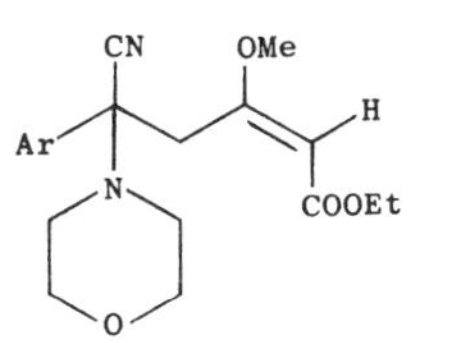

(14)

Further studies of the rhodium carbenoid induced tandem cyclisation-cycloaddition route to oxygen heterocycles has highlighted the ability of a ketone function to trap the carbenoid, the resulting oxonium ylide undergoing a 2,3-sigmatropic shift. A competing C-H insertion is significant where the trapping carbonyl is an ester, but this alternative route is negligible in aromatic systems (89JOC817). Application to the diazo-compounds (13, X=O or S) indicates that ylide formation is much more significant than C-H insertion for the sulphur compound, suggesting an enhanced ability of sulphur to coordinate to the carbene.

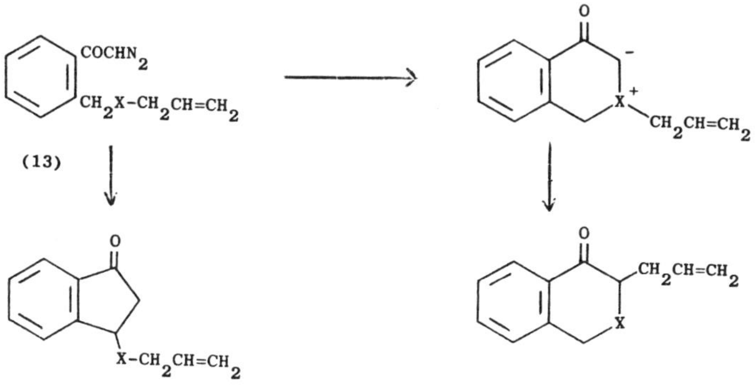

X = O 1 : 1.2

X = S 1 : 9

α-(Dialkylamino)nitriles (14) can act as masked acyl anions and provide a synthesis of the 2H-pyran-2-one ring system. The methodology has been applied to naturally occurring 6-substituted 4-methoxypyran-2-ones (89JOC3985).

(R)-O-Benzylglycidol is the starting point for a chiral synthesis of (4R,6S)-4-hydroxy-6-hydroxymethyltetrahydropyran-2-one of potential value for the synthesis of the compactin species (89S539). The (S)-enantiomer is a source of the spiroacetal unit of the macrolide antibiotics and lactone derived natural products (89H445).

Fused derivatives of pyran-2-ones (15) are produced stereoselectively in a Lewis acid catalysed intramolecular Diels Alder reaction of esters derived from dienyl cyclohexanols (89T6203) and an intramolecular Diels-Alder is involved in the biosynthesis of the phytotoxin solanapyrone A (89CC1282, 1284).

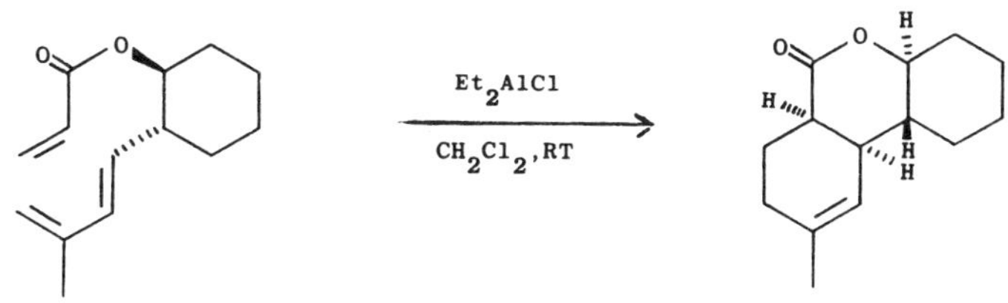

(15)

Interest in forskolin has been maintained because of its potential medicinal value. Variants on the forskolin structure include the synthesis of the 12-halogeno derivatives (89JCS(P1)1544), the regioselective acylations of 7-desacetylforskolin (89JOC2972) and the isolation and synthesis of the 1-deoxy analogue (89T763). Several groups have sought improved routes to key intermediates (89CC512, 1028, 89JOC2505, 89T5215, 89TL7297), whilst others have sought alternative routes to the AB ring system (89T661, 89TL1853) and thence to model tricyclic systems (89T1517).

A study of the C=^{17}O chemical shifts in a series of 1-heteracyclohexan-4-ones has revealed a correlation involving the electronegativity of the heteroatom X and the C-X bond lengths. Both sulphur and selenium cause significant downfield shifts of δ ^{17}O, but a hetero oxygen atom has very little effect (89JOC4758). The ^{17}O nmr spectra of a range of six-membered lactones have been reported (89H301).

5.4.1.5 Coumarins

Several naturally occurring 3,4-fused 5-methylcoumarins have been synthesised from 4-hydroxy-5-methylcoumarin via the quinone methide (89LA295) and 3,4-chromonocoumarins have been isolated from natural sources for the first time (89HCA1455). The 4-phenylsulphinyl and 4-phenylsulphonyl derivatives of coumarin show a greater tendency to react as dienophiles or dipolarophiles than coumarin itself, affording good yields of 3,4-fused coumarins (89T7587). The direct arylation of 4-hydroxycoumarins at C-3 occurs on reaction with aryl-lead triacetates (89TL1539) and organocerium compounds are of value in the synthesis of 4-methylcoumarins (89H703).

The incorporation of foreign substances can modify the crystal of coumarin and so facilitate its photodimerisation in the solid state. These crystal lattice controlling substances presumably alter the packing of coumarin in a way that is favourable for photocycloaddition (89T6979). The syn- and anti-coumarin dimers have been resolved and the absolute configurations assigned on the basis of the CD spectra (89JA4387).

5.4.1.6 Chromones

^{17}O nmr spectral data for a range of chromones and chromanones have been published (89MRC170), supplementing the previous data on coumarins (87MRC489). The chemical shifts of O-1 and O-4 are affected by both steric and electronic properties of substituents, whilst conformational changes and intramolecular hydrogen bonding also exert an influence.

5.4.1.7 Flavonoids

A compilation of the ^{13}C nmr spectra of flavonoids has been published (89MI1) and developments in flavonoid chemistry from 1980 have been reviewed (88MI1).

Dehydrogenation of flavanones to flavones can be achieved photochemically in the presence of triphenylpyrylium tetrafluoroborate, involving a single electron transfer from the 2-aryl ring to the excited state of the pyrylium salt (89H115). Hydrogen transfer reduction is a convenient way of converting isoflavones to isoflavanones (89H183) and the latter have also been obtained by the direct Pd-catalysed arylation of chroman-4-ones through the intermediacy of the enol acetate (89H15).

The cyclisation of 1-(2-hydroxyphenyl)-3-arylprop-2-

yn-1-ones (16) yields only flavones under acidic conditions, the ease of ring closure being related to the electron richness of the 3-aryl ring. However, under basic conditions both flavones and aurones result, indicating competition between 5-exo-dig and 6-endo-dig modes, nucleophilic attack occurring at either end of the triple bond (89JCS(P2)957). The acid catalysed cyclisation of the isomeric 1-phenyl-3-(2-hydroxyphenyl)prop-2-yn-3-ones also yields flavones (89HCA447).

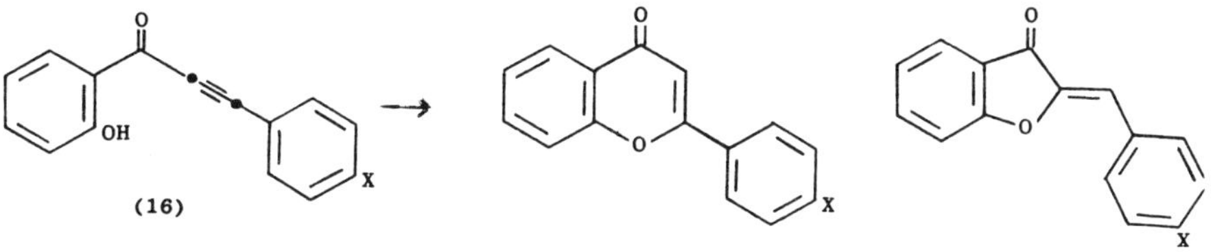

(16)

5.4.1.8 Xanthones

The xanthone ring system is produced in high yield by reaction of chromones with α,β-unsaturated ketones in the presence of t-butyldimethylsilyl triflate. A benzopyrylium (17) derived from the chromone is the reactive species (89H35).

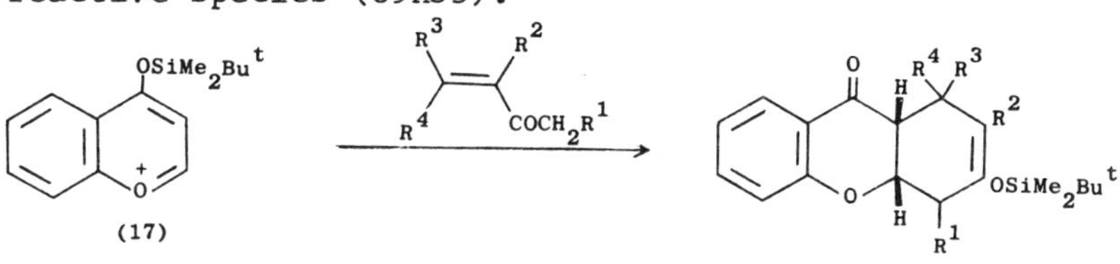

(17)

The base catalysed ring closure of 2,2'-dioxygenated benzophenones to xanthones seems to be a reversible process since treatment of the 1,3,5,8-tetraoxygenated xanthone (18) with hot aqueous morpholine leads to the 1,3,7,8-derivative. Ring opening and demethylation is followed by cyclisation at the 8-hydroxy group (89TL7265).

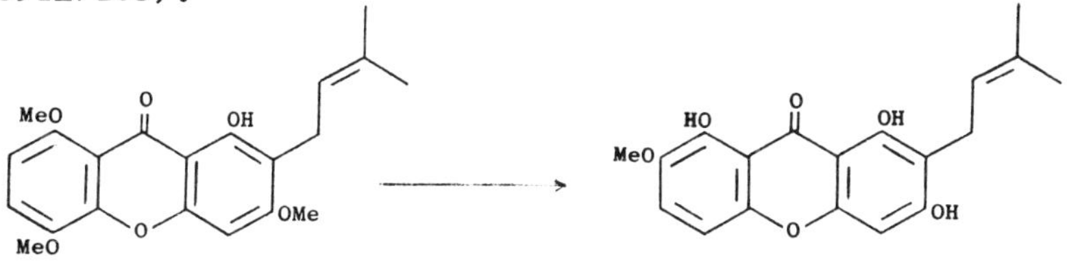

(18)

5.4.1.9 Pyrylium Salts

Fully optimised molecular geometries for both the ground and excited states of 2-phenylbenzopyrylium (89M715) and 5,6-dihydrobenzo[c]xanthylium ions (89M721) have been published.

The synthetic utility of pyrylium salts has been extended through a study of the reaction of 2,4,6-triarylpyrylium salts with 1,2-diones. Acyclic diketones afford the cyclopenta[b]pyran-2-ones (19), which undergo electrophilic attack initially in the alicyclic ring (89JPC293). The reaction with cycloalkan-1,2-diones is sensitive to the reaction conditions employed; the use of morpholine acetate or triethylamine in acetic acid leads to partially reduced benzopyran-8-ones (20). Further reaction with sodium acetate yields benzocycloalkenones, whilst treatment with perchloric acid affords a tetrahydrobenzopyrylium salt (89JPC303).

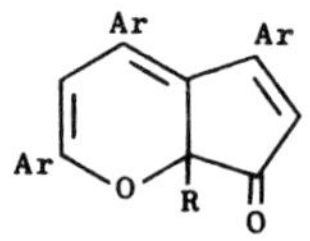

(19)

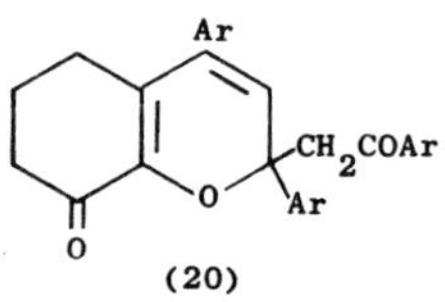

(20)

5.4.2. HETEROCYCLES CONTAINING ONE SULPHUR ATOM

5.4.2.1 Thiopyrans and analogues

Thiophene β-keto-esters and -imines (21) undergo a free radical ring expansion to the thiopyranone (89TL6129).

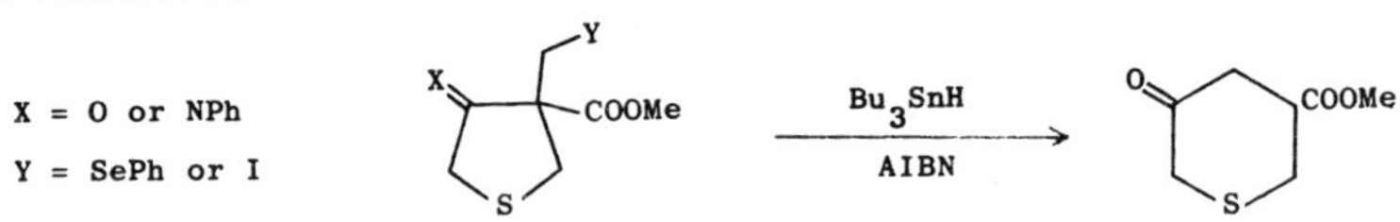

(21)

The palladium catalysed deallyloxycarbonylation of the readily available 3-alkyl-3-allyloxycarbonylthiopyran-4-ones provides a good route to 3-alkylthiopyran-4-

ones and their 1,1-dioxides (89S767).

Selenoketones, prepared directly from ketones, have been trapped by a range of dienes giving a variety of dihydroselenopyrans (89TL2095) and chiral dihydrothiopyrans have been obtained from monosaccharide O-thioformates and 1,1-dithiooxalates by the same methodology (89H887). Similarly, thiopyrano[3,2-c]benzothiopyrans have been prepared from 3-formyl-4-mercapto-2H-thiochromene (89ZC57).

Whereas 1H-2-benzothiopyran 2-oxide undergoes attack by activated methylene compounds at the 1-position, the corresponding 1-aryl derivatives ring contract to the benzo[c]thiophene. Attack at the 1-position is presumably sterically prevented (89JCS(P1)1611).

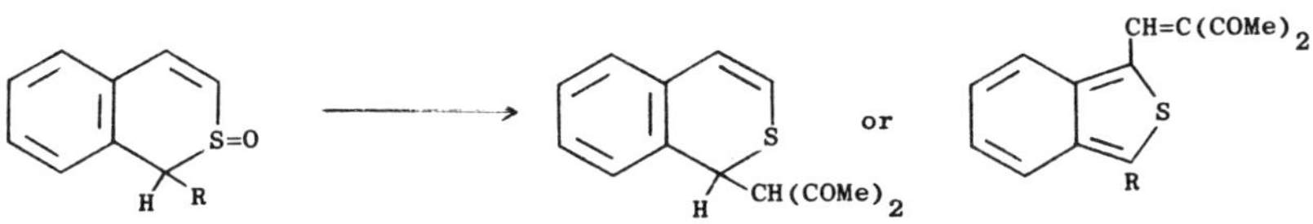

A further example (22) of the dithiaperylene ring system has been synthesised; the charge transfer complex of 1,7-dithiaperylene with nitrate is highly conducting (89CC1553).

Conformation and stereochemical aspects of thioxanthene and thianthrene and their derivatives have been reviewed (89MI2). Analysis of ^{13}C nmr data for a range of substituted thioxanthones indicates that the substituent alters the electron density in the unsubstituted benzene ring by transmission of its electronic effects by the heteroatom (89JCS(P2)325). Furthermore, the ^{17}O chemical shifts of the carbonyl oxygen correlate well with σ and σ_R^+ values again pointing to involvement of the sulphur atom (89JCS(P2)1657).

A significant development in the synthesis of selenoxanthones, and indeed applicable to thioxanthones, involves the use of trimethylsilyl polyphosphate as the catalyst in the Friedel Crafts cyclisation of phenylselenobenzoic acids (89JOC5642).

The reaction of lithium 2-lithiobenzenethiolate with thioxanthone leads to the triarylcarbinol (23) which possesses two ortho-bridging sulphur units. Complete conversion to the deep red cation occurs in CF_3COOH containing triflate, but it is suggested that the long C-S bonds reduce the resonance stabilisation of the ion (89JA654).

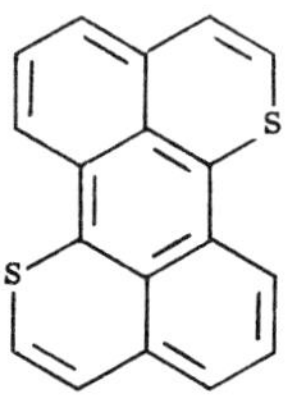

(22)

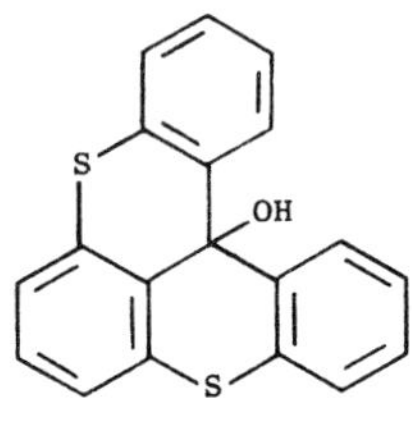

(23)

5.4.3 HETEROCYCLES CONTAINING TWO OR MORE OXYGEN ATOMS

5.4.3.1 Dioxins

Derivatives of 1,3-dioxan continue to attract attention both from conformational (89JOC5963, 89M725, 89TL4585) and synthetic viewpoints. In an extension of the synthetic applications of 1,3-dioxin-4-ones (86CPB4577), the 5-carboxyl analogues (24) have been prepared and shown to undergo thermal cycloreversion to the alkoxycarbonylformylketene, which may be trapped for example with isocyanates to yield oxazine derivates (89CPB665). A detailed kinetic study has confirmed that acetylketene is generated during the thermolysis of 2,2,6-trimethyl-1,3-dioxin-4-one (89JA2186). The decomposition of the 6-aryl analogues in the presence of 1,3-dicarbonyl compounds affords 3-acylpyran-4-ones via the aroyl ketene (89CHE126).

Not only is 2-t-butyl-6-methyl-1,3-dioxin-4-one (25) useful in the synthesis of enantiomerically pure compounds (88JA4763), but the derived dienolate is a chiral synthetic equivalent of acetoacetic ester dianion which shows a range of regio- and diastereoselectivity (89AGE472). Enantiomerically pure cyclopropane derivates are available as a consequence of the diastereofacial selectivity shown by spirocyclic derivatives of 5-arylmethylene-1,3-dioxan-

4,6-diones (26) and associated with the boat conformation adopted by the heterocyclic moiety (89TL5281). Application of the SAMP/RAMP methodology to 2,2-dimethyl-1,3-dioxan-5-one provides a chiral 1,3-dihydroxyacetone enolate equivalent and allows enantioselective alkylations in good yields (89S493).

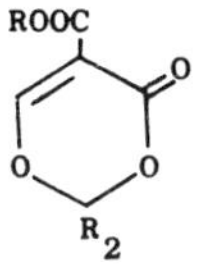
(24)

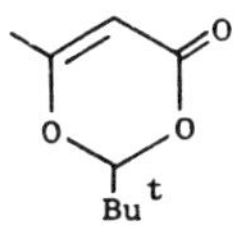
(25)

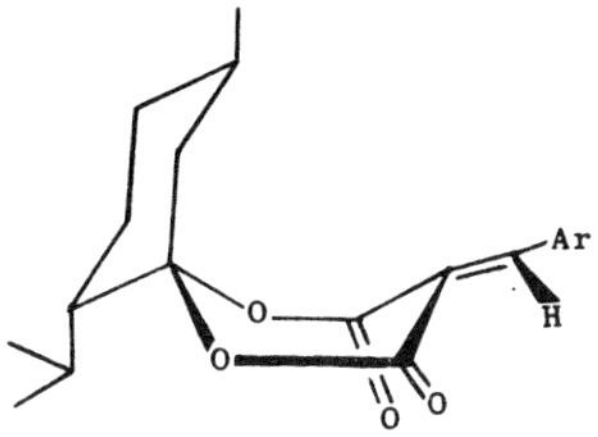
(26)

The four stereoisomers of 1,3-dimethyl-2,9-dioxabicyclo[3.3.1]nonane, the pheromone of the beetle Trypodendron lineatum, have been synthesised by cyclisation of the diastereoisomers of non-8-en-2,4-diol obtained by microbial reduction of the 2,4-dione. In the natural product, the dioxan ring adopts a boat conformation whilst the tetrahydropyran ring has a chair conformation (89JOC2238).

Although direct metallation of 1,4-benzodioxin occurs at the 2-position, prior formation of the tricarbonylchromium complex directs attack to C-5. Further lithiation occurs at C-8 (89TL5519). 3-Substituted 2-methylene-1,4-benzodioxins are available from 2-hydroxymethyl-1,4-benzodioxin by a zinc-mediated Mitsonubu substitution (89TL1637).

The chemistry of 2,3-dihydro-1,4-dioxin (89H521), and of 1,3-benzodioxin-4-ones (89ACS213) have been reviewed.

5.4.3.2. Trioxins

The search for new antimalarial agents has led to developments in the chemistry of artemisinin or qinghaosu (27). The pharmacological significance of the peroxide unit is uncertain and various derivatives have been prepared and assayed (89T7287, 89TL5973). The desmethyl analogue, synthesised from pyrrolidinocyclohexene and cis-1,4-dichlorobut-2-ene, retains antimalarial activity (89JOC1792), but activity is absent from the isochromanone sub-structure (89JOC1789). Some aspects of the chemistry of qinghaosu have been reported (89H421) and reviewed (89MI3).

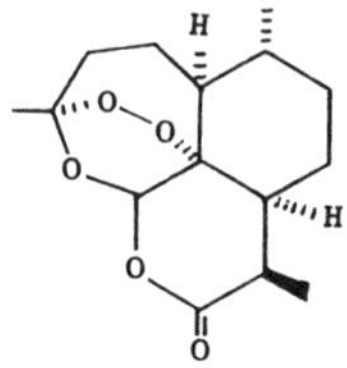

(27)

A range of 1,2,4-trioxanes have been obtained by the tungsten(VI) oxide and chlorosulphonic acid catalysed reaction of peroxides with epoxides and their reactions with a variety of nucleophiles investigated (89JCS(P1)1031).

5.4.4 HETEROCYCLES CONTAINING TWO OR MORE SULPHUR ATOMS

5.4.4.1 Dithiins

The synthesis, chemistry and pharmacological properties of 1,2-dithiins have been reviewed (89SR205). Since its introduction as an acyl anion equivalent by Corey and Seebach in 1965, the applications of 1,3-dithiane have blossomed. The more signficiant developments in the synthesis and uses of 1,3-dithiane during the period 1977 to 1988 have been reviewed (89T7643).

The enolates derived from 2-acyl-1,3-dithiane 1-oxides can exhibit high diastereoselective alkylation, the extent being related to the ability to achieve a chelated transition state involving the two oxygen atoms and the metal ion. The easy removal of the dithiane moiety suggests that this system has potential in the stereocontrolled alkylation of ketones (89JCS(P1)185,1158,2441).

A bicyclic sulphonium salt is postulated as an intermediate in the reaction of 2-(2-hydroxyethyl-1,3-

dithianes (28) with halogenating agents in order to account for the varied nature of the products. As the size of the substituent Z increases, the amount of the ring expanded product, a dithiacyclooctene (29), increases, whilst benzylic alcohols also yield a styrene derivative (89JCS(P1)1853).

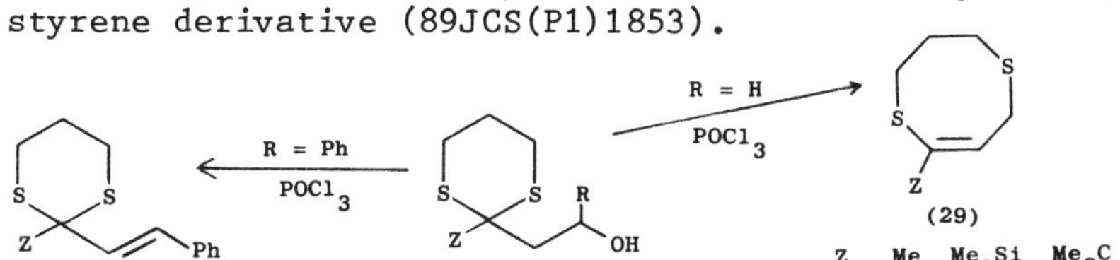

(28) (29)

Z	Me	Me_3Si	Me_3C
%	39	70	100

The anion derived from 2-(cyclohexanecarbonyl)-2-(methoxycarbonylmethyl)-1,3-dithiane undergoes spontaneous fragmentation. Recyclisation and lactonisation of the intermediate unsaturated thiolate affords the novel [1,4]dithiepino[2,3-c]furan system (30) (89SC973).

Sterically crowded diketo sulphides afford a mixture of 1,4-dithiins and the trithiabicycloheptane (31) on treatment with Lawesson's reagent (89H391).

The synthesis of 3,7,9-trithiabicyclo[3.3.1]nonane (32) has been achieved. It assumes the twin chair conformation and in concentrated sulphuric acid is converted into the orange dication which nmr indicates to exist as a rapidly equilibrating mixture of the twin chair and chair-boat conformations. The cation is stabilised by transannular S-S interaction (89TL4399).

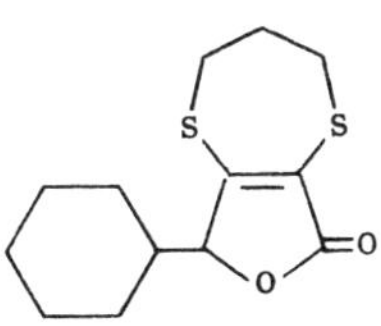

(30)

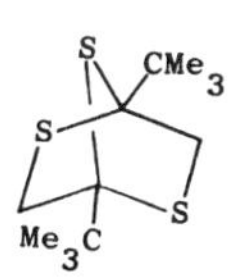

(31)

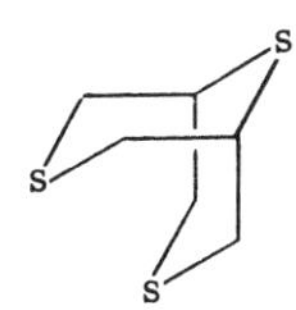

(32)

A ^{13}C and ^{77}Se nmr study of the conformational equilibria in a number of 1,3-diselenanes provides evidence for a significant anomeric effect involving selenium (89H389).

5.4.4.2 Trithiins

The molecular structure of 1,3,5-trithiane has been determined by electron diffraction, confirming the chair conformation and complementing the information for thiane and 1,3-dithiane (88ACS332, 89ACS953).

A 1,2,3-trithiane derivative (33) has been isolated from the cherry red New Zealand marine organism Aplidium sp. D. In weakly basic solution it inverts at C-4 to the trans isomer (89TL3703).

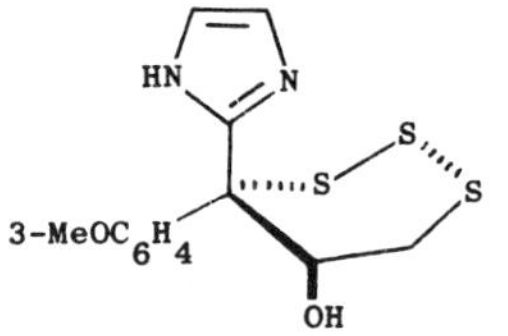

(33)

5.4.5 HETEROCYCLES CONTAINING BOTH OXYGEN AND SULPHUR IN THE SAME RING

5.4.5.1 1,2-Oxathiins

The novel 2,3-oxathiino[5,4-*c*][1]benzopyran system (34) is produced along with two benzoxathiinopyridines by the reaction of phenyl 4-chromone-3-sulphonate with methyl 3-aminoprop-2-enoate (89JHC753).

The vinyl sulphonates (35) undergo a highly diastereoselective intramolecular Diels Alder reaction to give the sultones (36), the configuration of which followed from ^{1}H nmr spectra backed up in one case by X-ray determination (89AGE202).

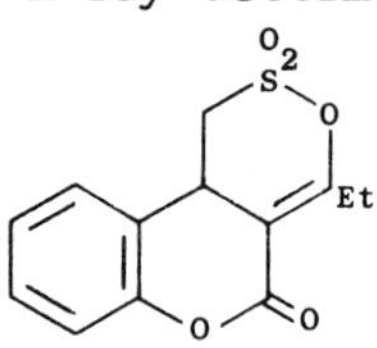

(34)

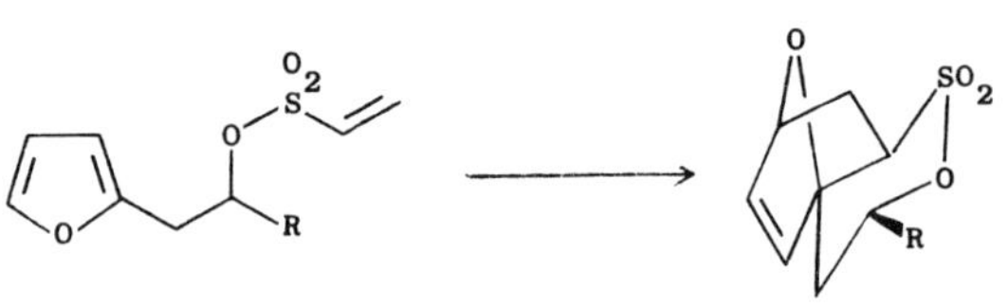

(35) (36)

5.4 REFERENCES

78S886 H.J. Kabbe, Synthesis, 1978, 886.

83JCS(P1)827 P.J. Brogden and J.D. Hepworth, J. Chem. Soc. Perkin Trans., 1, 1983, 827.

86CPB4577 M. Sato, N. Yoneda and C. Kaneko, Chem. Pharm. Bull., 1986, 34, 4577.

87MRC489 H. Duddeck, D. Rosenbaum, M.H.A. Elgamal and N.M.M. Shalaby, Mag. Res. Chem., 1987, 25, 489.

88ACS332 G. Schultz, A. Kucsman and I. Hargittai, Acta Chem. Scand. Ser. A, 1988, 42, 332.

88JA4763 D. Seebach, J. Zimmermann, U. Gysel, R. Ziegler and T.-K. Ha, J. Am. Chem. Soc., 1988, 110, 4763.

88MI1 'Flavonoids: Advances in Research', J.B. Harborne, Chapman and Hall, London, 1988, Vol. 2.

88T7063 F. Matsuda, N. Tomiyushi, M. Yanagiya, and T. Matsumoto, Tetrahedron, 1988, 44, 7063.

89ACH149 T. Timar, J.C. Jaszberenyi and S. Hosztafi, Acta Chim. Hung., 1989, 126, 149.

89ACS213 M. Ankersen, K.K. Nielsen and A. Senning, Acta Chem. Scand., 1989, 43, 213.

89ACS953 Z. Bencze, A. Kucsman, G. Schultz and I. Hargittai, Acta Chem. Scand., 1989, 43, 953.

89AGE202 E. Bovenschulte, P. Metz and G. Henkel, Angew. Chem. Int. Ed. Engl., 1989, 28, 202.

89AGE472 D. Seebach, U. Misslitz and P. Uhlmann, Angew. Chem. Int. Ed. Engl., 1989, 28, 472.

89CB643 L.F. Tietze, H. Meier and H. Nutt, Chem. Ber., 1989, 122, 643.

89CB1179 M. Buback, W. Tost, T. Hibsh, E. Voss and L.F. Tietze, Chem. Ber., 1989, 122, 1179.

89CC512 K.C. Nicolaou, S. Kubato and W.S. Li, J. Chem. Soc. Chem. Commun., 1989, 512.

89CC1028 K. Kanematsu and S. Nagashima, J. Chem. Soc. Chem. Commun., 1989, 1028.

89CC1151 W. Keim, W. Meltzow, A. Koehnes and T. Roethel, J. Chem. Soc. Chem. Commun., 1989, 1151.

89CC1250 E.R. Parmee, P.G. Steel and E.J. Thomas, J. Chem. Soc. Chem. Commun., 1989, 1250.

89CC1258 C.S.V. Houge-Frydrych, A. Marsh and I.L. Pinto, J. Chem. Soc. Chem. Commun., 1989, 1258.

89CC1282 H. Oikawa, T. Yokota, T. Abe, A. Ichihara, S. Sakamura, Y. Yoshizawa and J.C. Vederas, J. Chem. Soc. Chem. Commun., 1989, 1282.

89CC1284 H. Oikawa, T. Yokota, A. Ichihara and S. Sakamura, J. Chem. Soc. Chem. Commun., 1989, 1284.

89CC1505 L.Y. Foo, J. Chem. Soc. Chem. Commun., 1989, 1505.

89CC1553 K. Nakasuji, A. Oda, I. Murata, K. Imaeda and H. Inokuchi, J. Chem. Soc. Chem. Commun., 1989, 1553.

89CC1560 F. Camps, J.M. Moreto, S. Ricart, J.M. Vinas, E. Molins and C. Miravitlles, J. Chem. Soc. Chem. Commun., 1989, 1560.

89CHE126 Yu S. Andreichikov, O.V. Vinokurova and V.L. Gein, Chem. Hetercycl. Compd. (Engl. Transl.), 1989, 25, 126.

89CPB665 M. Sato, N. Katagiri, K. Takayama, M. Hirose and C. Kaneko, Chem. Pharm. Bull., 1989, 37, 665.

89CR1617 F. Perron and K.F. Albizati, Chem. Rev., 1989, 89, 1617.

89G481 G. Pastorini, P. Rodighiero, P. Manzini, M.T. Conconi, A. Chilin and A. Guiotto, Gazz. Chim. Ital., 1989, 119, 481.

89H15 D.M.X. Donnelly, J.-P. Finet and P.H. Stenson, Heterocycles, 1989, 28, 15.

89H35 Y. Lee, H. Iwasaki, Y. Yamamoto, K. Ohkata and K. Akiba, Heterocycles, 1989, 28, 35.

89H55 E. Negishi, T. Nguyen, B. O'Connor, J.M. Evans and A. Silveira, Jr., Heterocycles, 1989, 28, 55.

89H115 M.J. Climent, H. Garcia, S. Iborra, M.A. Miranda and J. Primo, Heterocycles, 1989, 28, 115.

89H183 W. Wähälä and T.A. Hase, Heterocycles, 1989, 28, 183.

89H301 D.W. Boykin, D.W. Sullins and E.J. Eisenbraun, Heterocycles, 1989, 28, 301.

89H389 B.M. Pinto, B.D. Johnston and R. Nagelkerke, Heterocycles, 1989, 28, 389.

89H391 J. Nakayama, K.S. Choi, S. Yamaoka and M. Hoshino, Heterocycles, 1989, 28, 391.

89H421 X. Shang, C.-H. He, Q.-T. Zheng, J.-J. Yang and X.-T. Liang, Heterocycles, 1989, 28, 421.

89H445 S. Takano, Y. Sekiguchi and K. Ogasawara, Heterocycles, 1989, 28, 445.

89H521 M. Fétizon, P. Goulaouic and I. Hanna, Heterocycles, 1989, 28, 521.

89H583 M.D. Bachi and D. Denenmark, Heterocycles, 1989, 28, 583.

89H703 K. Nagasawa and K. Ito, Heterocycles, 1989, 28, 703.

89H887 P. Herczegh, M. Zsély, L. Szilágyi and R. Bognár, Heterocycles, 1989, 28, 887.

89HCA447 D. Obrecht, Helv. Chim. Acta, 1989, 72, 447.

89HCA482 T. Bauer, C. Chapuis, J. Kozak and J. Jurczak, Helv. Chim. Acta, 1989, 72, 482.

89HCA1455 E.R. Di Paolo, M.O. Hamburger, H. Stoeckli-Evans, C. Rogers and K. Hostettmann, Helv. Chim. Acta, 1989, 72, 1455.

89HCA1649 W. Brade and A. Vasella, Helv. Chim. Acta, 1989, 72, 1649.

89JA654 G.D. Figuly, C.K. Loop and J.C. Martin, J. Am. Chem. Soc., 1989, 111, 654.

89JA1886 M.D. Bachi and D. Denenmark, J. Am. Chem. Soc., 1989, 111, 1886.

89JA2186 R.J. Clemens and J.S. Witzeman, J. Am. Chem. Soc., 1989, 111, 2186.

89JA2967 S.J. Danishefsky, D.M. Armistead, F.E. Wincott, H.G. Selnick and R. Hungate, J. Am. Chem. Soc., 1989, 111, 2967.

89JA4387 A. Hallberg, R. Isaksson, A.R. Martin and J. Sandström, J. Am. Chem. Soc., 1989, 111, 4387.

89JA5330 K.C. Nicolaou, C.V.C. Prasad, P.K. Somers and C.-K. Hwang, J. Am. Chem. Soc., 1989, 111, 5330.

89JA6666 K.C. Nicolaou, M.E. Duggan and C.-K. Hwang, J. Am. Chem. Soc., 1989, 111, 6666.

89JA6676 K.C. Nicolaou, M.E. Duggan and C.-K. Hwang, J. Am. Chem. Soc., 1989, 111, 6676.

89JA6682 K.C. Nicolaou, C.-K. Hwang and M.E. Duggan, J. Am. Chem. Soc., 1989, 111, 6682.

89JA7525 R.W. Armstrong et al., J. Am. Chem. Soc., 1989, 111, 7525.

89JA7530 R.W. Armstrong et al., J. Am. Chem. Soc., 1989, 111, 7530.

89JCR(S)36 P. Anastasis, P.E. Brown, Q. Islam and W.Y. Marcus, J. Chem. Res. (S), 1989, 36.

89JCS(P1)185 P.C. Bulman Page, A.M.Z. Slawin, D. Westwood and D.J. Williams, J. Chem. Soc. Perkin Trans. 1, 1989, 185.

89JCS(P1)204 M.J. Begley, L. Crombie, A.H. bin A. Hadi and J.L. Josephs, J. Chem. Soc. Perkin Trans. 1, 1989, 204.

89JCS(P1)207 F.M. Martin and T.J. Simpson, J. Chem. Soc. Perkin Trans. 1, 1989, 207.

89JCS(P1)1031 T. Fujisaka, M. Muira, M. Nojima and S. Kusabayashi, J. Chem. Soc. Perkin Trans. 1, 1989, 1031.

89JCS(P1)1158 P.C. Bulman Page, D. Westwood, A.M.Z. Slawin and D.J. Williams, J. Chem. Soc. Perkin Trans. 1, 1989, 1158.

89JCS(P1)1544 G.M. Shutske, J. Chem. Perkin Trans. 1, 1989, 1544.

89JCS(P1)1611 M. Hori, T. Kataoka, H. Shimizu, J. Hongo and M. Kido, J. Chem. Soc. Perkin Trans. 1, 1989, 1611.

89JCS(P1)1853 A.E. Davey, A.F. Parsons and R.J.K. Taylor, J. Chem. Soc. Perkin Trans. 1, 1989, 1853.

89JCS(P1)1901 K.C. Majumdar and R. De, J. Chem. Soc. Perkin Trans. 1, 1989, 1901.

89JCS(P1)2047 M.J. Crimmin, P.J. O'Hanlon, N.H. Rodgers and G. Walker, J. Chem. Soc. Perkin Trans. 1, 1989, 2047.

89JCS(P1)2329 D.N. Nicolaides, D.A. Lefkaditis, P.S. Lianis and K.E. Litinas, J. Chem. Soc. Perkin Trans. 1, 1989, 2329.

89JCS(P1)2441 P.C. Bulman Page, S.S. Klair and D. Westwood, J. Chem. Soc. Perkin Trans. 1, 1989, 2441.

89JCS(P2)325 J.S. Harwood, D.S. Marynick and A.L. Ternay, Jr., J. Chem. Soc. Perkin Trans., 2, 1989, 325.

89JCS(P2)957 C.M. Brennan, C.D. Johnson and P.D. McDonnell, J. Chem. Soc. Perkin Trans. 2, 1989, 957.

89JCS(P2)1657 J.S. Harwood and A.L. Ternay, Jr., J. Chem. Soc. Perkin Trans. 2, 1989, 1657.

89JHC753 W. Löwe and A. Kradepohl, J. Heterocycl. Chem., 1989, 26, 753.

89JOC560 S. Brownstein, G.W. Burton, L. Hughes and K.U. Ingold, J. Org. Chem., 1989, 54, 560.

89JOC817 A. Padwa, S.F. Hornbuckle, G.E. Fryxell and P.D. Stull, J. Org. Chem., 1989, 54, 817.

89JOC1157 M.T. Crimmins and R. O'Mahony, J. Org. Chem., 1989, 54, 1157.

89JOC1562 R.G. Salomon, S.M. Mazza and K. Lal, J. Org. Chem., 1989, 54, 1562.

89JOC1789 M.A. Avery, C. Jennings-White and W.K.M. Chong, J. Org. Chem., 1989, 54, 1789.

89JOC1792 M.A. Avery, C. Jennings-White and W.K.M. Chong, J. Org. Chem., 1989, 54, 1792.

89JOC1987 K. Omura, J. Org. Chem., 1989, 54, 1987.

89JOC2044 F. Perron and K.F. Albizati, J. Org. Chem., 1989, 54, 2044.

89JOC2238 G. Dauphin, A. Fauve and H. Veschambre, J. Org. Chem., 1989, 54, 2238.

89JOC2505 H. Venkataraman and J.K. Cha, J. Org. Chem., 1989, 54, 2505.

89JOC2972 R.W. Kosley, Jr. and R.J. Cherill, J. Org. Chem., 1989, 54, 2972.

89JOC3282 N. Cohen, B. Shaer, G. Saucy, R. Borer, L. Todaro and A.-M. Chiu, J. Org. Chem., 1989, 54, 3282.

89JOC3985 D.D. Dhavale, I.S. Aidhen and M. Shafique, J. Org. Chem., 1989, 54, 3985.

89JOC4723 A.G.M. Barrett, J.A. Flygare and C.D. Spilling, J. Org. Chem., 1989, 54, 4723.

89JOC4758 S.V. Mulekar and K.D. Berlin, J. Org. Chem., 1989, 54, 4758.

89JOC5642 E.M. Berman and H.D.H. Showalter, J. Org. Chem., 1989, 54, 5642.

89JOC5768 Z.Y. Wei, D. Wang, J.S. Li and T.H. Chan, J. Org. Chem., 1989, 54, 5768.

89JOC5963 E. Juaristi, B. Gordillo, R. Martinez and R.A. Toscano, J. Org. Chem., 1989, 54, 5963.

89JPC293 T. Zimmermann, G.W. Fischer and L. Kutschabksy, J. Prakt. Chem., 1989, 331, 293.

89JPC303 T. Zimmermann, G.W. Fischer, B. Olk and M. Findeisen, J. Prakt. Chem., 1989, 331, 303.

89LA295 C. Wolfrum and F. Bohlmann, Liebigs Ann. Chem., 1989, 295.

89M715 K. Gustav, R. Colditz and D. Steen, Monatsh. Chem., 1989, 120, 715.

89M721	K. Gustav, R. Colditz and D. Steen, Monatsh. Chem., 1989, 120, 721.
89M725	E. Kleinpeter, C.-P. Maschmeier and H. Matschiner, Monatsh. Chem., 1989, 120, 725.
89MI1	'Carbon-13 nmr of Flavonoids' ed. P.K. Agrawal, Elsevier, Amsterdam, 1989.
89MI2	S.A. Evans, Jr. in 'The Conformational Analysis of Cyclohexenes, Cyclohexadienes, and Related Hydroaromatic Compounds", ed. P.W. Rabideau, VCH Publishers, New York, 1989, 169.
89MI3	W.-S. Zhou and X.-X. Xu in 'Studies in Natural Products Chemistry' ed. A-u-Rahman, Elsevier, Amsterdam, 1989, vol. 3B, 495.
89MRC170	H. Duddeck, J.C. Jászberényi, A. Lévai, T. Timár and M.H.A. Elgamal, Mag. Res. Chem., 1989, 27, 170.
89RCR479	E.A. Markaryan and A.G. Samodurova, Russian Chem. Rev., 1989, 58, 479.
89S69	L. Rene, Synthesis, 1989, 69.
89S493	D. Enders and B. Bockstiegel, Synthesis, 1989, 493.
89S539	S. Takano, Y. Shimazaki, Y. Sekiguchi and K. Ogasawara, Synthesis, 1989, 539.
89S767	G. Casy, A.G. Sutherland, R.J.K. Taylor and P.G. Urben, Synthesis, 1989, 767.
89SC973	M. Fetizon, P. Goulaounic, I. Hanna and T. Prange, Synth. Commun., 1989, 19, 973.
89SR205	F. Freeman, D.S.H.L. Kim and E. Rodriguez, Sulphur Reports, 1989, 9, 205.
89T661	D. Neunert, H. Klein and P. Welzel, Tetrahedron, 1989, 45, 661.
89T763	Y. Khandelwal, B.R. Jotwani, P.K. Inamdar, N.J. de Souza and R.H. Rupp, Tetrahedron, 1989, 45, 763.
89T1517	P.G. Baraldi, A. Barco, S. Benetti, V. Ferretti, G.P. Pollini, E. Polo and V. Zanirato, Tetrahedron, 1989, 45, 1517.
89T5215	M.J. Begley, D.R. Cheshire, T. Harrison, J.H. Hutchinson, P.L. Myers and G. Pattenden, Tetrahedron, 1989, 45, 5215.
89T5895	S.M.F. Lai, J.J.A. Orchison and D.A. Whiting, Tetrahedron, 1989, 45, 5895.
89T6203	A. Alexakis and D. Jachiet, Tetrahedron, 1989, 45, 6203.
89T6875	J. Gawronski, K. Gawronska and A. Katrusiak, Tetrahedron, 1989, 45, 6875.
89T6979	J. Meng, D. Fu, X.-K. Yao, R.-J. Wang and T. Matsuura, Tetrahedron, 1989, 45, 6979.
89T7161	S.V. Ley et al., Tetrahedron, 1989, 45, 7161.
89T7287	B. Ye and Y.-L. Wu, Tetrahedron, 1989, 45, 7287.
89T7587	R. Grigg and D. Vipond, Tetrahedron, 1989, 45, 7587.
89T7643	P.C. Bulman Page, M.B. van Niel and J.C. Prodger, Tetrahedron, 1989, 45, 7643.
89TL253	J. Al-Dulayymi and M.S. Baird, Tetrahedron Lett., 1989, 253.
89TL1539	D.H.R. Barton, D.M.X. Donnelly, J.-P. Finet and P.J. Guiry, Tetrahedron Lett., 1539.
89TL1637	N. Ruiz and P. Rollin, Tetrahedron Lett., 1989, 1637.
89TL1853	A. Mukhopadhyay, S.M. Ali, M. Husain, S.N. Suryawanshi and D.S. Bhakuni, Tetrahedron Lett., 1989, 1853.
89TL2095	M. Segi, T. Koyama, T. Nakajima, S. Suga, S. Murai and N. Sonoda, Tetrahedron Lett., 1989, 2095.
89TL2297	R.S. Subramanian and K.K. Balasubramanian, Tetrahedron Lett., 1989, 2297.
89TL2437	P. Cox, M.F. Mahon, K.C. Molloy, S. Lister and T. Gallagher, Tetrahedron Lett., 1989, 2437.
89TL3703	B.R. Copp, J.W. Blunt and M.H.G. Munro, Tetrahedron Lett., 1989, 3703.

89TL4065	T.-H. Chan and P. Arya, Tetrahedron Lett., 1989, 4065.
89TL4399	H. Fujihara, R. Akaishi and N. Furukawa, Tetrahedron Lett., 1989, 4399.
89TL4585	C.A. Carr, S.L.R. Ellison and M.J.T. Robinson, Tetrahedron Lett., 1989, 4585.
89TL4731	A.J. Walkington and D.A. Whiting, Tetrahedron Lett., 1989, 4731.
89TL4863	J.v.d. Louw, G.J.J. Out, J.L.v.d. Baan, F.J.J. de Kanter, F.W. Bickelhaupt and G.W. Klumpp, Tetrahedron Lett., 1989, 4863.
89TL5281	M. Sato, H. Hisamichi and C. Kaneko, Tetrahedron Lett., 1989, 5281.
89TL5519	T.V. Lee, A.J. Leigh and C.B. Chapleo, Tetrahedron Lett., 1989, 5519.
89TL5973	M. Jung, X. Li, D.A. Bustos, H.N. Elsohly and J.D. McChesney, Tetrahedron Lett., 1989, 5973.
89TL5993	M.T. Crimmins and R. O'Mahony, Tetrahedron Lett., 1989, 5993.
89TL6129	P. Dowd and S.-C. Choi, Tetrahedron Lett., 1989, 6129.
89TL6573	A.I.D. Alanine, C.W.G. Fishwick and C. Szantay, Jr., Tetrahedron Lett., 1989, 6573.
89TL6653	M. Majewski and G. Bantle, Tetrahedron Lett., 1989, 6653.
89TL7265	G.J. Bennett and H.-H. Lee, Tetrahedron Lett., 1989, 7265.
89TL7297	E.J. Corey and P.D.S. Jardine, Tetrahedron Lett., 1989, 7297.
89ZC57	M. Pulst, D. Greif, A. Czerwonatis and M. Weissenfels, Z. Chem., 1989, 29, 57.
89ZOK155	A.V. Koblik and K.F. Suzdalev, Zh. Org. Khim. (Engl. Trans.), 1989, 25, 155.

CHAPTER 6
Seven-Membered Rings

JOHN M KANE AND NORTON P PEET
Merrell Dow Research Institute, Cincinnati, OH, USA

INTRODUCTION

Interesting reports on the construction of 7-membered heterocyclic compounds appeared in 1989. Although rearrangement chemistry, which was highlighted last year, was again utilized synthetically, a more resounding theme in the past year was the total synthesis of complex natural products, and it is this topic which is emphasized in this review.

This chapter includes 7-membered rings containing nitrogen, oxygen and sulfur. The emphasis on natural products limits the field to 12 classes, from a larger field of classes where investigations were reported. Coverage is by atom type beginning with azepines.

In the context of natural product synthesis, three topics are addressed, which are as follows: (1) total syntheses and total formal syntheses; (2) studies directed toward the synthesis of natural products; and (3) synthesis of analogs of natural products. Subtopics in these categories include (a) syntheses where the final product is a 7-membered ring heterocycle, and (b) syntheses where a 7-membered ring was utilized somewhere in the synthetic scheme.

In addition to synthetic studies, we highlight new 7-membered ring heterocyclic natural products whose structures have been established in the past year, usually by spectroscopic means. It is felt that this inclusion will be helpful to those seeking to identify future targets for total synthesis, especially since Heterocycles is no longer highlighting new natural products.

Although several syntheses of complex natural products were reported in 1989, one in particular stands out because of its complexity and the sheer magnitude of its undertaking. This is, of course, the total syntheses of palytoxin carboxylic acid and palytoxin amide <89JA(111)7525><89JA(111)-7530> which were conducted in the laboratories of Professor Yoshito Kishi at Harvard University during the past seven years.

6.1 7-Membered Rings with Nitrogen

6.1.1 Azepines

The total syntheses of a diverse array of azepine-containing natural products were reported in 1989. Williams <89JA(111)1923> described the total syntheses of (+)-Croomine, a pyrrolo[1,2-a]azepine derivative containing a spirocyclic lactone. The azepine ring was formed via the intramolecular addition of an iminophosphorane to an aldehyde followed by reduction of the resultant imine with sodium borohydride. The total synthesis of iboxyphylline <89JOC-(54)4553>, a pentacyclic indole alkaloid, was accomplished utilizing an intramolecular Mannich reaction for construction of the azepine moiety. Danishefsky <89TL(30)2747> outlined the total synthesis of chilenine (1) via a novel carbenoid-monothiophthalimide coupling reaction (Eq. 1) of 2 yielding 3.

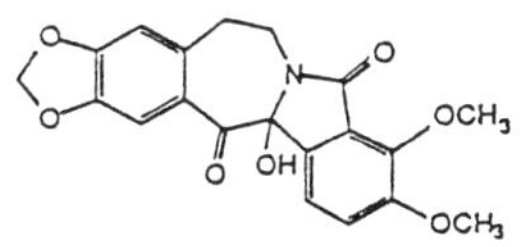

1

(Eq. 1)

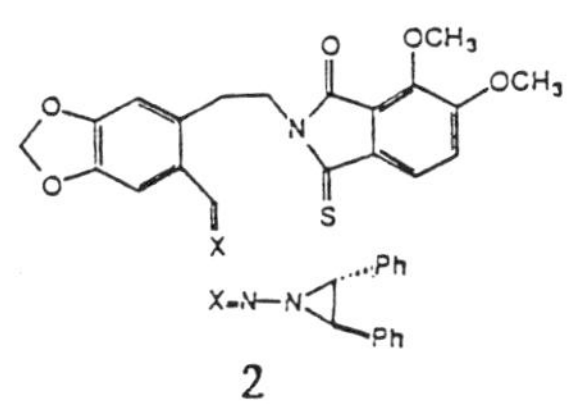

2

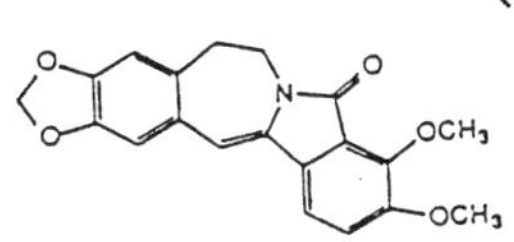

3

Bosch <89H(29)381> reported the total syntheses of both ngouniensine and epingouniensine, tetracyclic alkaloids possessing the octahydropyrido-[1′,2′:1,2]azepino[4,5-b]indole moiety. Construction of the azepine rings was accomplished via an intramolecular alkylation at the 3-position of the indole. Also reported was the synthesis of galanthamine <89T(45)3329> via an oxidative coupling of an amide tethered bis-phenol derivative.

The syntheses of a number of bridged azepine natural products were also described last year. Thus, Heathcock <89JOC(54)1548> reported a 13 step synthesis of (±)-fawcettimine, a 2-indanone derivative containing a 1,3-bridged azepine moiety. The azepine was introduced in the last step of the synthesis via intramolecular carbinolamine formation. The synthesis of (+)-aphanorphine <89CC1591>, a 2,6-methano-bridged azepine, was reported. The azepine-forming step involved an intramolecular amine alkylation. Speckamp <89JA(111)2588> disclosed the synthesis and absolute configuration of peduncularine, an indole alkaloid containing a 2,6-methano-bridged azepine. Construction of the azepine ring was accomplished via a silicon-assisted N-acyliminium ion cyclization. Jacobi <89TL(30)7173> described an interesting application of the intramolecular Diels-Alder reaction in the syntheses of both (+)- and (-)-norsecurinine, a tetracyclic, 2,6-methano-bridged azepine derivative which acts as an inhibitor of the acetylcholinesterase system. Finally, Rapoport <89JOC(54)4654> improved his existing synthesis of (+)-anatoxin, a 2,7-bridged azepine. Formation of the azepine ring involved the decarbonylative cyclization of a proline derivative.

In addition to the above completed total syntheses, several approaches were recorded to various azepine-containing natural products. Mariano <89TL-(30)4185> reported the synthesis of the pentacyclic skeleton of the harringtonine alkaloids based upon a photolytically-induced single electron transfer process to the spirocyclic D,E-rings. Formation of the azepine moiety was subsequently accomplished via an intramolecular amine alkylation. An interesting approach to the 3,8-diazabicyclo[3.2.1]octane moiety contained in the alkaloids naphthyridinomycin and quinocarcin employed an intramolecular azomethine ylide cycloaddition <89JOC(54)2041>. An intramolecular Mannich reaction onto an aromatic ring activated toward electrophilic substitution by three alkoxy groups was used for the azepine-forming step in an approach to the tetracyclic skeleton of the amaryllidaceae alkaloids <89TL(30)6223>.

The syntheses of some analogs of naturally occurring azepines were also noted during the past year. Thus, Bryce <89CC1162> reported the synthesis of some functionalized analogs of cephalotaxine via intramolecular carbinolamine formation. In addition, photolysis of (-)-quebrachamine afforded a 16-hydroxy derivative of an ibogamine-like ring system <89JOC(54)3174>.

Azepines have been utilized as key intermediates in the syntheses of natural products which do not themselves contain a 7-membered heterocycle. For example, Kuehne <89JOC(54)3407> described the syntheses of the indole alkaloids 20′-deoxyvinblastine, 20′-deoxyleurosidine, and 20′-deoxyvincovaline while Das <89TL(30)1959> reported the synthesis of 20-epi-ibophyllidine. All of these syntheses employed indolo[2,3-*d*]azepine derivatives which were subsequently fragmented to give products which underwent intramolecular Diels-Alder reactions. An interesting synthesis of the pyrrolo[1,4]benzodiazepines prothracarcin and DC-81 (see Section 7.1.2) utilized pyrrolobenzazepines as key intermediates. These materials were prepared via the photochemical ring expansion of appropriately substituted N-pentenylphthalimide derivatives <89JOC(54)5746>. Perhydrohistrionicotoxin, a fully hydrogenated analog of the natural histrionicotoxin, was prepared using an azepine derivative which was subsequently ring contracted to a substituted piperidine <89JOC(54)4419>.

Finally, the structures of several new, azepine-containing natural products were reported in 1989 (Eq. 2). Thus the bengamides C-F and isobengamide E (4) were isolated from a previously undescribed Jaspidae sponge <89JA(111)647> while (+)-acacialactam (5) was isolated from the seeds of *Acacia concinna* <89CPB(37)3164>. Also isolated was the benzo[c]azepine 1,6-epoxycomosivine <89P(28)1298>. Lastly, the isolation of the structurally complex, polycyclic alkaloid zoanthaminone (6) <89TL(30)6825A> should provide synthetic chemists with a challenging target.

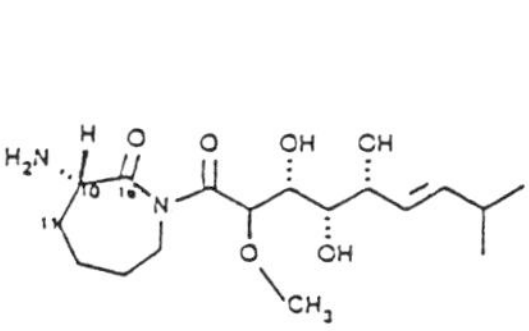

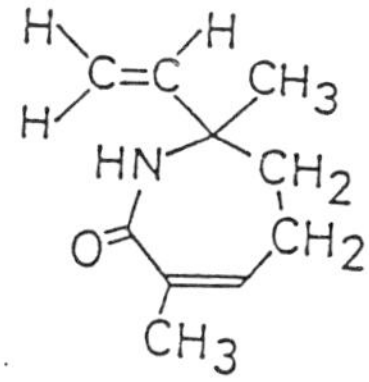

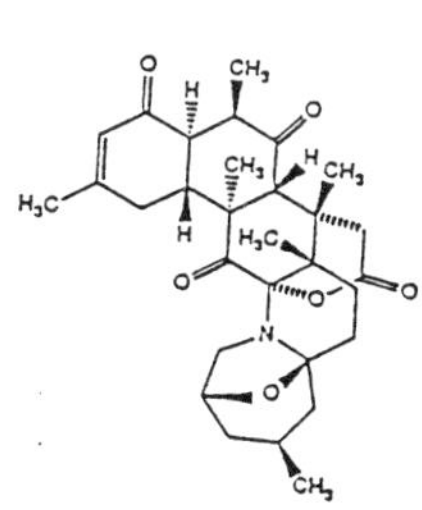

(Eq. 2)

4 5 6

6.1.2 1,4-Diazepines

Synthetic studies on the pyrrolo[1,4]benzodiazepine family of antitumor antibiotics continued in 1989 with several total syntheses being reported. Thus, Nagasaka described the total synthesis of tilivalline (7) <89TL(30)1871> and Stille reported the total synthesis of anthramycin (8) <89JA(111)5417. Both syntheses employed a known approach to the tricyclic nucleus, namely, the reaction of derivatives of isatoic anhydride with either proline or 4-hydroxyproline. The synthesis of both prothracarcin (9) and DC-81 (10) (Eq. 3) were also reported <89JOC(54)5746>. The 1,4-benzodiazepine rings in both syntheses were prepared via hemiaminal formation of the corresponding aminoaldehyde derivatives. During the past year a number of analogs of this family of natural products were described. The syntheses of 11-keto-11a-carboethoxyprothracarcin <89JHC(26)1023> as well as derivatives of both the 5*H*-1,2,3-triazolo[5,1-*c*][1,4]benzodiazepine <89JHC(26)1605> and the 1,2,3,6-tetrahydropyrido[1,6-*d*][1,4]benzodiazepine <89H(29)1721> ring systems were recorded. The benzodiazepine rings in these three instances were prepared by closure of the 10,11-bond via either amide or imine formation. A synthetic study directed toward another family of 1,4-benzodiazepine-containing natural products was reported last year. Thus, Okada described the synthesis of the pentacyclic skeleton of neosurugatoxin <89CL703>. The benzodiazepine ring was closed via imine formation of an aminoketone derivative.

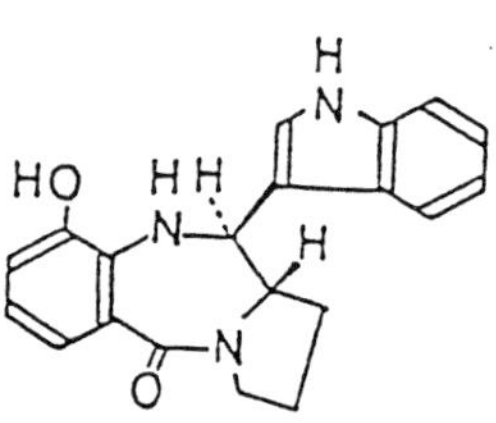

7

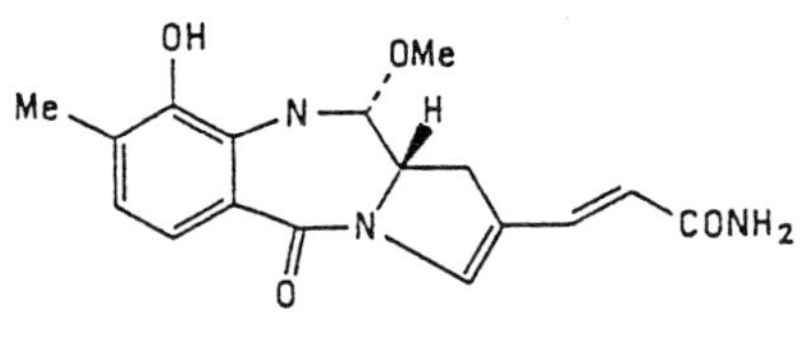

8

(Eq. 3)

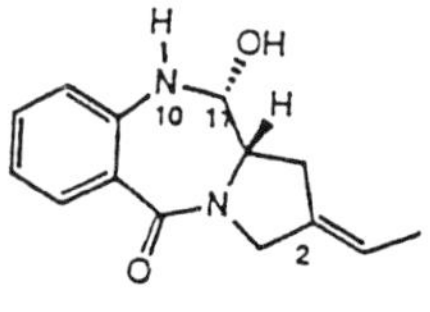

9

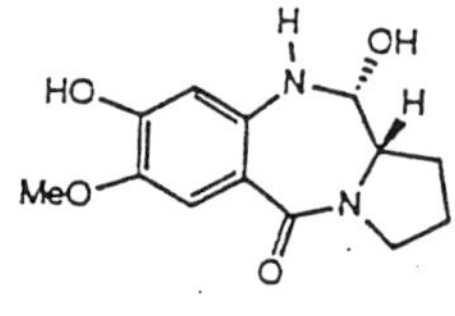

10

6.2 7-Membered Rings with Oxygen

6.2.1 Oxepanes

Oxepanes comprised the largest class of 7-membered heterocyclic natural products whose total or partial syntheses were reported in 1989. New methodology was used in the construction of many of these interesting natural products.

The brevicomins were the most studied oxepanes. Acid-catalyzed cyclization of chiral, protected dihydroxy ketone 11 gave (+)-endo-brevicomin (12), (Eq. 4) the aggregation pheromone of several pine beetle species; (-)-exo-brevicomin (13) (Eq. 4) was similarly prepared from an analogous cyclization <89T(45)7709>. Chirality was introduced via reduction of an acyl organostannane. In other syntheses of brevicomins, chirality was introduced by employing diethyl D-tartrate <89TL(30)3999> and by Baker's yeast reduction of a 2-acyl-1,3-dithiane <89CL1755>. Another synthesis of racemic exo- and endo-brevicomins employed a rhodium acetate catalyzed cycloaddition reaction of 1-diazo-2,5-hexanedione as the key step <89TL(30)1491>. In addition, a trialkylaluminum reduction of 6-methyl-2-propionyl-3,4-dihydro-2H-pyran gave a mixture of racemic exo- and endo-brevicomins <89SC(19)2915>.

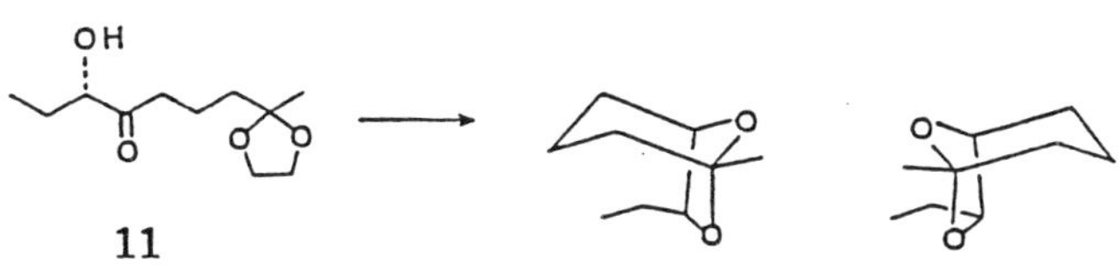

11 12 13 (Eq. 4)

Several lactones were prepared, many of which are biologically active targets. Pleurotonin is a fungal metabolite which displays antibiotic and antitumor activity. A synthesis of dl-pleurotonin (14) (Eq. 5) in which the trans perhydroindan ring was prepared by a stereoselective free radical cyclization was reported <89JA(111)7507>. Syntheses of (-)-picrotoxinin (15) (Eq. 5) and the related (-)-picrotin, both of which contain cis-fused hydrindanes, also appeared <89JA(111)3728>. Picrotoxinin is a specific GABA antagonist and a chloride channel blocker, and is an important pharmacological tool. (-)-Zeylena (16) (Eq. 5) was synthesized enantioselectively from styrene, by employing a microbial oxidation which produced a single cis-diol <89JOC(54)4239>. (-)-Echinosporin, an antibiotic-antitumor agent, was prepared from L-ascorbic acid. The key step in this sequence was an intramolecular Mitsunobu lactonization <89JA-(111)8039>.

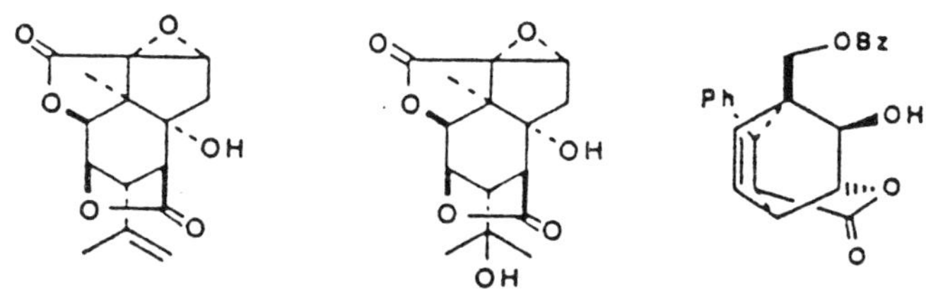

14 15 16 (Eq. 5)

Brassinolide and brassinosteroids are steroids with 7-membered ring lactone B rings. A synthesis of the plant growth regulator brassinolide (17) (Eq. 6) involving novel side chain construction via a pyranone intermediate was reported <89TL(30)-3141>. Another synthesis of 17, involving side chain construction by lactonization, was used for homobrassinolide as well <89CC612>. Brassinosteroids 18 and 19, which were active in the rice-lamina inclination test, were synthesized from ergosterol (Eq. 6) <89S426>.

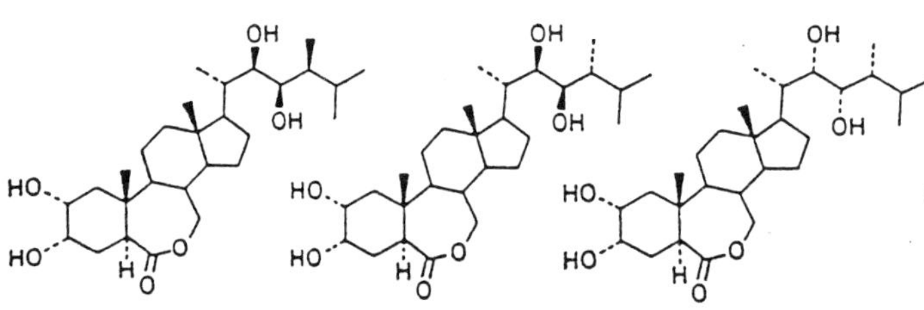

(Eq. 6)

17 18 19

An efficient synthesis of the fragrance chemical (+)-ambreinolide (20) (Eq. 7) from *l*-abietic acid was reported <89JOC(54)5123>. Ambreinolide is actually an odorless compound, but has fixative properties and is therefore valuable as a perfume base. (±)-Senoxepin (21) (Eq. 7), the first naturally occurring anti-Huckel oxepine, was prepared by an intramolecular [4+2] cycloaddition and a subsequent ring expansion of the epoxidized cyclohexadiene <89TL(30)-1241>. Stereocontrolled syntheses of erythronolides A (22a) and B (22b) (Eq. 7), which are aglycones of macrolide antibiotics, were accomplished starting from anhydro sugar levoglucosan <89T(45)5109>.

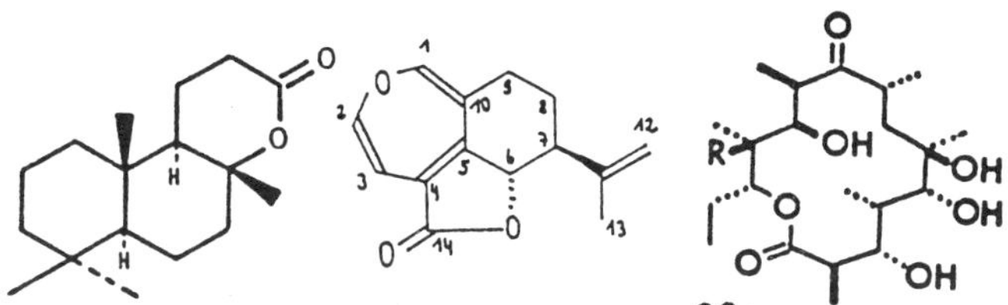

20 21 22a R=OH 22b R=H (Eq. 7)

Four naturally occurring piperidine derivatives were reported which involved oxepane intermediates in their construction. (+)-Conhydrine, a poisonous alkaloid constituent of the hemlock, whose extracts were used in ancient Greece for execution of criminals, was prepared from diethyl (S,S)-tartrate <89TL(30)6395>. Glycohydrolase inhibitor 1-deoxynojirimycin was prepared from a glucose derivative via a bridged anhydro sugar <89LA423>. Teneraic acid, a recently isolated marine natural product, is *trans*-piperidine-2,6-dicarboxylic acid. A stereoselective synthesis of (±)-teneraic acid, employing an oxatropone intermediate, was reported <89TL(30)6077>. Indole alkaloid (+)-quebrachamine was synthesized by reaction of tryptamine with a 7-membered ring lactol <89H(29)243>.

Several 7-membered ring ethers were reported. The syntheses of palytoxin carboxylic acid and palytoxin amide were accomplished by coupling eight different building blocks, each of which was a major synthetic effort <89JA(111)7525> <89JA(111)7530>. A stereocontrolled route to (±)-trichodermol (23), a sesquiterpene with diverse biological activities, was reported which used a tin enolate addition to a cyclohexadienyl-iron complex as the key step (Eq. 8) <89JA(111)1499><89JOC(54)4663>, (±)-Filiformin (24)

(Eq. 8), a bromine-containing sesquiterpene, was prepared using the Sheradsky rearrangement as the key step <89TL(30)2229>. (±)-Zoapatanol (25) (Eq. 8) was synthesized using a nickel-catalyzed coupling of methyl magnesium bromide with a dihydrofuran as a key step <89T(45)3839>. Zoapatanol is a constituent of Mexican tea, which contains other diterpenes as well. The tea has been used to induce labor and menses and to terminate early pregnancy. (-)-Phaseic acid is a metabolite of abscisic acid. This cyclohexenone derivative was synthesized with good stereocontrol from ethyl (1R, 2S)-5,5-ethylenedioxy-2-hydroxycyclohexanecarboxylate, which in turn was prepared in optically active form by reduction of the corresponding keto ester with Baker's yeast <89T(45)6387>. 7,7-Dimethyl-6,8-dioxabicyclo[3.2.1]octane, a relatively simple tricyclic ether which is a constituent of Japanese hops, was synthesized from acrolein dimer in racemic form in poor yield <89SC(19)2851>.

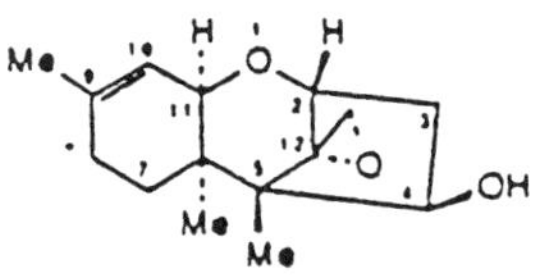

23

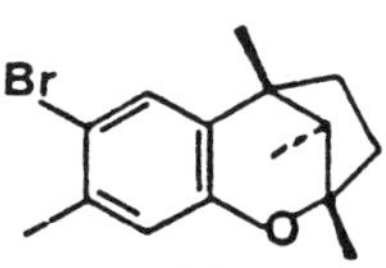

24

(Eq. 8)

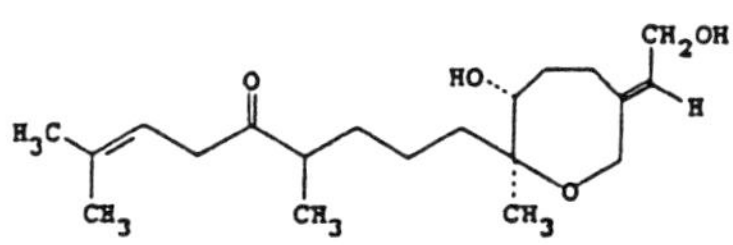

25

In addition, a few miscellaneous total syntheses were reported which utilized oxepanes as intermediates. (-)-Pulo'upone (26) (Eq. 9), an uncommon pyridine derivative whose structure had previously been determined only by spectral analysis, is a Hawaiian mollusk metabolite. A 7-membered ring lactone intermediate is produced during chiral auxiliary removal in this synthesis <89TL(30)1821>. Novel application of the Staudinger reaction allowed the preparation of mycosporins 27a and 27b (fungal metabolites) (Eq. 9) from D-(-)-quinic acid lactone <89JA(111)8970>. (+)-Phyllanthocin (28) (Eq. 9), the aglycone of an antineoplastic glycoside, was synthesized by an intermolecular nitrile oxide cycloaddition with (1R, 5R)-7-oxabicyclo[3.2.1]oct-2-en-6-one as a key step <89JOC(54)2209>.

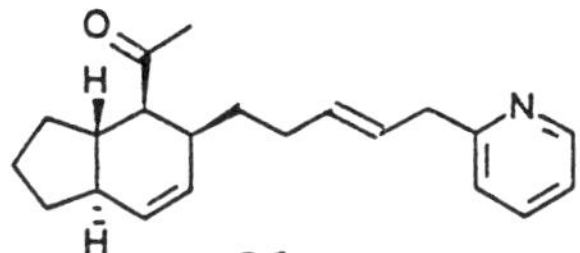

26

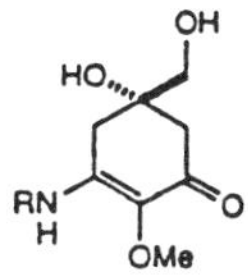

27a R = CH_2CO_2H (Mycosporin-Gly)
27b R = $CH(CH_2OH)_2$ (Mycosporin I)

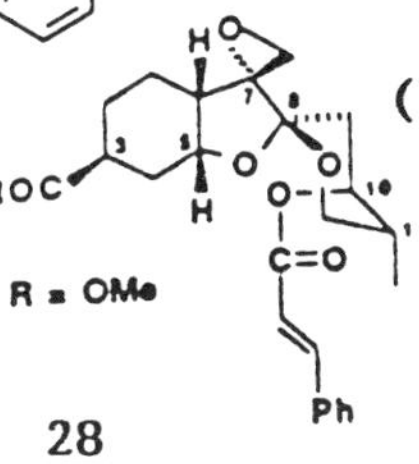

(Eq. 9)

28

Several interim studies appeared in which progress toward a natural product, either containing an oxepane ring or involving oxepanes as intermediates, was reported. Synthetic studies on the dioxepane region of the marine toxin brevetoxin B produced a model compound for the CDEF region, compound 29 (Eq. 10), which involved new methodology for the construction of oxepanes from hydroxy ketones <89JA(111)4136>. A stereoselective transformation of garderine to compound 30 (Eq. 10), which contains the humantenine skeleton, was reported <89TL(45)1327>. A new synthesis of Manske's ketone (31b) and its isomer 31a (Eq. 10) will allow a better route to isocularine alkaloids <89TL(30)6927>.

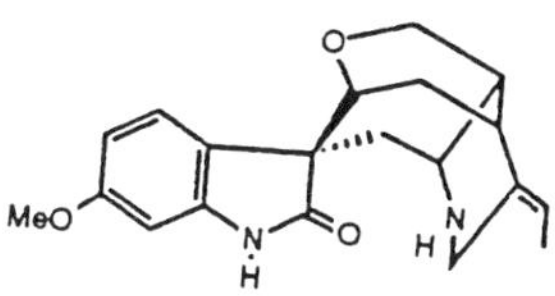

30

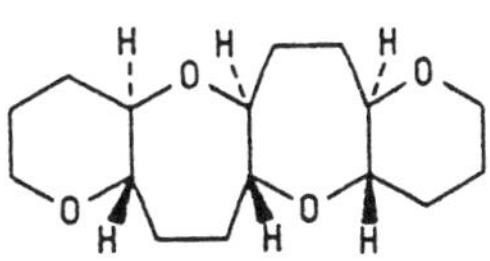

29

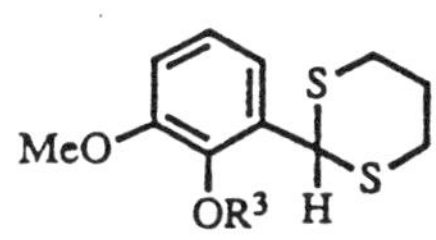

(Eq. 10)

31a R^3=$PhCH_2$
31b R^3=$MeOCH_2$

Two reports described synthetic methods for the construction of 7-membered cyclic ethers, as strategies directed toward isolaurepinnacin (32) (Eq. 11) <89JOC(54)5153><89JOC(54)5695>. An acetal unit, which is a possible intermediate for the total synthesis of (-)-azadirachtin (33) (Eq. 11), a highly oxygenated terpenoid with insect antifeedant and ecdysis inhibitory activity, was synthesized <89JOC-(54)3354>. Interestingly, a possible biosynthetic precursor to 33, which contains the furooxepane ring system, was isolated and characterized <89TL(30)1797>.

Gibberellin glycoside 34 (Eq. 11) was obtained by partial synthesis and shown to be identical to natural 34 <89T(45)1355>.

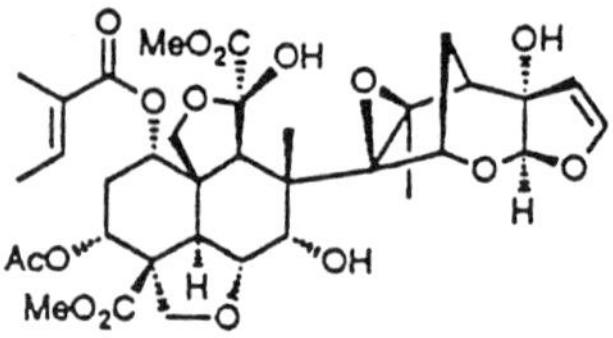

(Eq. 11)

33

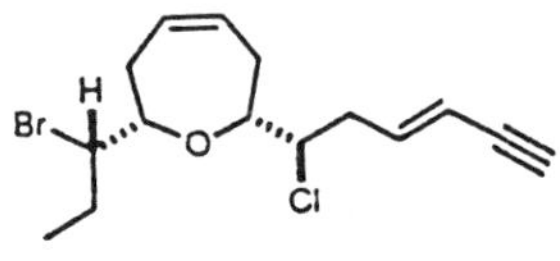

32

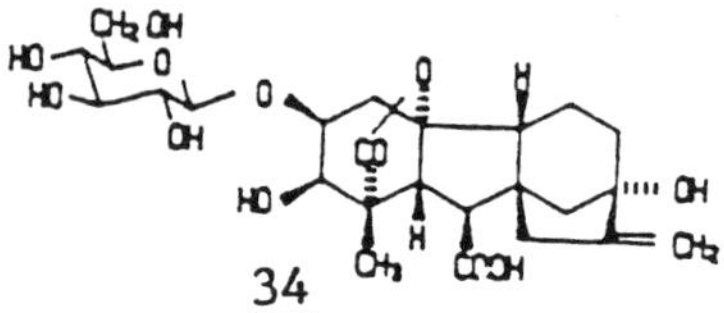

34

In addition, a 7-membered lactone was synthesized as a possible precursor to tirandomycin <89TL(30)1029>, and a bridged 7-membered lactone was prepared as an intermediate in the synthesis of the C-24 to C-34 fragment of the FK-506 family of immunosuppressants <89TL(30)6963>. An advanced intermediate which embodies the B, C and D rings of the penitrams, a small family of tremorgenic mycotoxins, was synthesized <89T(45)2431>. A 7-membered lactone was prepared in a projected general route to C-4 and C-10 functionalized abietane diterpenoids <89T(45)3547>, and another 7-membered bicyclic lactone was prepared as an intermeidate in an improved reserpine synthesis, based on Woodward's classical method <89SC(19)2987>. X-ray analysis confirmed the structure of a bridged oxepane intermediate in a route to germacranes, an ubiquitous class of cytotoxic sesquiterpenoids <89JOC(54)91>.

Analogs of several natural products were prepared in efforts to explore structure-activity relationships and enhance desired biological activities. Brassinolide analogs with a modified C-17 side chain <89CA(111)78497x> and lacking the C-17 side chain <89CCC(54)1019> were prepared, as well as 22,23-epoxybrassinosteroids <89ABC(53)263>. A key step in the synthesis of 11-oxaprogesterone was the homolytic cleavage of a C ring 7-membered lactol <89CA(111)-78493t>. A large number of trichothecene analogs were prepared and their cytotoxicities to mouse lymphoma cells were measured <89JMC(32)555>. A 3,8-dioxabicyclo[3.2.1]octane analog of diterpenoid zoapatanol (35) (Eq. 12) was prepared <89SC(19)2891>. Isocarbacyclins, e.g. 36 (Eq. 12), which are potent and chemically stable carbon analogs of prostacyclin, were prepared using either an intramolecular thermal ene reaction or an intramolecular aldol condensation

<89CPB(37)586>. Analogs of the aforementioned azadirachtin, an insect antifeedant and growth-disruption agent, and related limonoids were prepared <89T(45)-5175>. Tricyclic, relatively simple analogs of arteannuin, an unusual sesquiterpene with antimalarial properties, were prepared by photooxygenation of appropriate methoxymethylidine ketones <89TL(30)4485>. Simpler and less polar analogs of ginkolide B, a potent antagonist of platelet activating factor, were prepared in an attempt to increase potency and oral bioavailability <89TL(30)6959>. And cartenolide 37, 3-O-methyl uzarigenin (Eq. 12), was prepared from an androstane derivative <89T(45)2223>.

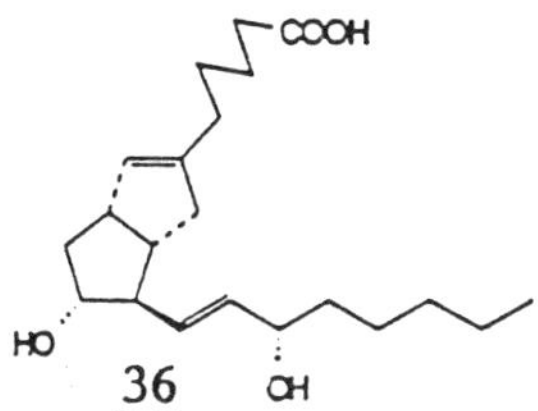

36

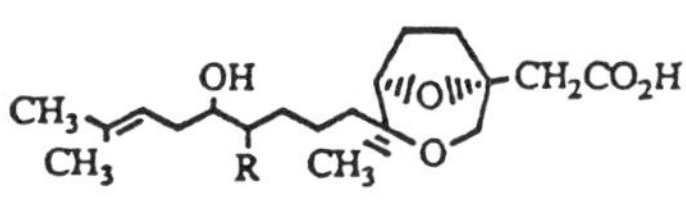

35 R = CH_3

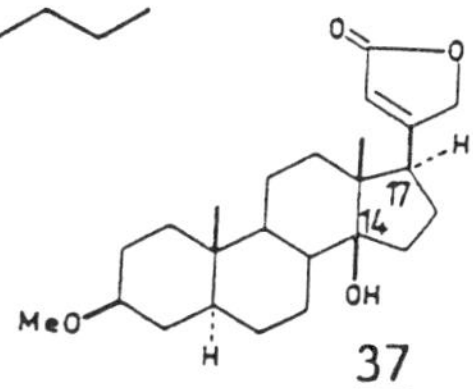

37

(Eq. 12)

In addition, a podophyliotoxin analog, an aglycone analog of the semisynthetic cytotoxic agent etoposide, was reported <89TL(30)5537>. The key step in this synthesis was the novel trimethylsilyl iodide mediated conversion of bromoisoxazoline 38 to bridged 7-membered lactone 39 (Eq. 13).

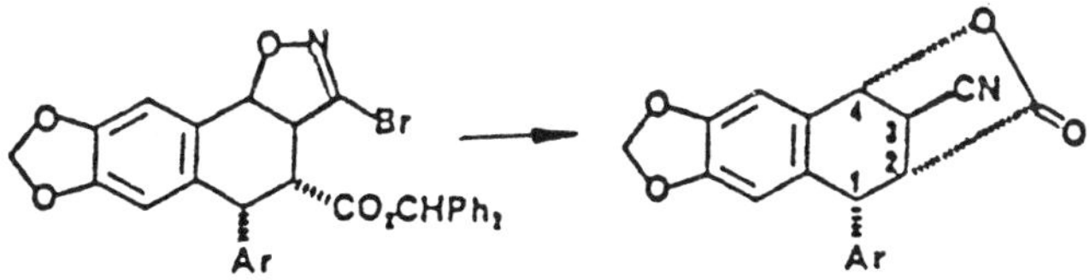

38 39

(Eq. 13)

A number of reports appeared where structure assignments were made for novel and diverse oxepanes which were not accompanied by synthetic efforts. The linear polycyclic ethers were the most complex group. Hemibrevetoxin-B, a biosynthetic precursor of the neurotoxin from the red tide organism, was reported <89JA(111)6476>. Ciguatoxin is structurally related, and is also a marine toxin which comes from the moray eel. The structure of ciguatoxin and a congener were reported <89TL(30)3793><89JA(111)8929>. Other polycyclic ethers containing piperazine rings and indole nuclei are fungal metabolite emethallicins A, B and C, which are inhibitors of histamine release from mast cells <89CPB(37)2592><89CC951>.

Ptilomycin A (40) (Eq. 14), a marine sponge natural product, possesses a unique polycyclic guanidine structure and displays antitumor, antiviral and antifungal activity <89JA(111)8925>. Nimbolicin is a new tetranortriterpene isolated from Azidirachta indica root bark <89H(29)729>. Ginsenoside la is a new saponin isolated from Panax ginseng leaves <89CPB-(37)1966>. Two new bicyclic ether lignans, kadsulignans A and B, were isolated from the seeds of K. coccinea <89CJC(67)682>. A new bicyclic ether alkaloid, rhazizine (41) (Eq. 14), was isolated from a Pakistani plant. Other natural products from this plant have displayed anticancer and antineoplastic activities <89T(45)3507>. Secospatane-like diterpene 42 (Eq. 14), isolated from algae, has shown feeding-deterrent activity against a young abalone strain <89TL(30)1567>.

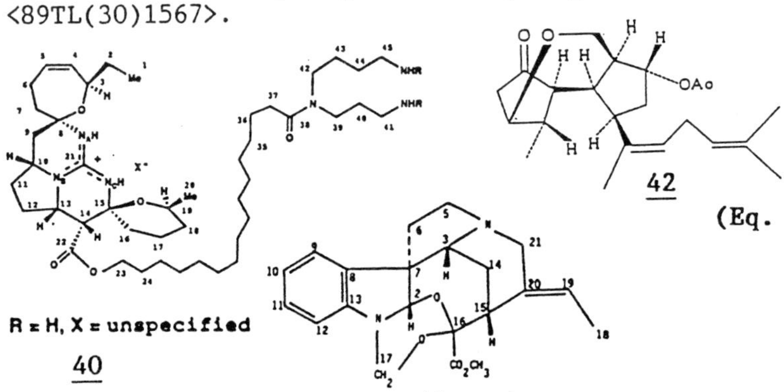

R = H, X = unspecified

40

41

42

(Eq. 14)

Lactones comprised the largest group of oxepane natural products whose structures were reported. Picrodendrins A (43) (Eq. 15), C and D, are new bicyclic sesquiterpene lactones isolated from Indonesian plants <89CPB(37)1805> <89CPB(37)2988>. Chromodorolide A (44) (Eq. 15), a rearranged diterpene lactone, is a putative repellent of the nudibranch, a shell-less mollusk <89JA(111)2712>. Two new limonids, clausenarin <89JNP(52)832> and dysoxylone <89CJC(67)-257> were isolated from C. anisata and D. richii, respectively. And pinthunamide (45) (Eq. 15), a new tricyclic sesquiterpene amide, is a fungal metabolite with plant growth stimulating properties <89TL(30)-1267>.

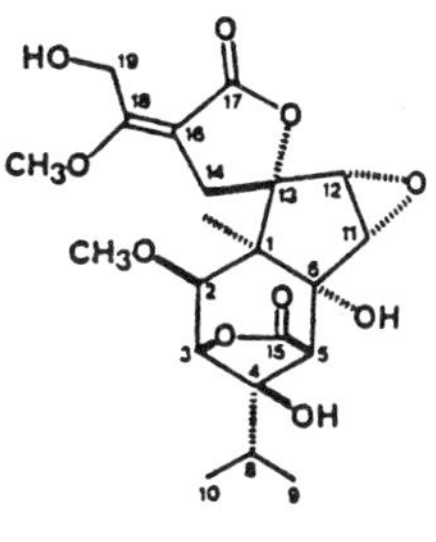
43

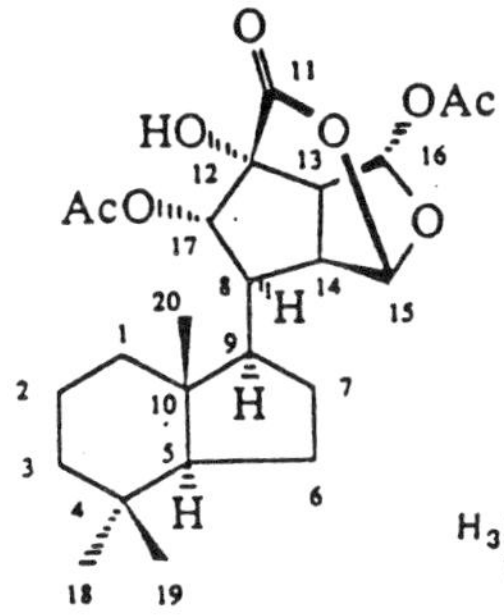
44

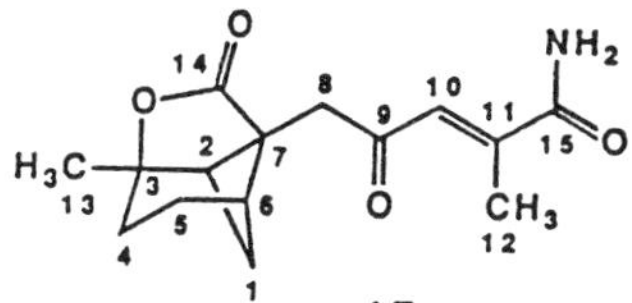
45

(Eq. 15)

6.2.2 1,2-Dioxepanes

The unusual structure of (+)-artemisinin (46) coupled with its antimalarial activity against chloroquine-resistant *P. falciparum* <85SCI(228)1049> has stimulated the interest of synthetic chemists, which resulted in three total syntheses <83JA(105)624> <86T(42)818> <87TL(28)4629>. From the perspective of this review, the structure of 46 is interesting in that it contains both oxepane and 1,2-dioxepane rings. During 1989 a number of analogs of 46 were reported. Avery described the syntheses of both a C,D-ring portion of 46 <89JOC(54)1789> and 6,9-desmethylartemisinin <89JOC(54)1792>. The methodology used to construct the 7-membered rings of the desmethyl analog paralleled that employed in their previously published total synthesis of 46. The lactone moiety in 46 was reduced to a lactol which, when treated with a variety of alcohols and BF_3, afforded ether analogs of 46 <89JMC(32)1249>. An unsuccessful synthetic approach to similar ethers was also disclosed <89TL(30)1125>. Jung reported syntheses of (+)-homodeoxoartemisinin (48) <89H(29)2273>, (+)-deoxoartemisinin, and (-)-deoxodesoxyartemisinin <89TL(30)5973>. Cyclization of 47 was noteworthy in that three 7-membered heterocycles were formed (Eq. 16). The syntheses of carba-analogs of 46 in which the peroxide oxygen adjacent to the central, quaternary carbon (C_6) was replaced by either a carbonyl group or a methylene group were also reported <89T(45)7287>.

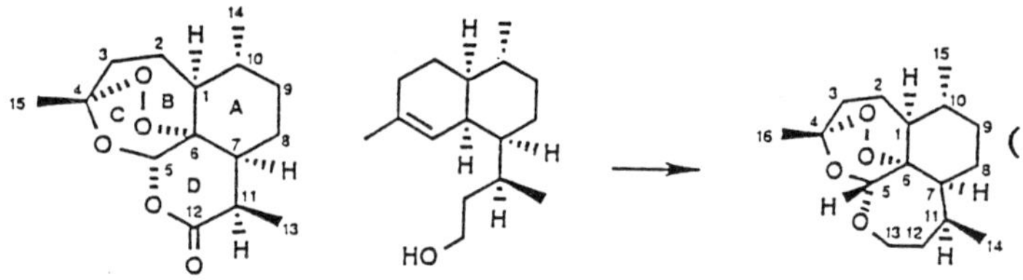 (Eq. 16)

46 47 48

6.2.3 1,3-Dioxepanes

1,3-Dioxepanes were utilized as synthetic intermediates in the syntheses of several natural products. As part of a study on the biosynthesis of the aflatoxins, both versicolorin A and 6-deoxyversicolorin were prepared <89JA(111)8306>. A key transformation in these syntheses involved the conversion of aldehyde 49 to ring-expanded bicyclic acetal 50 using triisopropylsilyl (TIPS) triflate and diethylaluminum chloride (Eq. 17). Both Nicolaou <89CC512> and Rúveda <89JOC(54)1539> employed 51 in their syntheses of advanced forskolin intermediates. The cyclic ketal was prepared in both instances from the corresponding diol via standard methodology. Finally, the new 1,3-dioxepine 52 (dracooxepine) was isolated from the fruits of Daemonorops draco Blume <89H(29)1119>.

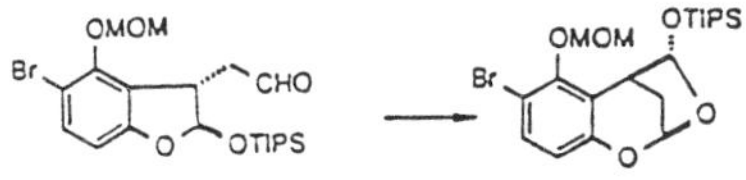

49 50

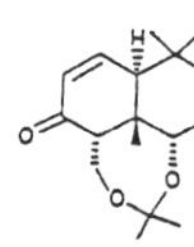

51

(Eq. 17)

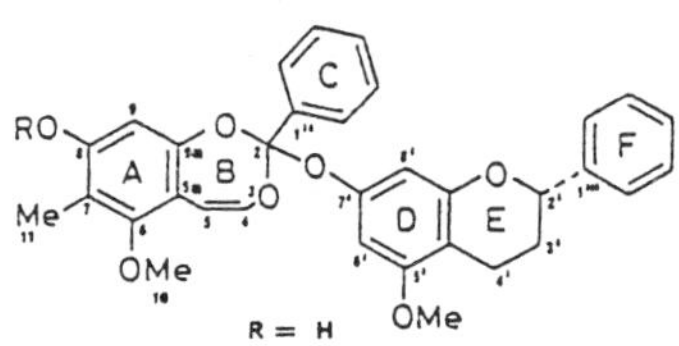

52

6.2.4 1,4-Dioxepines

Blizzard reported several chemical modifications of paraherquamide (53) <89JOC(54)2657>.

Treatment of 53 with two equivalents of bromine followed by treatment of the resulting tribromo derivative with zinc dust afforded 5-bromoparaherquamide. Reaction of this intermediate with phosgene followed by hydroxide gave alcohol 54 (Eq. 18). The absolute stereochemistry of 54 was then determined by x-ray analysis, which also defined the absolute stereochemistry of 53.

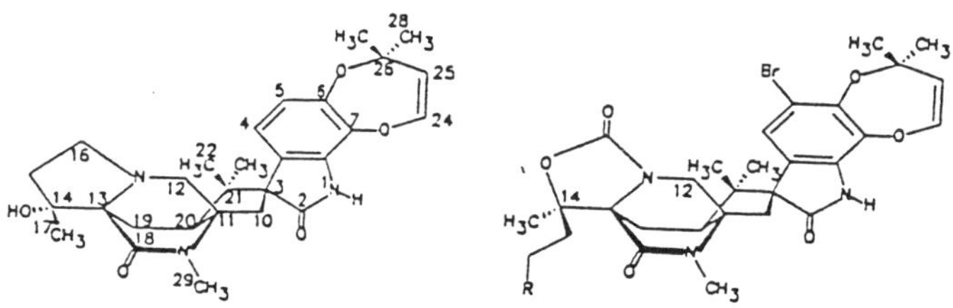

R = OH

53 54

(Eq. 18)

6.3 7-Membered Rings with Heteroatom Combinations

6.3.1 1,3-Oxazepines

Ueda <89CPB(37)660> recently prepared several carbon-bridged cyclonucleosides as conformational probes for the stereochemical investigation of nucleosides and nucleotides. The key 7-membered ring-forming step involved the intramolecular glycosylation of 55 using stannic chloride (Eq. 19). The imino ether of the resulting cyclonucleoside 56 was then manipulated to yield cyclouridine, cyclocytidine and cyclo-4-thiouridine.

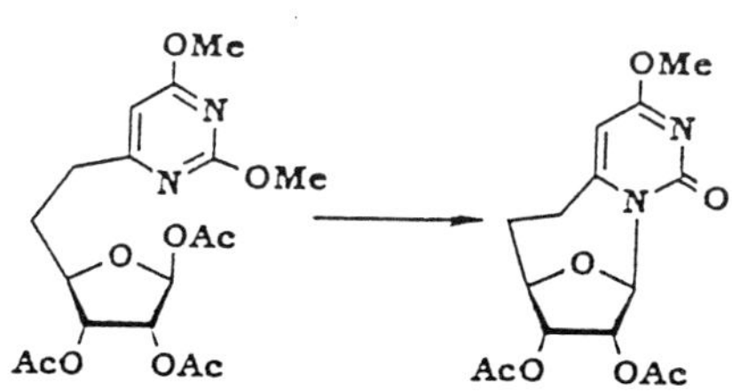

(Eq. 19)

55 56

6.3.2 1,3-Thiazepines

Baldwin and coworkers continued their mechanistic studies on the enzyme isopenicillin N synthase (IPNS) <89CC758><89CC802><89CC978>. Interestingly, in a number of instances, incubation of selected peptides with IPNS afforded ring-fused 1,3-thiazepine derivatives. For example, incubation of tripeptide 57 with IPNS gave 1,3-thiazepine 58 (Eq. 20) as a major product whereas the corresponding penam derivative was isolated as only a very minor product.

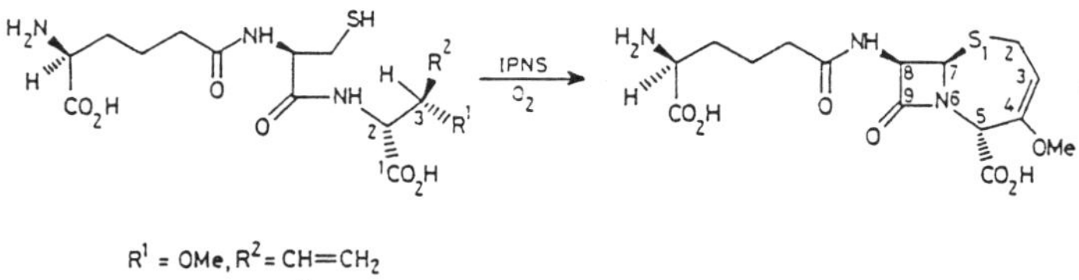

(Eq. 20)

57 58

6.3.3 1,3,7-Oxathiazepines

Further synthetic studies on the eudistomin alkaloids were reported in 1989. Cyclization of 59 with NCS followed by removal of the Boc group completed the total synthesis of (-)-eudistomin L (60) <89JA(111)2721> (Eq. 21). A similar synthetic strategy involving a sila-Pummerer rearrangement afforded the corresponding acetamide of 60 <89TL(30)1041>. Two syntheses of the unsubstituted tetracyclic skeleton of the eudistomin alkaloids were also reported and these both involved Pictet-Spengler ring closures in the key 7-membered ring-forming step <89TL(30)5009> <89TL(30)-6809>. Also reported was the isolation of the debromoderivative of 60 <89AJC(42)1201> from the New Zealand ascidian *R. sigillinoides*.

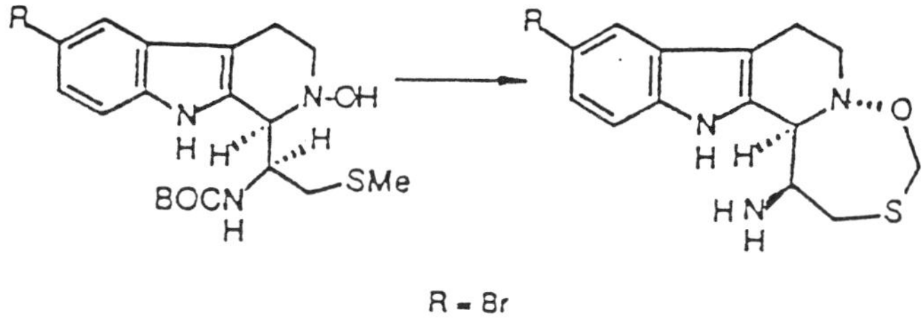

(Eq. 21)

59 60

6.3.4 1,3,6-Oxadiazepines

Three new staurosporin-related, protein kinase C inhibitors were recently isolated. Thus, UNC-01 (61a) and its stereoisomer, UNC-02 (61b) (Eq. 22) were isolated from the N-126 strain of *Streptomyces* <89JAB(42)564><89JAB(42)571> while SF2370 (62) (Eq. 22) was isolated from the culture broth of *Actinomadura* <89LS(44)1351>. In addition, several analogs of staurosporin were synthesized and evaluated as inhibitors of protein kinases <89IJC(43)851>.

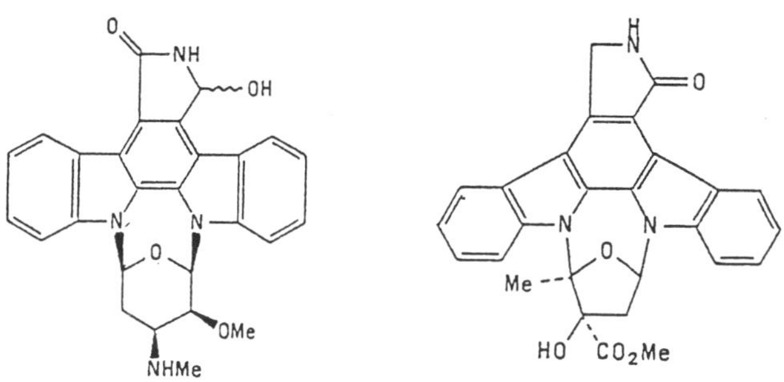

(Eq. 22)

61a,b 62

6.3.5 1,2,5-Thiadiazepines

Interest in the pyrrolo[1,4]benzodiazepine family of antitumor antibiotics (see Section 7.1.2) has stimulated the search for biologically active, totally synthetic analogs. During the past year, the synthesis of tricyclic 1,2,5-thiadiazepine 64 <89H-(29)1529> was reported. Formation of the 7-membered ring was accomplished by stirring aminoaldehyde 63 in dichloromethane in the presence of an alcohol and trifluoroacetic acid (Eq. 23).

(Eq. 23)

63 64

6.3.6 1,4-Oxathiapanes

The novel withanolide, withaperuvin H (65) (Eq. 24) was recently isolated from the roots of *P. peruviana* <89CC628>.

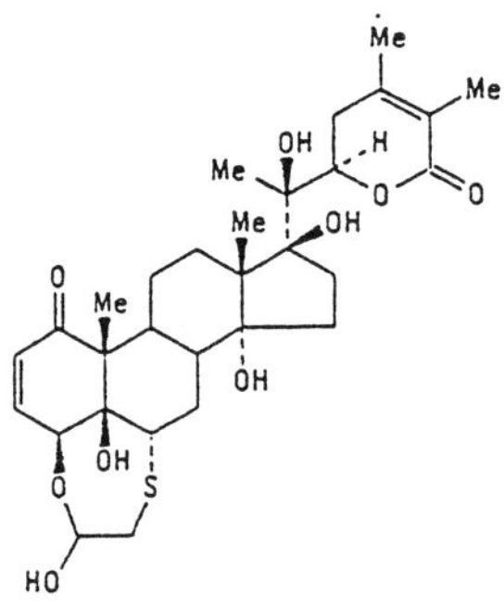

(Eq. 24)

65

VOLUME	REFERENCES
83JA(105)624	G. Schmid and W. Hofheinz; J. Am. Chem. Soc., 1983, 105, 624.
85SCI(228)1049	D.L. Klayman; Science, 1985, 228, 1049.
86T(42)818	X.-X. Xu, J. Zhu, D.-Z. Huang and W.-S. Zhou; Tetrahedron, 1986, 42, 818.
87TL(28)4629	M.A. Avery, C. Jennings-White and W.K.M. Chong; Tetrahedron Lett., 1987, 28, 4629.
89ABC(53)263	S. Takatsuto, F. Futatsuya, K. Kobayashi and H. Satoh; Agric. Biol. Chem., 1989, 53, 263.
89AJC(42)1201	R.J. Lake, J.W. Blunt and M.H.G. Munro; Aust. J. Chem., 1989, 42, 1201.
89BCJ(62)193	H. Suginome and J.B. Wang; Bull. Chem. Soc. Jpn., 1989, 62, 193.
89CA(111)78497X	W. Zhou, B. Jiang and X. Pan; Chem. Abstr., 1989, 111, 78497X.
89CC512	K.C. Nicolaou, S. Kubota and W.S. Li; J. Chem. Soc., Chem. Commun., 1989, 512.
89CC612	Z. Wei-Shan, B. Jiang and X. Pan; J. Chem. Soc., Chem. Commun., 1989, 612.
89CC628	Y. Oshima, H. Hikino, M. Sahai and A.B. Ray; J. Chem. Soc., Chem. Commun., 1989, 628.
89CC758	J.E. Baldwin, R.M. Adlington, C.J. Schofield and H.-H. Ting; J. Chem. Soc., Chem. Commun., 1989, 758.
89CC802	J.E. Baldwin, R.M. Adlington, A. Basak, P. Imming, K. Ponnamperuma, H. Ronneberg, C.J. Schofield, H.-H. Ting and N.J. Turner; J. Chem. Soc., Chem. Commun., 1989, 802.
89CC978	J.E. Baldwin, R.M. Adlington, M. Bradley, N.J. Turner and A.R. Pitt; J. Chem. Soc., Chem. Commun., 1989, 978.
89CC1162	M. Bryce and J.M. Gardiner; J. Chem. Soc., Chem. Commun., 1989, 1162.
89CC1591	S. Takano, K. Inomata, T. Sato and K. Ogasawara; J. Chem. Soc., Chem. Commun., 1989, 1591.
89CCC(54)1019	L. Kohout and M. Strnad; Collect. Czech. Chem. Commun., 1989, 54, 1019.
89CJC(67)257	M.K. Jogia and R.J. Andersen; Can. J. Chem., 1989, 67, 257.
89CJC(67)682	J. Liu, L. Li and H. Yu; Can. J. Chem., 1989, 67, 682.
89CL703	K. Okada, Y. Mizuno, H. Tanino, H. Kakoi and S. Inoue; Chem. Lett., 1989, 703.
89CL1755	Y. Noda and M. Kikuchi; Chem. Lett., 1989, 1755.
89CPB(37)586	M. Sodeoka, Y. Ogawa, T. Mase and M. Shibasaki; Chem. Pharm. Bull., 1989, 37, 586.
89CPB(37)660	Y. Yoshimura, A. Matsuda and T. Ueda; Chem. Pharm. Bull., 1989, 37, 660.

89CPB(37)1805 T. Ohmoto, K. Koike, H. Fukuda, K. Mitsunaga, K. Kagei, T. Kawai and T. Sato; Chem. Pharm. Bull., 1989, 37, 1805.

89CPB(37)1966 S. Zhang, X. Yao, Y. Chen, C. Cui, Y. Tezuka and T. Kikuchi; Chem. Pharm. Bull., 1989, 37, 1966.

89CPB(37)2592 N. Kawahara, S. Nakajima, M. Yamazaki and K. Kawai; Chem. Pharm. Bull., 1989, 37, 2592.

89CPB(37)2988 T. Ohmoto, K. Koike, H. Fukuda, K. Mitsunaga, K. Ogata and K. Kagei; Chem. Pharm. Bull., 1989, 37, 2988.

89CPB(37)3164 T. Sekine, J. Arita, K. Saito, F. Ikegami, S. Okonogi and I. Murakoshi; Chem. Pharm. Bull., 1989, 37, 3164.

89H(28)103 C.B. Rao, D.V. Rao, V.S.N. Raju, B.W. Sullivan and D.J. Faulkner; Heterocycles, 1989, 28, 103.

89H(29)243 M. Asaoka and H. Takei; Heterocycles, 1989, 29, 243.

89H(29)381 M.-L. Bennasar, E. Zulaica, R. Vila and J. Bosch; Heterocycles, 1989, 29, 381.

89H(29)729 I. Ara, B.S. Siddiqui, S. Faizi and S. Siddiqui; Heterocycles, 1989, 29, 729.

89H(29)1119 A. Arnone, G. Nasini and L. Merlini; Heterocycles, 1989, 29, 1119.

89H(29)1529 N. Langlois and R.Z. Andriamialisoa; Heterocycles, 1989, 29, 1529.

89H(29)1721 D.J. Wustrow and W. Wierenga; Heterocycles, 1989, 29, 1721.

89H(29)2273 D.A. Bustos, M. Jung, H.N. ElSohly and J.D. McChesney; Heterocycles, 1989, 29, 2273.

89IJC(43)851 T. Meyer, U. Regenass, D. Fabbro, E. Alteri, J. Roesel, M. Mueller, G. Caravatti and A. Matter; Int. J. Cancer, 1989, 43, 851.

89JA(111)647 M. Adamczeski, E. Quiñoá and P. Crews; J. Am. Chem. Soc., 1989, 111, 647.

89JA(111)1499 M.K. O'Brien, A.J. Pearson, A.A. Pinkerton, W. Schmidt and K. Willman; J. Am. Chem. Soc., 1989, 111, 1499.

89JA(111)1923 D.R. Williams, D.L. Brown and J.W. Benbow; J. Am. Chem. Soc., 1989, 111, 1923.

89JA(111)2588 W.J. Klaver, H. Hiemstra and W.N. Speckamp; J. Am. Chem. Soc., 1989, 111, 2588.

89JA(111)2712 E.J. Dundei, E.D. de Silva and R.J. Andersen; J. Am. Chem. Soc., 1989, 111, 2712.

89JA(111)2721 M. Nakagawa, J.-J. Liu and T. Hino; J. Am. Chem. Soc., 1989, 111, 2721.

89JA(111)3728 M. Miyashita, T. Suzuki and A. Yoshikoshi; J. Am. Chem. Soc., 1989, 111, 3728.

89JA(111)4136 K.C. Nicolau, C.-K. Huang and D.A. Nugiel; J. Am. Chem. Soc., 1989, 111, 4136.

89JA(111)5417 M.R. Peña and J.K. Stille; J. Am. Chem. Soc., 1989, 111, 5417.

89JA(111)6476 A.V.K. Prasad and Y. Shimizu; J. Am. Chem. Soc., 1989, 111, 6476.

89JA(111)7507 D.J. Hart, H.-C. Huang, R. Krishnamurthy and T. Schwartz; J. Am. Chem. Soc., 1989, 111, 7507.

89JA(111)7525 R.W. Armstrong, J.-M. Beau, S.H. Cheon, W.J. Christ, H. Fujioka, W.-H. Ham, L.D. Hawkins, H. Jin, S.H. Kang, Y. Kishi, M.J. Martinelli, W.W. McWhorter, M. Mizuno, M. Nakata, A.E. Stutz, F.X. Talamas, M. Taniguchi, J.A. Tino, K. Ueda, J. Uenishi, J.B. White and M. Yonaga; J. Am. Chem. Soc., 1989, 111, 7525.

89JA(111)7530 R.W. Armstrong, J.-M. Beau, S.H. Cheon, W.J. Christ, H. Fujioka, W.-H. Ham, L.D. Hawkins, H. Jin, S.H. Kang, Y. Kishi, M.J. Martinelli, W.W. McWhorter, M. Mizuno, M. Nakata, A.E. Stutz, F.X. Talamas, M. Taniguchi, J.A. Tino, K. Ueda, J. Uenishi, J.B. White and M. Yonaga; J. Am. Chem. Soc., 1989, 111, 7530.

89JA(111)8039 A.B. Smith, G.A. Sulikowski and K. Fujimoto; J. Am. Chem. Soc., 1989, 111, 8039.

89JA(111)8306 T.L. Graybill, K. Pal, S.M. McGuire, S.W. Brobst and C.A. Townsend; J. Am. Chem. Soc., 1989, 111, 8306.

89JA(111)8925 Y. Kashman and S. Hirsh; J. Am. Chem. Soc., 1989, 111, 8925.

89JA(111)8929 M. Murata, A.M. Legrand, Y. Ishibashi and T. Yasumoto; J. Am. Chem. Soc., 1989, 111, 8929.

89JA(111)8970 J.D. White, J.H. Cammack and K. Sakuma; J. Am. Chem. Soc., 1989, 111, 8970.

89JAB(42)564 I. Takahashi, K. Asano, I. Kawamoto, T. Tamaoki and H. Nakano; J. Antibiot., 1989, 42, 564.

89JAB(42)571 I. Takahashi, Y. Saitoh, M. Yoshida, H. Sano, H. Nakano, M. Morimoto and T. Tamaoki; J. Antibiot., 1989, 42, 571.

89JHC(26)1023 P. DeCaprariis, G. DeMartino, E. Abignente, P. Avara and L. Mayol; J. Heterocycl. Chem., 1989, 26, 1023.

89JHC(26)1605 F. Melani, L. Cecchi, V. Colotta and G. Filacchioni; J. Heterocycl. Chem., 1989, 26, 1605.

89JMC(32)555 D.W. Anderson, R.M. Black, C.G. Lee, C. Pottage, R.L. Rickard, M.S. Sandford, T.D. Webber and N.E. Williams; J. Med. Chem., 1989, 32, 555.

89JMC(32)1249 A.J. Lin, M. Lee and D.L. Klayman; J. Med. Chem., 1989, 32, 1249.

89JNP(52)832 B.V. Ngadjui, J.F. Ayafor, B.L. Sondergam and J.D. Connolly; J. Nat. Prod., 1989, 52, 832.

89JOC(54)91 P.G. McDougal, Y.-I. Oh and D. VanDerveer; J. Org. Chem., 1989, 54, 91.

89JOC(54)1539 C. Somoza, J. Darias and E.A. Rúveda; J. Org. Chem., 1989, 54, 1539.

89JOC(54)1548 C.H. Heathcock, T.A. Blumenkopf and K.M. Smith; J. Org. Chem., 1989, 54, 1548.

89JOC(54)1789 M.A. Avery, C. Jennings-White and W.K.M. Chong; J. Org. Chem., 1989, 54, 1789.

89JOC(54)1792 M.A. Avery, C. Jennings-White and W.K.M. Chong; J. Org. Chem., 1989, 54, 1792.

89JOC(54)2041 P. Garner, K. Sunitka, W.-B. Ho, W.J. Youngs, V.O. Kennedy and A. Djebli; J. Org. Chem., 1989, 54, 2041.

89JOC(54)2209 S.F. Martin, M.S. Dappen, B. Dupre, C.J. Murphy and J.A. Colapret; J. Org. Chem., 1989, 54, 2209.

89JOC(54)2657 T.A. Blizzard, G. Marino, H. Mrozik, M.H. Fisher, K. Hoogsteen and J.P. Springer; J. Org. Chem., 1989, 54, 2657.

89JOC(54)3174 J.A. Postigo and R. Erra-Balsells; J. Org. Chem., 1989, 54, 3174.

89JOC(54)3354 Y. Nishikimi, T. Timori, M. Sodeoka and M. Shibasaki; J. Org. Chem., 1989, 54, 3354.

89JOC(54)3407 M.E. Kuehne and W.G. Bornmann; J. Org. Chem., 1989, 54, 3407.

89JOC(54)4239 T. Hudlicky, G. Seoane and T. Pettus; J. Org. Chem., 1989, 54, 4239.

89JOC(54)4419 P. Duhamel, M. Kotera, T. Monteil, B. Marabout and D. Davoust; J. Org. Chem., 1989, 54, 4419.

89JOC(54)4553 M.E. Kuehne and J.B. Pitner; J. Org. Chem., 1989, 54, 4553.

89JOC(54)4654 F.J. Sardina, M.H. Howard, A.M.P. Koskinen and H. Rapoport; J. Org. Chem., 1989, 54, 4654.

89JOC(54)4663 A.J. Pearson and M.K. O'Brien; J. Org. Chem., 1989, 54, 4663.

89JOC(54)5123 A. Abad, C. Agullo, M. Arno, A.C. Cunat and R.J. Zaragoza; J. Org. Chem., 1989, 54, 5123.

89JOC(54)5153 H. Kotsuki, Y. Ushio, I. Kadota and M. Ochi; J. Org. Chem., 1989, 54, 5153.

89JOC(54)5695 A. Castaneda, D.J. Kucera and L.E. Overman; J. Org. Chem., 1989, 54, 5695.

89JOC(54)5746 M.A. Weidner-Wells, A. DeCamp and P.H. Mazzocchi; J. Org. Chem., 1989, 54, 5746.

89LA423 R.R. Schmidt, J. Michel and E. Rucker; Liebigs Ann. Chem., 1989, 423.

89LS(44)1351 M. Hachisu, T. Hiranuma, M. Koyama and M. Sezaki; Life Sci., 1989, 44, 1351.

89P(28)1298 J. Pusset, S. La Barre, N. Langlois and J. Hamon; Phytochemistry, 1989, 28, 1298.

89S426 W.-S. Zhou, Y.-P. Zhou and B. Jiang; Synthesis, 1989, 426.

89SC(19)2851 R.R. Kopp and B.P. Mundy; Syn. Commun., 1989, 19, 2851.

89SC(19)2891 Z.G. Hajos, M.P. Wachter, R.E. Adams and H.M. Werblood; Syn. Commun., 1989, 19, 2891.

89SC(19)2915 K.E. Bartelt and B.P. Mundy; Syn. Commun., 1989, 19, 2915.

89SC(19)2987 L. Gottlieb, D. Kellner and H.J.E. Loewenthal; Syn. Commun., 1989, 19, 2987.

89T(45)1327 H. Takayama, K. Masubuchi, M. Kitajima, N. Aimi and S. Sakai; Tetrahedron, 1989, 45, 1327.

89T(45)1355 G. Schneider, G. Sembdner, K. Schreiber and B.O. Phinney; Tetrahedron, 1989, 45, 1355.

89T(45)2223 G. Groszek, A. Kurek-Tyrlik and J. Wicha; Tetrahedron, 1989, 45, 2223.

89T(45)2431 A.B. Smith, J.N. Haseltine and M. Visnick; Tetrahedron, 1989, 45, 2431.

89T(45)3329 R. Vlahov, D. Krikorian, G. Spassov, M. Chinova, I. Vlahov, S. Parushev, G. Snatzke, L. Ernst, K. Kieslich, W.-R. Abrahan and W.S. Sheldrick; Tetrahedron, 1989, 45, 3329.

89T(45)3507 A. Rahman, T. Fatima and S. Khanum; Tetrahedron, 1989, 45, 3507.

89T(45)3547 B.K. Banik and U.R. Ghatak; Tetrahedron, 1989, 45, 3547.

89T(45)3839 P.J. Kocienski, C.L. Love, R.J. Whitby, G. Costello and D.A. Roberts; Tetrahedron, 1989, 45, 3839.

89T(45)5109 N.K. Kochetkov, A.F. Sviridov, M.S. Ermolenko, D.V. Yashunsky and V.S. Borodkin; Tetrahedron, 1989, 45, 5109.

89T(45)5175 S.V. Ley, J.C. Anderson, W.M. Blaney, P.S. Jones, Z. Lidert, E.D. Morgan, N.G. Robinson, D. Santafianos, M.S.J. Simmonds and P.L. Toogood; Tetrahedron, 1989, 45, 5175.

89T(45)6387 T. Kitahara, K. Touhara, H. Watanabe and K. Mori; Tetrahedron, 1989, 45, 6387.

89T(45)7287 B. Ye and Y.-L. Wu; Tetrahedron, 1989, 45, 7287.

89T(45)7709 J.M. Chong and E.K. Mar; Tetrahedron, 1989, 45, 7709.

89TL(30)1029 M. Taschner and A.S. Aminbhavi; Tetrahedron Lett., 1989, 30, 1029.

89TL(30)1041 I.W.J. Still, J.R. Strautmanis; Tetrahedron Lett., 1989, 30, 1041.

89TL(30)1125 F. Binns and T.W. Wallace; Tetrahedron Lett., 1989, 30, 1125.

89TL(30)1241 A. Cleve and F. Bohlmann; Tetrahedron Lett., 1989, 30, 1241.

89TL(30)1267 Y. Kimura, H. Nakajima, T. Hamasaki, F. Sugawara, L. Parkanyi and J. Clardy; Tetrahedron Lett., 1989, 30, 1267.

89TL(30)1491 A. Padwa, R.L. Chinn and L. Zhi; Tetrahedron Lett., 1989, 30, 1491.

89TL(30)1567 K. Kurata, K. Taniguchi, K. Shiraishi and M. Suzuki; Tetrahedron Lett., 1989, 30, 1567.

89TL(30)1797 W. Kraus, H. Gutzeit and M. Bokel; Tetrahedron Lett., 1989, 30, 1797.

89TL(30)1821 T. Sugahara, T. Iwata, M. Yamaoka and S. Takano; Tetrahedron Lett., 1989, 30, 1821.

89TL(30)1871 T. Nagasaka, Y. Koseki and F. Hamaguchi; Tetrahedron Lett., 1989, 30, 1871.

89TL(30)1959 S. Jegham, J.-L. Fourrey and B.C. Das; Tetrahedron Lett., 1989, 30, 1959.

89TL(30)2229 J.-Y. Laronze, R.E. Boukili, D. Cartier, J. Laronze and J. Levy; Tetrahedron Lett., 1989, 30, 2229.

89TL(30)2747 F.G. Fang and S.I. Danishefsky; Tetrahedron Lett., 1989, 30, 2747.

89TL(30)3141 T. Kametani, K. Keino, M. Kigawa, M. Tsubuki and T. Honda; *Tetrahedron Lett.*, 1989, 30, 3141.

89TL(30)3793 M. Murata, A.M. Legrand and T. Yasumoto; *Tetrahedron Lett.*, 1989, 30, 3793.

89TL(30)3999 H. Kotsuki, I. Kadota and M. Ochi; *Tetrahedron Lett.*, 1989, 30, 3999.

89TL(30)4185 R.W. Kavash and P.S. Mariano; *Tetrahedron Lett.*, 1989, 30, 4185.

89TL(30)4485 C.W. Jefford, J. Velarde and G. Bernardinelli; *Tetrahedron Lett.*, 1989, 30, 4485.

89TL(30)5009 P.H.H. Hermkens, J.H. v. Maarseveen, C.G. Kruse and H.W. Scheeven; *Tetrahedron Lett.*, 1989, 30, 5009.

89TL(30)5537 A. Haber; *Tetrahedron Lett.*, 1989, 30, 5537.

89TL(30)5973 M. Jung, X. Li, D.A. Bustos, H.N. ElSohly and J.D. McChesney; *Tetrahedron Lett.*, 1989, 30, 5973.

89TL(30)6077 H.H. Wasserman, K. Rodriques and R. Kucharczyk; *Tetrahedron Lett.*, 1989, 30, 6077.

89TL(30)6223 C.W. Bird, A.L. Brown, C.C. Chan and A. Lewis; *Tetrahedron Lett.*, 1989, 30, 6223.

89TL(30)6395 Y. Masaki, T. Imaeda, K. Nagata, H. Oda and A. Ito; *Tetrahedron Lett.*, 1989, 30, 6395.

89TL(30)6809 M.P. Kirkup, B.B. Shankar, S. McCombie, A.K. Ganguly and A.T. McPhail; *Tetrahedron Lett.*, 1989, 30, 6809.

89TL(30)6825 Atta-ur-Rahman, K.A. Alvi, S.A. Abbas, M.I. Choudhary and J. Clardy; *Tetrahedron Lett.*, 1989, 30, 6825.

89TL(30)6927 C. Lamas, A. Garcia, L. Castedo and D. Dominguez; *Tetrahedron Lett.*, 1989, 30, 6927.

89TL(30)6959 E.J. Corey and A.V. Gavai; *Tetrahedron Lett.*, 1989, 30, 6959.

89TL(30)6963 A.B. Smith, K.J. Hale, L.M. Laakso, K. Chen and A. Riera; *Tetrahedron Lett.*, 1989, 30, 6963.

89TL(30)7173 P.A. Jacobi, C.A. Blum, R.W. DeSimone and U.E.S. Udodong; *Tetrahedron Lett.*, 1989, 30, 7173.

CHAPTER 7

Eight-Membered and Larger Rings

GEORGE R NEWKOME & CLAUS D WEIS
University of South Florida, Tampa, FL, USA

7.1 INTRODUCTION

In 1989, the rapid development in the field of supramolecular chemistry was highlighted in three texts. The first outlined the progress in crown ethers, cryptands, podands, spherands, and their profound effect on host-guest chemistry, liquid crystals and clatherates <89MI:(3)>. The second described the "beauty of distinct organic molecules", such as: pagodanes, dodecahedrane, and cubanes, as well as hetero-circulene, -helicene, -cyclophanes, superphanes, superferroceneophane, porphycene, and octavinylogue[34]porphyrine <89MI:(2)>. The third overviewed the rapid development in host-guest chemistry of macrocyclic ligands upon chemistry and biochemistry <89MI:(1)>.

The coordination chemistry of bi- and poly-nuclear manganese complexes, which contain *O,N*-donors and a variety of bridging ligands, has been presented <89AG(E)1153>. Complex stability and selectivity for different metal ions of preorganized ligands, such as cryptands, spherands and rigid ligands with pendant donor groups were reviewed <89CR1875>. Oxo-bridged compounds as versatile ligands <89MI219>, synthesis of macrobicyclic lanthanide complexes <89MI938>, photochemical reactions of cage-type polypyridine complexes of Eu^{+3} and Ru^{+3} and $[Co(sepulchrate)]^{+3}$ (**1**) <89G311>, inclusion complexes of carboxylic hosts <89MI12>, molecular recognition in host-guest chemistry <89MI493> and future developments <89MI583><89MI484> in particular cyclodextrins <89MI29> have been reviewed. Molecular belts, strips, carcerands (hollow container-like molecules) with nanometer scale dimensions have been surveyed <89AG(E)1103>.

Supramolecular chemistry <89MI63> and chiral recognition <89CR929> (**2**) <89CR1581><89CR347> are the development areas in hetero-macrocyclic chemistry of the nineties. Effects of molecular strain on the basicities of large ring bridgehead amines, such as 1,6-diazabicyclo-

[4.4.4]-tetradecane, with expanded C-N-C angles has been discussed <89CR1215>. Elucidation of the active site structure and reactivity of binuclear copper proteins, hemocyanine and tyrosinase has been summarized <89T3>.

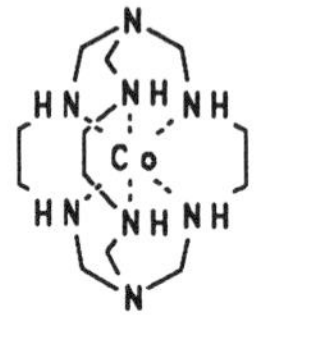

1 $[Co(sep)]^{+3}$

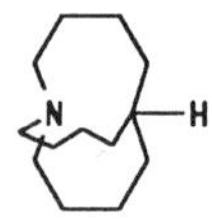

2

Because of spacial limitations, only macrocycles possessing heteroatoms and subheterocyclic rings were reviewed; lactams, lactones, and cyclic imines were excluded.

7.2 CARBON-OXYGEN RINGS

The synthesis of new chiral macrocycles continues at rapid pace affording enantiomer-selective complexes with amines in liquid membrane-type electrode systems <89CC864>, partial separation of free amino acids <89TL4231>, racemic substrate differentiation <89TL3849> for the creation of enzyme mimics and analogues, chiral azophenolic acerands (chromo-acerands) which were used to judge the absolute configurations of chiral amines <89JA1881><89JA742>, asymmetric syntheses <89TL3605><89CB1113>, and enantioselective phase transfer catalysis <89LA181>.

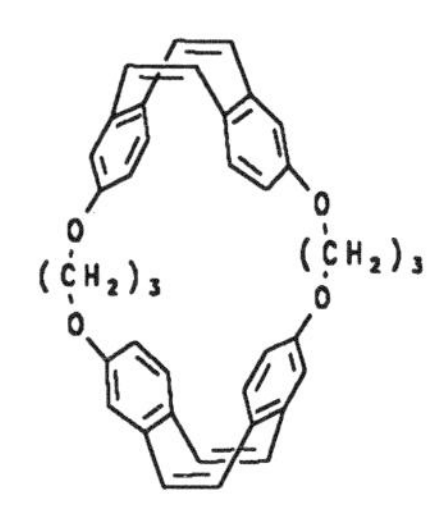

3

The synthesis of crowned macrocycles has attracted specific interest because of their multiple functions and their highly selective applications; thus a thiazolium macrocycle <89TL2759>, furophanes <89CB2147>, thiocrown ethers, containing 3,4-thiophenediyl*bis*-(methylene) subunit <89BCJ838>, crown with propeller units <89JCR(S)64>, crown ethers with inward-facing methoxy groups <89JOC861>, piperazino-30-crown-10-ethers <89TL4333>, a macrocyclic ethereal host bearing *bis*-acetal ether functionality <89T2743>, and a macrocycle incorporating cyclooctatetraene units (**3**) <89TL159> have been described. Considerable effort has been made to synthesize phthalocyanines with pendant crown ethers (**4**) <89CB291>, porphyrines bridged by large functionalized heterorings (**5**) <89JA7006>, and molecular belts and collars incorporating macrohetero-rings <89PAC1581>.

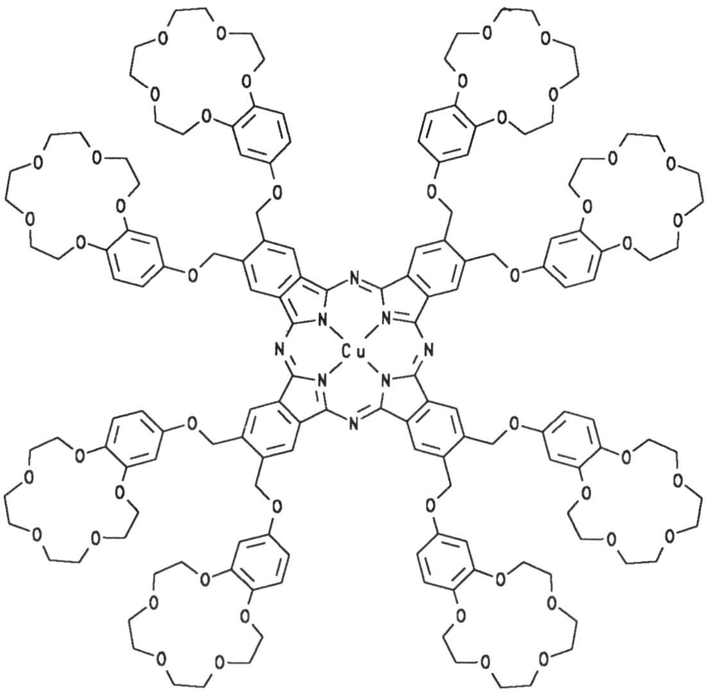

4

The quest for molecular recognition with ethereal crowns gave rise to numerous papers dealing with host-guest interactions and the inherent binding sites. The central role of solvation for polar host-guest interactions <89PAC1523> (**6**) <89JA8438><89JOC1000><89JOC496> has been recognized. Anion effects were reported for the complexation of ammonium ions by simple macrocyclic polyethers <89JOC4712>.

Aside from the synthetic efforts, mechanistic considerations and kinetic studies have appeared. Kinetic studies in the presence of crown ethers <89JOC2009> and proton transfer from a crown ether to a water-soluble molecule complexed in the molecular cavity <89JA700> have been presented. The thermodynamics of complex formation involving crown ethers <89DOK1374>, ion pair formation <89JA7622>, and the photo-

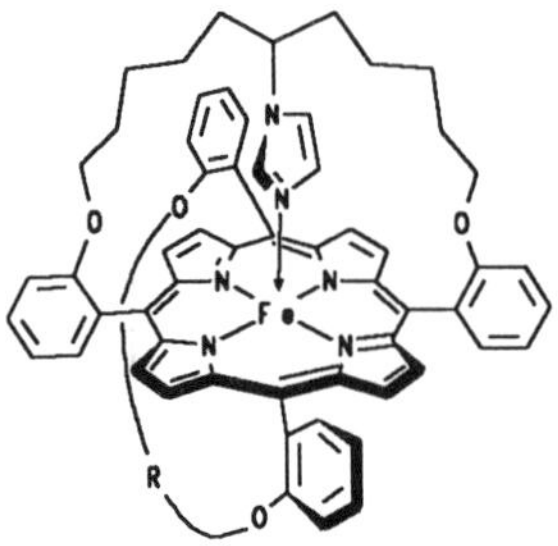

5 $R = (CH_2)_{12}$

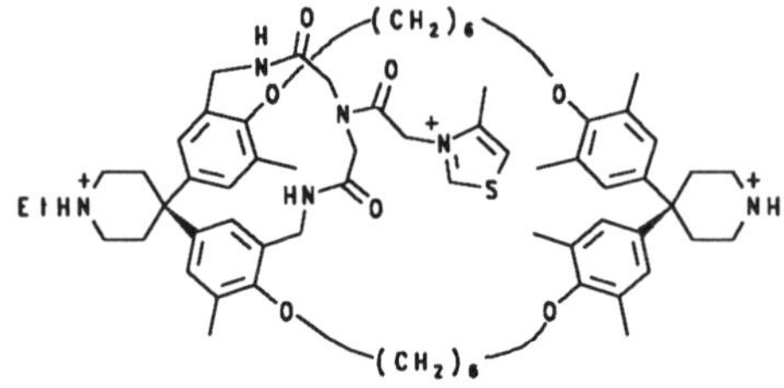

6

chemical and photophysical behavior of host-guest species <89JA7072> have been described. Results of simulations on spherands and crown ethers <89PAC593> using computer graphics, distance geometry, molecular dynamics <89JA3011><89JA2152> have been presented. X-ray crystal structures are playing a more pivatol role in elucidating complexes between host-guest molecules; crown ether derivatives of phthalic acid and urea <89JHC981>, a cryptophane-E/$CHCl_3$ <89AG(E)1246>, a donor-acceptor stack made of paraquat and crown ether <89AG(E)1394>, an inclusion complex of CH_2Cl_2 in a molecular cavity (**7**) <89JCSP(1)212>, a complex of sodium perchlorate and bicyclic polyether <89MI171>, complexation studies <89IC2638><89PAC1587>, biomimetic ion channel structure, and its activity involving a crown ethereal framework <89JA767> have been reported.

7.3 CARBON-NITROGEN RINGS

Syntheses of polyazamacrocycles <89CC580><89T219><89PAC545> and reports <89TL3983><89JCSP(1)2079><89JOC2990><89CC797><89HCA131> <89HCA343><89HCA1253><89HCA307><89JA4662> about their metal ion complexing abilities continue at a rapid pace. Their ability to function as cryptands <89TL4125><89CPB257><89JCSP(2)245><89CC701> and to form monolayers <89CC1016> have been explored. The incorporation of pyridine moieties into macrocycles <89AC95><89CB347><89CB343> <89CB1725> and their ruthenium complexes <89JA5442> <89JA4662> have been reported. Phthalocyanine analogues <89JCSP(2)797>, electrochemical studies of dipyrrolic compounds <89JOC1943>, and Vogel's porphycene (**8**) systems <89MI257> offer insight in things to come.

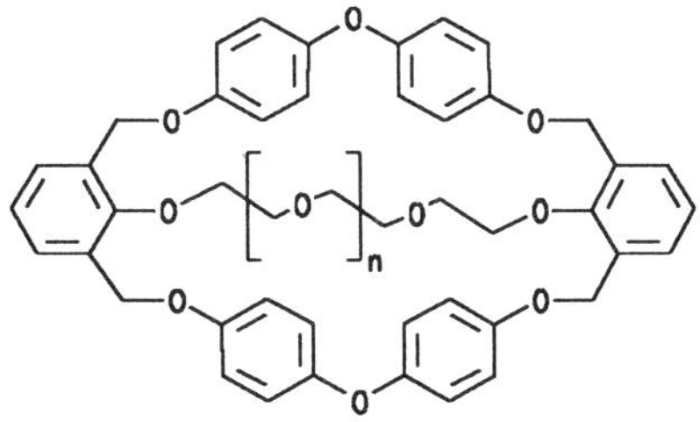

7

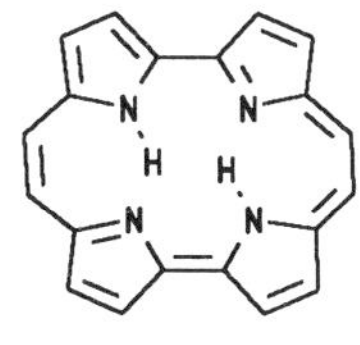

8

7.4 CARBON-SULFUR RINGS

The topic of sulfur macrocycles has not attracted the attention of its oxygen counterpart due to obvious reasons. Several papers report synthetic progress in this area. [3.3]Dithiaphane <89CB1013>, trithia-metacyclophanes (**9**) <89JA3007>, dithiacyclohexadecadiin <89S557>, 1-thia-2-cyclooctin <89LA435> have been synthesized and the X-ray crystal structures of *S*-macroheterocycles <89AX(E)1531><89IC3021> were reported. An inexpensive synthesis of new *O*/*S*-crown ethers has been reported <89TL2903>.

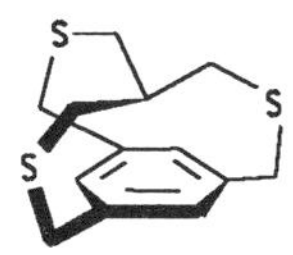

9

7.5 CARBON-OXYGEN-NITROGEN RINGS

The oxazamacrocycles are playing an increasing role in this field due to their increased ease of functionalization and diverse character in complexation. Interlocking catenates were prepared <89JA7791> and their solid state structures, electrochemical studies, electron reductive/transfer processes, and their use as phototransmitters have been studied. Calix-crowns and -spherands <89JA7567>, macrobicyclic cryptates <89TL2209>, polyazamacrocycles <89TL1323>, polyazacrown ethers <89JOC4061>, macro-bicyclic cryptates (**10**) incorporating *bis*thiazole, *bis*imidazole, and bipyrimidine subunits <89TL2209>, peptide side arm derivatives of bibrachial lariat ethers <89JOC937>, imidazol-crown ethers <89JOC1256>, dipyridino-crown ethers (**11**) <89JOC5105><89HCA1066><89CB959>

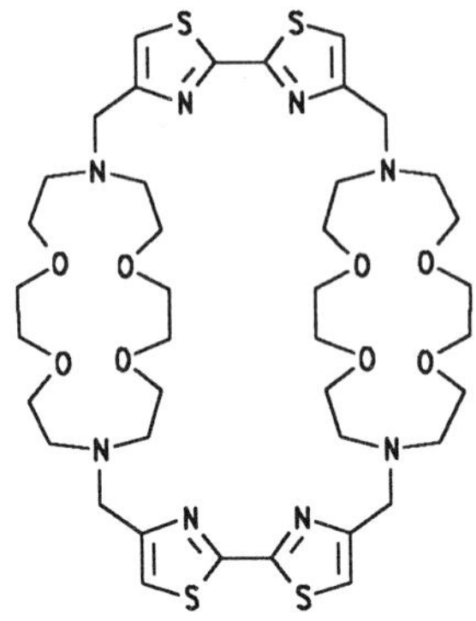

10

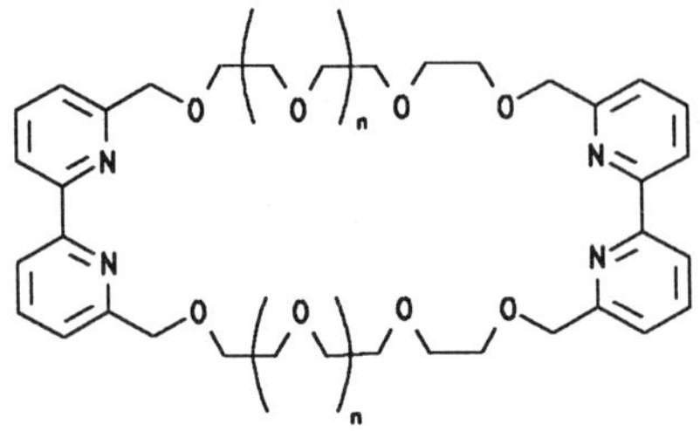

11

<89AG(E)1396>, macro bicyclic phenanthrolines <89LA367>, molecular trifoil knots **(12)** <89AG(E)189>, and crown capped metalloporphyrins <89PAC1547> exemplify the breadth of structural design.

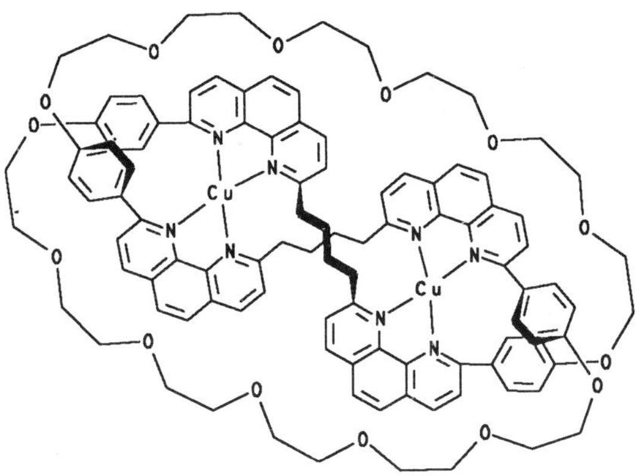

12

Numerous physicochemical studies of complexes of this class have been undertaken: ESR studies in steroidal lariat ethers <89CC578>, voltametric studies on the complexation of metal ions <89CC419>, and polyammonium macrocycles as enzyme mimics for phosphoryl transfer in ATP hydrolysis <89JA6330> have recently appeared. Transport phenomena on host-guest complexes in solutions <89JCSP(1)89> <89JOC1391><89JOC2988><89JA1409><89JA3075> and molecular dynamics of configurational changes in macrocycles <89PAC1593> have been reported. Silica gel bound macrocycles were found to separate heavy metal ions <89PAC1619>.

7.6 CARBON-SULFUR-NITROGEN RINGS

Thiadiazaparacyclophane <89CB1503> and diverse nickel and copper complexes <89IC655><89IC4065><89JA1494> with *N,S*-bonding site(s) were prepared. A kinetic analysis of azobenzenophenophanes <89JA4627> was presented. Topologically constrained macrocycles were used as DNA *bis*intercalators <89JA6805>.

7.7 CARBON-NITROGEN-OXYGEN-SULFUR RINGS

The synthesis of macrocyclic cages (**13**) <89TL941> and their various metal complexes <89MI41><89IC4593><89IC4279><89JCSP(2)1131> continue to open up new structural vistas.

7.8 CARBON-TIN RINGS

Crystalline complexes of macrocycles containing tin have been studied <89JA6294> (**14**) <89TL3501><89MI2755><89JOC4626>.

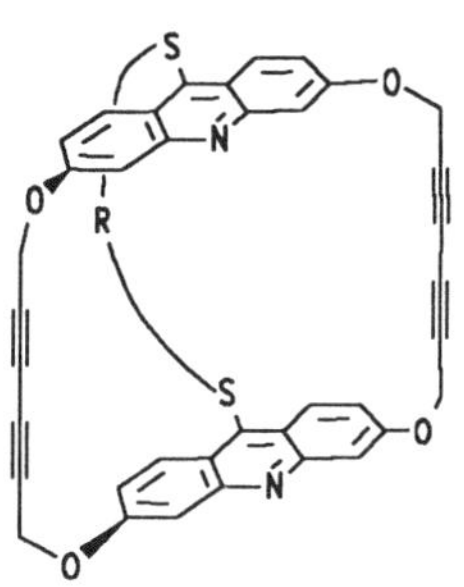

13 R = $(CH_2)_6$, $CH_2CH_2OCH_2CH_2$, $CH_2CH_2OCH_2CH_2OCH_2CH_2$

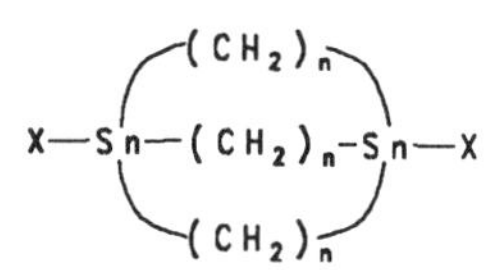

14 X = Cl, Br
n = 6, 7, 8, 10, 12

7.9 CARBON-SELENIUM RINGS

A series of novel selenium coronands <89JAC6582>, selenacyclophanes (**15**) <89T1611>, and selenapyridinophanes (**16**) <89JOC393> have been reported.

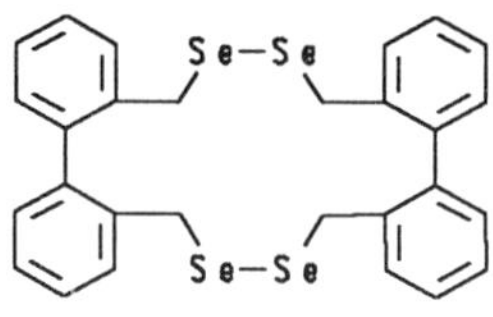

15

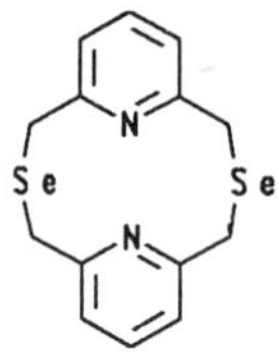

16

7.10 REFERENCES

89AC95 F. Ebmeyer, and F. Vögtle; *Angew. Chem.*, 1989, **101**, 95.

89AG(E)189 C. O. Dietrich-Buchecker, and J. P. Sauvage; *Angew. Chem., Int. Ed. Engl.*, 1989, **28**, 189.

89AG(E)1103 F. H. Kohnke, J. P. Mathias, and J. F. Stoddart; *Angew. Chem. Int. Ed. Adv. Mat.*, 1989, **28**, 1103.

89AG(E)1153 K. Wieghardt; *Angew. Chem. Int. Ed. Engl.*, 1989, **28**, 1153.

89AG(E)1246 J. Canceill, M. Cesario, A. Collet, J. Guilhem, L. Lacombe, B. Lozach, and C. Pascard; *Angew. Chem. Int. Ed. Engl.*, 1989, **28**, 1246.

89AG(E)1394 J.-Y. Ortholand, A. M. Z. Zlawin, N. Spenser, J. F. Stoddart, and D. J. Williams; *Angew. Chem. Int. Ed. Engl.*, 1989, **28**, 1394.

89AG(E)1396 P. R. Ashton, T. T. Goodnow, A. E. Kaifer, M. V. Reddington, A. M. Z. Slavin, N. Spencer, J. F. Stoddart, C. Vicent, and D. J. Williams; *Angew. Chem., Int. Ed. Engl.*, 1989, **28**, 1396.

89AX(C)1531 I. Zamaev, V. E. Shklover, Y. E. Ovchinnikov, Y. T. Struchkov, A. V. Astankov, V. I. Nedel'kin, and V. Sergeyev; *Acta Cryst.*, 1989, **C45**, 1531.

89BCJ838 T. Sone, K. Sato, and Y. Ohba; *Bull. Chem. Soc. Jpn.*, 1989, **62**, 838,

89CB291 S. Sarigul and Ö. Bekâroğlu; *Chem. Ber.*, 1989, **122**, 291.

89CB343 F. Vögtle, U. U. Koersgen, H. Puff, and H. Reuter; *Chem. Ber.*, 1989, **122**, 343.

89CB347 K. J. Przybilla and F. Vögtle; *Chem. Ber.*, 1989, **122**, 347.

89CB959 E. Weber, H. J. Koehler, and H. Reuter; *Chem. Ber.*, 1989, **122**, 959.

89CB1013 G. Bodwell, L. Ernst, and H. Hopf; *Chem. Ber.*, 1989, **122**, 1013.

89CB1113 S. Billen and F. Vögtle; *Chem. Ber.*, 1989, **122**, 1113.

89CB1503 U. Funke and H. F. Gruetzmacher; *Chem. Ber.*, 1989, **122**, 1503.

89CB1725 F. Ebmeyer and F. Vögtle; *Chem. Ber.*, 1989, **122**, 1725.

89CB2147 I. Heinze, K. Knoll, R. Muller, and W. Eherbach; *Chem. Ber.*, 1989, **122**, 2147.

89CC419 C. D. Hall, N. W. Sharpe, I. Danks, and Y. P. Sang; *J. Chem. Soc., Chem. Comm.*, 1989, 419.

89CC578 H. Fasoli, L. E. Echegoyen, J. C. Hernandez, G. W. Gokel, and L. Echegoyen; *J. Chem. Soc., Chem. Comm.*, 1989, 778.

89CC580 H. J. Schneider, T. Blatter, S. Simova, and I. Theis; *J. Chem. Soc., Chem. Comm.*, 1989, 580.

89CC701 A. Bencini, A. Bianchi, M. Ciampolini, E. Garcia-Espana, P. Dapporto, M. Micheloni, P. Paoli, J. A. Ramirez, and B. Valtancoli; *J. Chem. Soc., Chem. Comm.*, 1989, 701.

89CC797 J. P. L. Cox, K. J. Jankowski, R. Kataki, D. Parker, N. R. A. Beeley,

B. A. Boyce, M. A. W. Eaton, K. Millar, A. T. Millican, A. Harrison, and C. Walker; *J. Chem. Soc., Chem. Comm.*, 1989, 797.
89CC864 K. Maruyama, H. Sohmiya, and H. Tsukube; *J. Chem. Soc., Chem. Commun.*, 1989, 864.
89CC1016 J. Malthete, D. Poupinet, R. Vilanove, and J. M. Lehn; *J. Chem. Soc., Chem. Comm.*, 1989, 1016.
89CPB257 K. Odashima, H. Kawakami, A. Miwa, I. Sasaki, and K. Koga; *Chem. Pharm. Bull.*, 1989, **37**, 257.
89CR347 W. H. Pirkle, and T. C. Pochapsky; *Chem. Rev.*, 1989, **89**, 347.
89CR929 K. E. Krakowiak, J. S. Bradshaw, and D. J. Zamecka-Krakowiak; *Chem. Rev.*, 1989, **89**, 929.
89CR1215 R. W. Adler; *Chem. Rev.*, 1989, **89**, 1215.
89CR1581 J. K. Whitesell; *Chem. Rev.*, 1989, **89**, 1581.
89CR1875 R. D. Hancock, and A. E. Martell; *Chem. Rev.*, 1989, **89**, 1875.
89DOK1374 V. P. Solov'ev, N. N. Strakhova, O. A. Raevskii, V. E. Zubareva, and I. I. Bulgak; *Dok. Akad. Nauk SSSR*, 1989, **304**, 1374.
89G311 V. Balzani; *Gazz. Chim. Ital.*,1989, **119**, 311.
89HCA131 D. Tschudin, A. Riesen, and T. A. Kaden; *Helv. Chim. Acta*, 1989, **72**, 131.
89HCA343 D. Tschudin, A. Riesen, and T. A. Kaden; *Helv. Chim. Acta*, 1989, **72**, 343.
89HCA307 M. Studer, A. Riesen, and T. A. Kaden; *Helv. Chim. Acta*, 1989, **72**, 307.
89HCA1066 M. W. Hosseini, J. Comarmond, and J. M. Lehn; *Helv. Chim. Acta*, 1989, **72**, 1066.
89HCA1253 M. Studer, A. Riesen, and T. A. Kaden; *Helv. Chim. Acta*, 1989, **72**, 1253.
89IC655 D. G. Fortier and A. McAuley; *Inorg. Chem.*, 1989, **28**, 655.
89IC2638 J.-C. G. Bunzli and F. Pilloud; *Inorg. Chem.*, 1989, **28**, 2638.
89IC3021 H. Elias, G. Schmidt, H. J. Kueppers, M. Saher, K. Wieghardt, B. Nuber, and J. Weiss; *Inorg. Chem.*, 1989, **28**, 3021.
89IC4065 L. Latos-Grazynski, M. M. Olmstead, and A. L. Balch; *Inorg. Chem.*, 1989, **28**, 4065.
89IC4279 A. Bencini, A. Bianchi, A. Borselli, M. Ciampolini, E. Garcia-Espania, P. Dapporto, M. Micheloni, P. Paoli, J. A. Ramirez, B. Valtancoli; *Inorg. Chem.*, 1989, **28**, 4279.
89IC4593 R. Bonomo, F. Bottino, F. R. Fronczek, A. Mamo, and S. Pappalardo; *Inorg. Chem.*, 1989, **28**, 4593.
89JA700 J. V. Eerden, M. Skowronska-Ptasinka, P. D. J. Grootenhuis, S. Harkema, and D. N. Reinhoudt; *J. Am. Chem. Soc.*, 1989, **111**, 700.
89JA742 T. Kaneda, K. Hirose, and S. Misumi; *J. Am. Chem. Soc.*, 1989, **111**, 742.
89JA767 V. E. Carmichael, P. J. Dutton, T. M. Fyles, T. D. James, J. A. Swan, and M. Zojaji; *J. Am. Chem. Soc.*, 1989, **111**, 767.
89JA1409 H. Jahansouz, J. Jiang, R. H. Himes, M. P. Mertes, and K. B. Mertes; *J. Am. Chem. Soc.*, 1989, **111**, 1409.
89JA1494 Y. Agnus, J. P. Gisselbrecht, R. Louis, and B. Metz; *J. Am. Chem. Soc.*, 1989, **111**, 1494.
89JA1881 T. Kaneda, S'i. Umeda, Y. Ishizaki, H.-S. Kuo, S. Misumi, Y. Kai, N. Kanehisa, and N. Kasai; *J. Am. Chem. Soc.*, 1989, **111**, 1881.
89JA2151 P. D. J. Grootenhuis and P. A. Kollman; *J. Am. Chem. Soc.*, 1989, **111**, 2152.
89JA3007 R. A. Pascal, C. G. Winans, and D. Van Engen; *J. Am. Chem. Soc.*,

1989, **111**, 3007.

89JA3011 D. Gehin, P. A. Kollman, and G. Wipff; *J. Am. Chem. Soc.*, 1989, **111**, 3011.

89JA3075 K. T. Chapman and W. C. Still; *J. Am. Chem. Soc.*, 1989, **111**, 3075.

89JA4627 G. Ritter, G. Haeflinger, E. Luedecke, and H. Rau; *J. Am. Chem. Soc.*, 1989, **111**, 4627.

89JA4662 F. Brigelletti, L. De Cola, V. Balzani, P. Belser, A. von Zelewski, F. Vögtle, F. Ebmeyer, and S. Grammenudi; *J. Am. Chem. Soc.*, 1989, **111**, 4662.

89JA5442 J. T. Groves and L. A. Baron; *J. Am. Chem. Soc.*, 1989, **111**, 5442.

89JA6294 M. Newcomb, J. H. Hoener, M. T. Blanda, and P. J. Squattrito; *J. Am. Chem. Soc.*, 1989, **111**, 6294.

89JA6330 M. W. Hosseini, J. M. Lehn, K. C. Jones, K. E. Plute, K. B. Mertes, and M. P. Mertes; *J. Am. Chem. Soc.*, 1989, **111**, 6330.

89JA6582 R. J. Batchelor, F. W. B. Einstein, I. D. Gray, J. H. Gu, B. D. Johnston, and B. M. Pinto; *J. Am. Chem. Soc.*, 1989, **111**, 6582.

89JA6805 S. C. Zimmerman, C. R. Lamberson, M. Cory, and T. A. Fairley; *J. Am. Chem. Soc.*, **111**, 6805.

89JA7006 I. P. Gerothanassis, M. Momenteau, and B. Loock; *J. Am. Chem. Soc.*, 1989, **111**, 7006.

89JA7072 R. Ballardini, M. T. Gandolfi, L. Prodi, M. Ciano, V. Balzani, F. H. Kohnke, H. Shahriari-Zavareh, N. Spenser, and J. F. Stoddart; *J. Am. Chem. Soc.*, 1989, **111**, 7072.

89JA7567 P. J. Dijkstra, J. A. J. Brunik, K. E. Bugge, D. N. Reinhoudt, S. Harkema, R. Ungaro, F. Uggozzoli, and E. Ghidini; *J. Am. Chem. Soc.*, 1989, **111**, 7567.

89JA7622 H. Bock and H.F. Herrmann, *J. Am. Chem. Soc.*, 1989, **111**, 7622.

89JA7791 C. Dietrich-Buchecker, J. P. Sauvage, and J. M. Kern; *J. Am. Chem. Soc.*, 1989, **111**, 7791.

89JA8438 F. Diederich and H.-D. Lutter; *J. Am. Chem. Soc.*, 1989, **111**, 8438.

89JCR(S)64 K. I. Kinnear, J. C. Lockhart, C. T. Hewkin, and D. J. Rushton; *J. Chem. Res. (S)*, 1989, 64.

89JCSP(1)89 H. Tsukube, H. Adachi, and S. M. Orosawa; *J. Chem. Soc. Perkins Trans. I*, 1989, 89.

89JCSP(1)212 G. R. Bower, A. M. Z. Slawin, D, J. Williams, G. R. Brown, S. S. Chana, and J. F. Stoddart; *J. Chem. Soc. Perkin 1*, 1989, 212.

89JCSP(1)2079 I. M. Helps, D. Parker, K. J. Jankowski, J. Chapman, and P. E. Nicholson; *J. Chem. Soc., Perkin Trans.*, 1989, 2079.

89JCSP(2)245 S. Kumar and H. J. Schneider; *J. Chem. Soc., Perkin Trans. II*, 1989, 245.

89JCSP(2)797 F. Fernandez-Lazaro, J. de Mendoza O. Mo, S. Rodriguez-Morgade, T. Torres, and M. Yanez; *J. Chem. Soc., Perkin Trans. II*, 1989, 797.

89JCSP(2)1131 A. Bianchi, M. Ciampolini, M. Micheloni, N. Nardi, B. Valtancoli, S. S. Mangani, and J. A. Ramirez; *J. Chem. Soc., Perkin Trans. II*, 1989, 1131.

89JHC981 F. Benetollo, G. Bombieri, and M. R. Truter; *J. Heterocycl. Chem.*, 1989, **26**, 981.

89JOC393 S. Muralidharan, M. Hoijatie, M. Firestone, and H. Freiser; *J. Org. Chem.*, 1989, **54**, 393.

89JOC496 K. Laali and R. P. Lattimer; *J. Org. Chem.*, 1989, **54**, 496.

89JOC861 E. Chapoteau, B. P. Czech, A. Kumar, A. Pose, R. A. Bartsch, R. A. Holwerda, N. K. Dalley, B. E. Wilson, and J. Weining; *J. Org. Chem.*, 1989, **54**, 861.

89JOC937 B. D. White, J. Mallen, K. A. Arnold, F. R. Fronczek, R. D. Gandour, L. M. B. Gehrig, and G. W. Gokel; *J. Org. Chem.*, 1989, **54**, 937.

89JOC1000 T. B. Stolwijk, E. J. R. Sudholter, D. N. Reinhoudt, J. V. Eerden, and S. Harkema; *J. Org. Chem.*, 1989, **54**, 1000.

89JOC1256 S. C. Zimmerman, K. D. Cramer, and A. A. Galan; *J. Org. Chem.*, 1989, **54**, 1256.

89JOC1391 P. Navarro, M. I. Rodriguez-Franco, C. Foces-Foces, F. Cano, and A. Samat; *J. Org. Chem.*, 1989, **54**, 1391.

89JOC1943 H. D. Tabba, J. A. S. Cavaleiro, D. Jeyakumar, M. G. P. M. S. Neves, and K. M. Smith; *J. Org. Chem.*, 1989, **54**, 1943.

89JOC2009 J. Suh and B. S. Mun; *J. Org. Chem.*, 1989, **54**, 2009.

89JOC2988 Y. Nakatsuji, R. Wakita, Y. Harada, and M. Okahara; *J. Org. Chem.*, 1989, **54**, 2988.

89JOC2990 F. Chavez and A. D. Sherry; *J. Org. Chem.*, 1989, **54**, 2990.

89JOC4061 K. E. Krakowiak, J. S. Bradshaw, R. M. Izatt, and D. J. Zamecka-Krakowiak; *J. Org. Chem.*, 1989, **54**, 4061.

89JOC4626 M. T. Blanda, J. H. Horner, and M. Newcomb; *J. Org. Chem.*, 1989, **54**, 4626.

89JOC4712 K. M. Doxsee; *J. Org. Chem.*, 1989, **54**, 4712.

89JOC5105 G. R. Newkome, G. E. Kiefer, D. K. Kohli, Y. J. Xia, F. R. Fronczek, and G. R. Baker; *J. Org. Chem.*, 1989, **54**, 5105.

89LA181 E. V. Dehmlow and C. Sauerbier; *Liebigs Ann. Chem.*, 1989, 181.

89LA367 U. Luening and M. Mueller; *Liebigs Ann. Chem.*, 1989, 367.

89LA435 E. Stavridou, H. Schuhmacher, and H. Meier; *Liebigs Ann. Chem.*, 1989, 435.

89MI:(1) L. F. Lindoy; "The Chemistry of Macrocyclic Ligand Complexes." Cambridge University Press, Cambridge, 1989.

89MI:(2) F. Vögtle; "Reizvolle Moleküle der Organischen Chemie," Teubner, Stuttgart, 1989.

89MI:(3) F. Vögtle; "Supramolekulare Chemie," Teubner Stuttgart, 1989.

89MI12 E. Weber; *J. Mol. Graphics,* 1989, **7**, 12.

89MI29 Y. Yamamoto and Y. Inoue; *J. Carbohydr. Chem.*, 1989, **8**, 29.

89MI41 J. P. Collin, J. A. Osborn, and M. T. Youinou; *Inorg. Chim. Acta*, 1989, **165**, 41.

89MI63 P. Scrimin; *Chemicaoggi*, 1989, 63.

89MI171 M. R. Truter; *Inorg. Chim. Acta*, 1989, **160**, 171.

89MI219 B. O. West; *Polyhedron,* 1989, **8**, 219.

89MI257 E. Vogel, M. Koecher, J. Lex, and O. Ermer; *Israel J. Chem.*, 1989, **29**, 257.

89MI484 Y. Murakami and J.-I. Kikuchi; *J. Org. Org. Jpn.*, 1989, **47**, 484.

89MI493 S. Sasaki and K. Koga; *J. Syn. Org. Jpn.*, 1989, **47**, 493.

89MI583 E. Kimura; *J. Syn. Org. Chem. Jpn.*, 1989, **47**, 583.

89MI938 T. J. McMurry, K.N. Raymond, and P. H. Smith; *Science*, 1989, **244**, 938.

89MI2755 K. Jurgschat, H. G. Kuivila, S. Liu, and J. A. Zubieta; *J. Am. Chem. Soc.*, 1989, **8**, 2755.

89PAC545 K. N. Raymond, T. J. McMurry, and T. M. Garrett; *Pure & Appl. Chem.*, 1989, **60**, 545.

89PAC593 P. A. Kollman, P. D. J. Grootenhuis, and M. A. Lopez; *Pure & Appl. Chem.*, 1989, **61**, 593.

89PAC1523 S. B. Ferguson, E. M. Sewaard, E. M. Sanford, M. Hester, M. Uyeki, and F. Diederich; *Pure & Appl. Chem.*, 1989, **61**, 1523.

89PAC1547 I. O. Sutherland; *Pure & Appl. Chem.*, 1989, **61**, 1547.

89PAC1581 F. H. Kohnke and J. F. Stoddart; *Pure & Appl. Chem.*, 1989, **61**, 1581.
89PAC1587 J. Dale and S. B. Fredriksen; *Pure & Appl. Chem.*, 1989, **61**, 1587.
89PAC1593 L. J. Rodriguez, M. Xu, E. M. Eyring, and S. Petrucci; *Pure & Appl. Chem.*, 1989, **61**, 1593.
89PAC1619 J. S. Bradshaw, K. E. Krakowiak, B. J. Tarbet, R. L. Bruening, J. F. Biernat, M. Bochenska, R. M. Izatt, and J. J. Christensen; *Pure & Appl. Chem.*, 1989, **61**, 1619.
89S557 H. Schuhmacher and H. Meier; *Synthesis*, 1989, 557.
89T3 T. N. Sorrell; *Tetrahedron*, 1989, **45**, 3.
89T219 I. M. Helps, D. Parker, J. R. Morphy, and J. Chapman; *Tetrahedron*, 1989, **45**, 219.
89T1611 M. Hojjattie, S. Muralidharan, and H. Freiser; *Tetrahedron*, 1989, **45**, 1622.
89T2743 G. Mehta, K. S. Rao, N. Krishnamurthy, V. Srinivas, and D. Balasubramanian; *Tetrahedron*, 1989, **45**, 2743.
89TL159 W. Heinz, H.-J. Rader, and K. Mullen; *Tetrahedron Lett.*, 1989, **30**, 159.
89TL941 S. Claude, J. M. Lehn, and J. P. Vigneron; *Tetrahedron Lett.*, 1989, **30**, 941.
89TL1323 K. Gu, K. B. Mertes, and M. Mertes; *Tetrahedron Lett.*, 1989, **30**, 1323.
89TL2209 J. M. Lehn and J. B. Regnouf de Vains; *Tetrahedron Lett.*, 1989, **30**, 2209.
89TL2759 L. Jimenez and F. Diederich; *Tetrahedron Lett.*, 1989, **30**, 2759.
89TL2903 G. Maerkl and R. Vybiral; *Tetrahedron Lett.*, 1989, **30**, 2903.
89TL3501 M. T. Blanda and M. Newcomb; *Tetrahedron Lett.*, 1989, **30**, 3501.
89TL3605 C. Vicent, M. Martin-Lomas, and S. Penades; *Tetrahedron Lett.*, 1989, **30**, 3605.
89TL3849 J. Crosby, M. E. Fakley, C. Gemmeli, K. Martin, A. Quick, A. M. Z. Slawin, H. Shahriari-Zavareh, J. F. Stoddart, and D. J. Williams; *Tetrahedron Lett.*, 1989, **30**, 3849.
89TL3983 H. Tsukube, K. Yamashita, T. Iwachido, and M. Zenki; *Tetrahedron Lett.*, 1989, **30**, 3983.
89TL4125 B. Dietrich, J. M. Lehn, J. Guilhem, and C. Pascard; *Tetrahedron Lett.*, 1989, **30**, 4125.
89TL4231 J.-P. Joly and B. Gross; *Tetrahedron Lett.*, 1989, **30**, 4231.
89TL4333 H. D. H. Stover and C. Detellier; *Tetrahedron Lett.*, 1989, **30**, 4333.

SUBJECT INDEX